Statistical Signal Processing

Swagata Nandi · Debasis Kundu

Statistical Signal Processing

Frequency Estimation

Second Edition

 Springer

Swagata Nandi
Theoretical Statistics and Mathematics Unit
Indian Statistical Institute
New Delhi, Delhi, India

Debasis Kundu
Department of Mathematics and Statistics
Indian Institute of Technology Kanpur
Kanpur, Uttar Pradesh, India

ISBN 978-981-15-6282-2   ISBN 978-981-15-6280-8   (eBook)
https://doi.org/10.1007/978-981-15-6280-8

This Springer imprint is published by the registered company Springer Nature Singapore Pte Ltd.
The registered company address is: 152 Beach Road, #21-01/04 Gateway East, Singapore 189721, Singapore

To my mother and to the memory of my father
S. Nandi

To my mother and to the memory of my father
D. Kundu

Preface to the Second Edition

The first edition of this book was written in 2012 as a *Springer Brief*, and the main emphasis was on the sinusoidal model. It was written in response to an invitation from *Springer* to write an extended review on this particular topic targeting the Statistics community. The main idea was to introduce to the Statistics community the class of problems from the *Statistical Signal Processing* area which would be of interest to them. We had the intention to extend it to a full-fledged research monograph in the future.

Since the first edition has been published, an extensive amount of work has been done in some of the related areas. Two Ph.D. theses have been written at the Indian Institute of Technology, Kanpur, in the Department of Mathematics and Statistics dealing with different problems associated with chirp and some related models. Interesting connections have been made with some of the well-known number theoretic results. Many other interesting new developments have taken place in this area. Thus, we decided to write this second edition with an intention to include all these new materials in one place.

The second edition of this book has undergone a massive change, although the main structure and mathematical level remain the same. It still cannot be called a textbook. It can be used as a reference book in this area. We have revised most of the chapters and added three major chapters namely (i) Fundamental Frequency Model and its Generalizations, (ii) Chirp Model, and (iii) Random Amplitude Models. In addition to the major changes, many new comments and references have been included. Numerous errors have been taken care of, and several open problems have been proposed for future research.

The research for this second edition has been supported in part by a grant from the Science and Engineering Research Board, Government of India. Finally, we would like to thank Ms. Sagarika Ghosh and Ms. Nupur Singh from Springer for helping us to finish this monograph within a reasonable time.

Delhi, India   Swagata Nandi
Kanpur, India   Debasis Kundu
April 2020

Preface to the First Edition

Both of us have done our Ph.D. theses on *Statistical Signal Processing* although in a gap of almost 15 years. The first author was introduced to the area by his Ph.D. supervisor Prof. C.R. Rao, while the second author was introduced to this topic by the first author. It has been observed that the frequency estimation plays an important role in dealing with different problems in the area of *Statistical Signal Processing*, and both the authors have spent a significant amount of their research career dealing with this problem for different models associated with the Statistical Signal Processing.

Although an extensive amount of the literature is available in engineering dealing with the frequency estimation problem, not that much attention has been paid in the statistical literature. The book by Quinn and Hannan [15] is the only book dealing with the problem of frequency estimation written for the statistical community. Both of us were thinking about writing a review article on this topic for quite sometime. In this respect, the invitation from Springer to write a Springer Brief on this topic came as a pleasant surprise to both of us.

In this Springer brief, we provide a review of different methods available till date dealing with the problem of frequency estimation. We have not attempted an exhaustive survey of frequency estimation techniques. We believe that would require separate books on several topics themselves. Naturally, the choice of topics and examples is in favor of our own research interests. The list of references is also far from complete.

We have kept the mathematical level quite modest. Chapter 4 mainly deals with somewhat more demanding asymptotic theories, and this chapter can be avoided during the first reading without losing any continuity. Senior undergraduate-level mathematics should be sufficient to understand the rest of the chapters. Our basic goal to write this Springer Brief is to introduce the challenges of the frequency estimation problem to the statistical community, which are present in different areas of science and technology. We believe that statisticians can play a major role in solving several problems associated with frequency estimation. In Chap. 8, we have provided several related models, where there are several open issues which need to be answered by the scientific community.

Every book is written with a specific audience in mind. This book definitely cannot be called a textbook. This book has been written mainly for senior undergraduates and graduate students specializing in Mathematics or Statistics. We hope that this book will motivate students to pursue higher studies in the area of Statistical Signal Processing. This book will be helpful to young researchers who want to start their research career in the area of Statistical Signal Processing. We will consider our efforts to be of worth if the target audience finds this volume useful.

Kanpur, India   Debasis Kundu
Delhi, India   Swagata Nandi
January 2012

Acknowledgements

The authors would like to thank their respective Institutes and all who contributed directly and indirectly to the production of this monograph. The second author would like to thank his wife Ranjana for her continued support and encouragement.

Contents

About the Authors

Swagata Nandi is currently an Associate Professor at the Theoretical Statistics and Mathematics Unit of the Indian Statistical Institute, Delhi Center. She received her M.Sc. and Ph.D. from the Indian Institute of Technology, Kanpur. In her Ph. D, she worked under the guidance of Professor Debasis Kundu and Professor S. K. Iyer. Before joining Indian Statistical Institute in 2005, she was a Postdoctoral Fellow at the University of Heidelberg and at the University of Mannheim, Germany. Her research interests include statistical signal processing, nonlinear regression, analysis of surrogate data, EM algorithms and survival analysis. She has published more than 35 research papers in various national and international journals and co-authored a monograph on Statistical Signal Processing. She is the recipient of the Indian Science Congress Association Young Scientist award and the C.L. Chandana Award for Students.

Prof. Debasis Kundu received his B-Stat and M-Stat from the Indian Statistical Institute in 1982 and 1984, respectively, MA in Mathematics from the University of Pittsburgh in 1985 and Ph.D. from the Pennsylvania State University in 1989 under the guidance of Professor C.R. Rao. After fnishing his Ph.D. he joined The University of Texas at Dallas as an Assistant Professor before joining the Indian Institute of Technology Kanpur in 1990. He served as the Head of the Department of Mathematics and Statistics, IIT Kanpur from 2011 to 2014, currently he is the Dean of Faculty Affairs at IIT Kanpur. Professor Kundu works on different areas of Statistics. His major research interests are on Distribution Theory, Lifetime Data Analysis, Censoring, Statistical Signal Processing and Statistical Computing. He has published more than 300 research papers in different referred journals with close to 12,000 citations and H-index 55. He has supervised/ co-supervised 14 Ph.D. and 5 M.Tech students. He has co-authored two research monographs on Statistical Signal Processing and on Step-Stress Models, and co-edited one book on Statistical Computing. He is a Fellow of the National Academy of Sciences, India and a Fellow of the Indian Society of Probability and Statistics. He has received Chandana Award from the Canadian Mathematical Society, Distinguished Statistician Award from the Indian Society of Probability and Statistics, P.C.

Mahalanobis Award from the Indian Society of Operation Research and Distinguished Teacher's Award from IIT Kanpur. He is currently the Editor-in-Chief of the *Journal of the Indian Society of Probability and Statistics*, and in the editorial boards of *Communications in Statistics - Theory and Methods, Communications in Statistics - Simulation and Computation, Communications in Statistics - Case Studies and Sankhya, Ser. B*. He has served as members in the editorial boards of the *IEEE Transactions on Reliability, Journal of Statistical Theory and Practice and Journal of Distribution Theory*.

Acronyms

1-D	One-dimensional
2-D	Two-dimensional
3-D	Three-dimensional
AIC	Akaike's Information Criterion
ALSE(s)	Approximate Least Squares Estimator(s)
AM	Amplitude Modulated
AMLE(s)	Approximate Maximum Likelihood Estimator(s)
AR	AutoRegressive
ARCOS	AutoRegressive amplitude modulated COSinusoid
ARIMA	AutoRegressive Integrated Moving Average
ARMA	AutoRegressive Moving Average
BIC	Bayesian Information Criterion
CV	Cross Validation
ECG	Electro Cardio Graph
EDC	Efficient Detection Criterion
EM	Expectation Maximization
ESPRIT	Estimation of Signal Parameters via Rotational Invariance Technique
EVLP	EquiVariance Linear Prediction
FFM	Fundamental Frequency Model
GFFM	Generalized Fundamental Frequency Model
i.i.d.	independent and identically distributed
Im(z)	Imaginary Part of a complex number z
ITC	Information Theoretic Criterion
IQML	Iterative Quadratic Maximum Likelihood
KIC	Kundu's Information Criterion
LAD	Least Absolute Deviation
LS	Least Squares
LSE(s)	Least Squares Estimator(s)
MA	Moving Average

MCMC	Markov Chain Monte Carlo
MDL	Minimum Description Length
MFBLP	Modified Forward Backward Linear Prediction
MLE(s)	Maximum Likelihood Estimator(s)
MRI	Magnetic Resonance Imaging
MSE	Mean Squared Error
NSD	Noise Space Decomposition
PCE	Probability of Correct Estimate
PDF	Probability Density Function
PMF	Probability Mass Function
PPC	Polynomial Phase Chirp
PE(s)	Periodogram Estimator(s)
QIC	Quinn's Information Criterion
QT	Quinn and Thomson
Re(z)	Real Part of a complex number z
RGB	Red-Green-Blue
RSS	Residual Sum of Squares
SAR	Synthetic Aperture Radar
SIC	Sakai's Information Criterion
SVD	Singular Value Decomposition
TLS-ESPRIT	Total Least Squares Estimation of Signal Parameters via Rotational Invariance Technique
WIC	Wang's Information Criterion
WLSE(s)	Weighted Least Squares Estimator(s)

Symbols

a.e.	almost everywhere
$\arg(z)$	$\tan^{-1}(\theta)$ where $z = r\,e^{i\theta}$
$\mathbf{I}_n$	Identity matrix of order n
$\mathrm{rank}(A)$	Rank of matrix A
$\mathrm{tr}(A)$	Trace of matrix A
$\det(A)$	Determinant of matrix A
$\mathcal{N}(a, b^2)$	Univariate normal distribution with mean a and variance b^2
$\mathcal{N}_p(\mathbf{0}, \Sigma)$	p-variate normal distribution with mean vector $\mathbf{0}$ and dispersion matrix Σ
χ_p^2	Chi-square distribution with p degrees of freedom
$X_n \xrightarrow{d} X$	X_n converges to X in distribution
$X_n \xrightarrow{p} X$	X_n converges to X in probability
a.s.	almost surely
$X_n \to X$ a.s.	X_n converges to X almost surely

$X_n \overset{a.s.}{\to} X$	X_n converges to X almost surely
$X \overset{d}{=} Y$	X and Y have the same distribution
$o(a_n)$	$(1/a_n)o(a_n) \to 0$ as $n \to \infty$
$O(a_n)$	$(1/a_n)O(a_n)$ is bounded as $n \to \infty$
$X_n = o_p(a_n)$	$\lim_{n\to\infty} P(|X_n/a_n| \geq \varepsilon) = 0$ for every positive ε
$X_n = O_p(a_n)$	X_n/a_n is bounded in probability as $n \to \infty$
$\|\mathbf{x}\|$	$\sqrt{x_1^2 + \cdots + x_n^2}, \mathbf{x} = (x_1, x_2, \cdots, x_n)$
$\mathbb{R}$	set of real numbers
$\mathbb{R}^d$	d-dimensional Euclidean space
$\mathbb{C}$	set of complex numbers
$|\mathbf{A}|$	determinant of matrix $\mathbf{A}$

Chapter 1
Introduction

Signal processing may broadly be considered to involve the recovery of information from physical observations. The received signal is usually disturbed by thermal, electrical, atmospheric, or intentional interferences. Due to random nature of the signal, statistical techniques play important roles in analyzing it. Statistics is also used in the formulation of the appropriate models to describe the behavior of the system, the development of appropriate techniques for estimation of model parameters, and the assessment of the model performances. Statistical signal processing basically refers to the analysis of random signals using appropriate statistical techniques.

The main aim of this monograph is to introduce different signal processing models which have been used in analyzing periodic data, and different statistical and computational issues associated with them. We observe periodic phenomena everyday in our lives. The daily temperature of Delhi or the number of tourists visiting the famous Taj Mahal everyday, or the ECG signal of a normal human being clearly follow periodic nature. Sometimes, the observations/signals may not be exactly periodic because of different reasons, but they may be nearly periodic. It should be clear from the following examples, where the observations are obtained from different disciplines that they are nearly periodic. In Fig. 1.1, we provide the ECG signal of a healthy person. In Fig. 1.2, we present an astronomical data set which represents the daily brightness of a variable star on 600 successive midnights. Figure 1.3 represents a classical data set of the monthly international airline passengers during January 1953–December 1960, and are collected from the Time Series Data Library http://www.robhyndman.info/TDSL.

The simplest periodic function is the sinusoidal function, and it can be written in the following form

$$y(t) = A \cos(\omega t) + B \sin(\omega t); \tag{1.1}$$

here, $A^2 + B^2$ is known as the amplitude of $y(t)$, and ω is the frequency. In general, a *smooth* mean zero periodic function $y(t)$ can always be written in the form:

© Springer Nature Singapore Pte Ltd. 2020
S. Nandi and D. Kundu, *Statistical Signal Processing*,
https://doi.org/10.1007/978-981-15-6280-8_1

Fig. 1.1 ECG signal of a healthy person

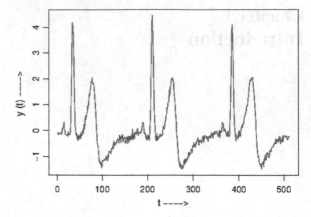

Fig. 1.2 Brightness of a variable star

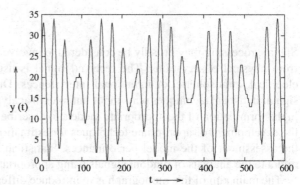

Fig. 1.3 Monthly international airline passengers during January 1953–December 1960

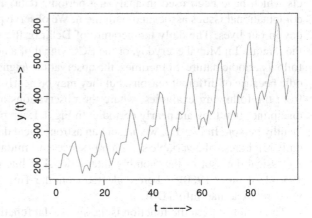

$$y(t) = \sum_{k=1}^{\infty} [A_k \cos(k\omega t) + B_k \sin(k\omega t)], \tag{1.2}$$

and it is well-known as the Fourier expansion of $y(t)$. From $y(t)$, A_k and B_k for $k = 1, \ldots, \infty$ can be obtained uniquely. But, in practice we hardly observe *smooth* $y(t)$. Most of the time, $y(t)$ is corrupted with noise as observed in the above three examples, hence it is quite natural to use the following model

$$y(t) = \sum_{k=1}^{\infty} [A_k \cos(k\omega t) + B_k \sin(k\omega t)] + X(t), \tag{1.3}$$

to analyze noisy periodic signals. Here $X(t)$ is the noise component. It is impossible to estimate infinite number of parameters in practice, hence (1.3) is approximated by

$$y(t) = \sum_{k=1}^{p} [A_k \cos(\omega_k t) + B_k \sin(\omega_k t)] + X(t), \tag{1.4}$$

for some $p < \infty$. Due to this reason, quite often, the main problem boils down to estimate p, and A_k, B_k, ω_k for $k = 1, \ldots, p$, from the observed signal $\{y(t); t = 1, \ldots, n\}$.

It should be mentioned that often instead of working with model (1.4), it might be more convenient to work with the associated complex-valued model. With the abuse of notation we use the corresponding complex-valued model as

$$y(t) = \sum_{k=1}^{p} \alpha_k e^{i\omega_k t} + X(t). \tag{1.5}$$

In model (1.5), $y(t)$s, α_ks, and $X(t)$s are all complex-valued, and $i = \sqrt{-1}$. Model (1.5) can be obtained by taking Hilbert transformation of (1.4). Therefore, the two models are equivalent. Although any observed signal is always real-valued, by taking the Hilbert transformation of the signal, the corresponding complex model can be used. Any analytical result or numerical procedure for model (1.5) can be used for model (1.4), and vice versa.

The problem of estimating the parameters of model (1.4) from the data $\{y(t); t = 1, \ldots, n\}$ becomes a classical problem. Starting with the work of Fisher [2], this problem has received considerable amount of attention because of its wide scale applicability. Brillinger [1] discussed some of the very important real-life applications from different areas and provided solutions using the sum of sinusoidal model. Interestingly, but not surprisingly, this model has been used quite extensively in the signal processing literature. Kay and Marple [4] wrote an excellent expository article from the signal processors point of views. More than three hundred list of references can be found in Stoica [9] on this particular problem till that time. See also two other review articles by Prasad, Chakraborty, and Parthasarathy [7] and Kundu [6] in this area. The monograph of Quinn and Hannan [8] is another important contribution in this area.

This problem has several different important and interesting aspects. Although model (1.4) can be viewed as a nonlinear regression model, this model does not satisfy the standard assumptions needed for different estimators to behave *nicely*. Therefore, deriving the properties of the estimators is an important problem. The usual consistency and asymptotic normality results do not follow from the general results. Moreover, finding estimators of these unknown parameters is well-known to be a numerically difficult problem. The problem becomes more complex if $p \geq 2$. Because of these reasons, this problem becomes a challenging problem both from the theoretical and computational point of view.

Model (1.4) is a nonlinear regression model, hence in the presence of independent and identically distributed (i.i.d.) error $\{X(t)\}$, the least squares method seems to be a reasonable choice for estimating the unknown parameters. But interestingly, instead of the least squares estimators (LSEs), traditionally the periodogram estimators (PEs) became more popular. The PEs of the frequencies can be obtained by finding the local maximums of the periodogram function $I(\omega)$, where

$$I(\omega) = \frac{1}{n} \left| \sum_{t=1}^{n} y(t) e^{-i\omega t} \right|^2.$$
(1.6)

Hannan [3] and Walker [10] independently first obtained the theoretical properties of the PEs. It is observed that the rate of convergence of the PEs of the frequencies are $O_p(n^{-3/2})$. Kundu [5] observed that the rate of convergence of the LSEs of the frequencies and amplitudes are $O_p(n^{-3/2})$ and $O_p(n^{-1/2})$, respectively. This unusual rate of convergence of the frequencies makes the model interesting from the theoretical point of view.

Finding the LSEs or the PEs is a computationally challenging problem. The problem is difficult because the least squares surface as well as the periodogram surface of the frequencies are highly nonlinear in nature. There are several local optimums in both the surfaces. Thus, very good (close enough to the true values) initial estimates are needed for any iterative process to work properly. It is also well-known that the standard methods like Newton–Raphson or Gauss–Newton do not work well for this problem. One of the common methods to find the initial guesses of the frequencies is to find the local maximums of the periodogram function $I(\omega)$, at the Fourier frequencies, that is, restricting the search space only at the discrete points $\{\omega_j = 2\pi j/n; \ j = 0, \ldots n - 1\}$. Asymptotically, the periodogram function has local maximums at the true frequencies. But, if two frequencies are very close to each other, then this method may not work properly.

Just to see the complexity of the problem, consider the periodogram function of the following synthesized signal:

$$y(t) = 3.0 \cos(0.20\pi t) + 3.0 \sin(0.20\pi t)$$
$$+ 0.25 \cos(0.19\pi t) + 0.25 \sin(0.19\pi t) + X(t); \ t = 1, \ldots, 75. \quad (1.7)$$

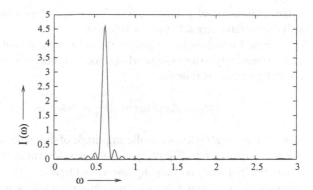

Fig. 1.4 Periodogram function of the synthesized data obtained from model (1.7)

Here $\{X(t); t = 1, \ldots, 75\}$ are i.i.d. normal random variables with mean 0 and variance 0.5. The periodogram function of the observed signal from model (1.7) is provided in Fig. 1.4. In this case clearly, two frequencies are not resolvable. It is not immediate how to choose initial estimates in this case to start any iterative process to find the LSEs or the PEs.

Because of this, several other techniques are available in practice, which attempted to find efficient estimators without using any iterative procedure. Most of the methods make use of the recurrence relation formulation (1.8) obtained from the celebrated Prony's equation (see Chap. 2):

$$\sum_{j=0}^{2p} c(j)y(t - j) = \sum_{j=0}^{2p} c(j)X(t - j); \quad t = 2p + 1, \ldots, n, \qquad (1.8)$$

where $c(0) = c(2p) = 1, c(j) = c(2p - j)$ for $j = 0, \ldots, 2p$. Relation (1.8) is formally equivalent to saying that a linear combination of p sinusoidal signals can be modeled as an ARMA$(2p, 2p)$ process. The coefficients $c(1), \ldots, c(2p - 1)$ depend only on the frequencies $\omega_1 \ldots, \omega_p$ and they can be obtained uniquely from $c(1), \ldots, c(2p - 1)$. Due to this relation, a variety of procedures have been evolved since 70s on estimating the coefficients $c(j)$ for $j = 1, \ldots, 2p - 1$ from the observed signal $\{y(t); t = 1, \ldots, n\}$. From the estimated $c(1), \ldots, c(2p - 1)$, the estimates of $\omega_1, \ldots, \omega_p$ can be easily obtained. Since all these methods are non-iterative in nature, they do not demand any initial guesses. But, the frequency estimates produced by these methods are mostly $O_p(n^{-1/2})$, not $O_p(n^{-3/2})$. Therefore, their efficiency is much lower than the LSEs or the PEs.

Another important problem is to estimate p, when it is unknown. Fisher [2] first treated this as a testing of the hypothesis problem. Later, several authors attempted different information theoretic criteria, namely AIC, BIC, EDC, etc. or their variants. But choosing the proper penalty function seems to be a really difficult problem. Cross-validation technique has also been used to estimate p. But computationally it is quite demanding, particularly if p is large, which may happen in practice quite often. Estimation of p for model (1.4) seems to be an open problem for which we still do not have any satisfactory solution.

A model which is very close to the sinusoidal model and which has been used quite extensively to analyze nearly periodic signals is known as the chirp model and it can be written as follows:

$$y(t) = A\cos(\alpha t + \beta t^2) + B\sin(\alpha t + \beta t^2) + X(t).$$

Here also $A^2 + B^2$ is known as the amplitude of the signal $y(t)$, α is the frequency, and β is the frequency rate. The chirp model can be seen as a frequency modulation model, where the frequency is changing linearly with time t. This model has been used quite extensively in several areas of engineering namely sonar, radar, communications, etc. A more general chirp model can be defined as follows:

$$y(t) = \sum_{k=1}^{p} \left\{ A_k \cos(\alpha_k t + \beta_k t^2) + B_k \sin(\alpha_k t + \beta_k t^2) \right\} + X(t). \tag{1.9}$$

The problem of estimating the unknown parameters of model (1.9) based on the observed signal $\{y(t); t = 1, \ldots, n\}$ is a challenging problem. This problem also has several important and interesting aspects. An extensive amount of work has taken place both in the Signal Processing and the Statistics literature in the last three decades dealing with theoretical and computational issues related to this model.

The main aim of this monograph is to provide a comprehensive review of different aspects of these problems mainly from statisticians' perspective, which is not available in the literature. It is observed that several related models are also available in the literature. We try to provide a brief account on those different models. Interestingly, natural two-dimensional (2-D) extension of this model has several applications in texture analysis and in spectrography. We provide a brief review on 2-D and three-dimensional (3-D) models also. For a better understanding of the different procedures discussed in this monograph, we present some real data analysis. Finally, we present some open and challenging problems in these areas.

Rest of the monograph is organized as follows. In Chap. 2, we provide the preliminaries. Different methods of estimation are discussed in Chap. 3. Theoretical properties of the different estimators are presented in Chap. 4. Different order estimation methods are reviewed in Chap. 5. Fundamental frequency model and generalized fundamental frequency model are discussed in Chap. 6. A few real data sets are analyzed in Chap. 7. Multidimensional models are introduced in Chap. 8. In Chap. 9, chirp signal model has been presented, and random amplitude sinusoidal and chirp models are provided in Chap. 10. Finally, we provide several related models in Chap. 11.

References

1. Brillinger, D. (1987). Fitting cosines: Some procedures and some physical examples. In I. B. MacNeill & G. J. Umphrey (Eds.), *Applied probability, stochastic process and sampling theory* (pp. 75–100). Dordrecht: Reidel.
2. Fisher, R. A. (1929). Tests of significance in harmonic analysis. *Proceedings of the Royal Society of London, Series A, 125*, 54–59.
3. Hannan, E. J. (1971). Non-linear time series regression. *Journal of Applied Probability, 8*, 767–780.
4. Kay, S. M., & Marple, S. L. (1981). Spectrum analysis - a modern perspective. *Proceedings of the IEEE, 69*, 1380–1419.
5. Kundu, D. (1997). Estimating the number of sinusoids in additive white noise. *Signal Processing, 56*, 103–110.
6. Kundu, D. (2002). Estimating parameters of sinusoidal frequency; some recent developments. *National Academy Science Letters, 25*, 53–73.
7. Prasad, S., Chakraborty, M., & Parthasarathy, H. (1995). The role of statistics in signal processing - a brief review and some emerging trends. *Indian Journal of Pure and Applied Mathematics, 26*, 547–578.
8. Quinn, B. G., & Hannan, E. J. (2001). *The estimation and tracking of frequency.* New York: Cambridge University Press.
9. Stoica, P. (1993). List of references on spectral analysis. *Signal Processing, 31*, 329–340.
10. Walker, A. M. (1971). On the estimation of a harmonic component in a time series with stationary independent residuals. *Biometrika, 58*, 21–36.

Chapter 2
Notations and Preliminaries

In this monograph, the scalar quantities are denoted by regular lower or uppercase letters. The lower and uppercase bold type faces of English alphabets are used for vectors and matrices, and for Greek alphabates, it should be clear from the context. For a real matrix, $\mathbf{A}$, $\mathbf{A}^T$ denotes its transpose. Similarly, for a complex matrix $\mathbf{A}$, $\mathbf{A}^H$ denotes its complex conjugate transpose. An $n \times n$ diagonal matrix, with diagonal elements, $\lambda_1, \ldots, \lambda_n$, is denoted by diag$\{\lambda_1, \ldots, \lambda_n\}$. If $\mathbf{A}$ is a real or complex square matrix, the projection matrix on the column space of $\mathbf{A}$ is denoted by $\mathbf{P}_\mathbf{A} = \mathbf{A}(\mathbf{A}^T\mathbf{A})^{-1}\mathbf{A}^T$ or $\mathbf{P}_\mathbf{A} = \mathbf{A}(\mathbf{A}^H\mathbf{A})^{-1}\mathbf{A}^H$. The following definition and matrix theory results may not be very familiar with the readers and, therefore, we are providing it for ease of reading. For more details, the readers are referred to Rao [21].

2.1 Matrix Theory-Related Results

Definition 2.1 An $n \times n$ matrix $\mathbf{J}$ is called a reflection or an exchange matrix if

$$\mathbf{J} = \begin{bmatrix} 0 & 0 & \ldots & 0 & 1 \\ 0 & 0 & \ldots & 1 & 0 \\ \vdots & \vdots & \ddots & \vdots & \vdots \\ 0 & 1 & \ldots & 0 & 0 \\ 1 & 0 & \ldots & 0 & 0 \end{bmatrix}. \tag{2.1}$$

© Springer Nature Singapore Pte Ltd. 2020
S. Nandi and D. Kundu, *Statistical Signal Processing*,
https://doi.org/10.1007/978-981-15-6280-8_2

2.1.1 Eigenvalues and Eigenvectors

Let $\mathbf{A}$ be an $n \times n$ matrix. The determinant $|\mathbf{A} - \lambda\mathbf{I}|$ is a polynomial in λ of degree n and is called the **characteristic polynomial** of $\mathbf{A}$. The equation

$$|\mathbf{A} - \lambda\mathbf{I}| = 0$$

is called the **characteristic equation** of $\mathbf{A}$. The roots of this equation are called the **eigenvalues** of $\mathbf{A}$. Corresponding to any root λ_i, there exists a non-null column vector $\mathbf{p}_i$, called the **eigenvector**, such that $\mathbf{A}\mathbf{p}_i = \lambda_i\mathbf{p}_i$, $\mathbf{p}_i^H\mathbf{p}_i = 1$.

2.1.2 Eigenvalues and Trace

1. When matrices are conformable, $\mathrm{tr}(\mathbf{A} + \mathbf{B}) = \mathrm{tr}(\mathbf{A}) + \mathrm{tr}(\mathbf{B})$ and $\mathrm{tr}(\mathbf{CD}) = \mathrm{tr}(\mathbf{DC})$.
2. Let λ_i, $i = 1, \ldots, n$, be eigenvalues of an $n \times n$ matrix $\mathbf{A}$, then

$$\mathrm{tr}(\mathbf{A}) = \sum_{i=1}^{n} \lambda_i, \qquad \det(\mathbf{A}) = \prod_{i=1}^{n} \lambda_i.$$

3. For a symmetric matrix $\mathbf{A}$, $\mathrm{tr}(\mathbf{A}^p) = \sum_{i=1}^{n} \lambda_i^p$, $p > 0$.
4. For a symmetric, non-singular matrix $\mathbf{A}$ of order $n \times n$ with eigenvalues $\lambda_1, \ldots, \lambda_n$, the eigenvalues of $\mathbf{A}^{-1}$ are λ_i^{-1}, $i = 1, \ldots, n$, and therefore, $\mathrm{tr}(\mathbf{A}^{-1}) = \sum_{i=1}^{n} \lambda_i^{-1}$.

2.1.3 Positive Semi-Definitive Matrix

Definition 2.2 A symmetric matrix is called **positive semi-definite** if and only if $\mathbf{x}^T\mathbf{A}\mathbf{x} \geq 0$ for all $\mathbf{x}$.

1. The eigenvalues of a positive semi-definite matrix are non-negative.
2. For a positive semi-definite matrix $\mathbf{A}$, $\mathrm{tr}(\mathbf{A}) \geq 0$.
3. An $n \times n$ matrix $\mathbf{A}$ of rank r is positive semi-definite if and only if there exists an $n \times n$ matrix $\mathbf{B}$ of rank r such that $\mathbf{A} = \mathbf{B}\mathbf{B}^T$.
4. If $\mathbf{A}$ is positive semi-definite, $\mathbf{X}^T\mathbf{A}\mathbf{X} = \mathbf{0} \Rightarrow \mathbf{A}\mathbf{X} = \mathbf{0}$.

2.1.4 Positive Definitive Matrix

Definition 2.3 A symmetric matrix is called **positive definite** if for all non-zero vectors $\mathbf{x}$, $\mathbf{x}^T \mathbf{A} \mathbf{x} > 0$. A positive definite matrix is also a positive semi-definite matrix.

1. The identity matrix $\mathbf{I}_n$ is clearly positive definite and so is a diagonal matrix with positive entries along the diagonal.
2. If $\mathbf{A}$ is positive definite, then it is non-singular.
3. The eigenvalues of a positive definite matrix are all positive.
4. $\mathbf{A}$ is positive definite if and only if there exists a non-singular matrix $\mathbf{B}$ such that $\mathbf{A} = \mathbf{B}\mathbf{B}^T$.
5. If $\mathbf{A}$ is positive definite, then any principal sub-matrix of $\mathbf{A}$ is also positive definite.
6. The diagonal elements of a positive definite matrix are all positive.
7. If $\mathbf{A}$ is positive definite, so is $\mathbf{A}^{-1}$.
8. If $\mathbf{A}$ is positive definite, then $\text{rank}(\mathbf{B}\mathbf{A}\mathbf{B}^T) = \text{rank}(\mathbf{B})$.
9. If $\mathbf{A}$ is an $n \times n$ positive definite matrix and $\mathbf{B}$ is a $p \times n$ matrix of rank p, then $\mathbf{B}\mathbf{A}\mathbf{B}^T$ is positive definite.
10. If $\mathbf{A}$ is $n \times p$ of rank p, then $\mathbf{A}^T\mathbf{A}$ is positive definite.
11. If $\mathbf{A}$ is positive definite, then for any vector $\mathbf{d}$,

$$\max_{\mathbf{h}:\mathbf{h}\neq 0} \left\{ \frac{(\mathbf{h}^T\mathbf{d})^2}{\mathbf{h}^T\mathbf{A}\mathbf{h}} \right\} = \mathbf{d}^T\mathbf{A}^{-1}\mathbf{d}.$$

2.1.5 Idempotent Matrix

Definition 2.4 A matrix $\mathbf{P}$ is **idempotent** if $\mathbf{P}^2 = \mathbf{P}$. A symmetric idempotent matrix is called a **projection matrix**.

1. If $\mathbf{P}$ is symmetric, then $\mathbf{P}$ is idempotent with rank r if and only if it has r eigenvalues equal to be 1 and $n - r$ eigenvalues equal to be 0.
2. If $\mathbf{P}$ is idempotent, so is $\mathbf{I} - \mathbf{P}$.
3. If $\mathbf{P}$ is a projection matrix, then $\text{tr}(\mathbf{P}) = \text{rank}(\mathbf{P})$.
4. With the exception of the identity matrix, an idempotent matrix is singular.
5. Projection matrices are positive semi-definite.
6. An idempotent matrix is always diagonalizable and has eigenvalues either 0 or 1.
7. The rank of an idempotent matrix is equal to its trace.
8. $\mathbf{P}^2 = \mathbf{P} \Leftrightarrow \text{rank}(\mathbf{P}) + \text{rank}(\mathbf{I} - \mathbf{P}) = n$, the order of the matrix.

2.1.6 Hermitian Matrix

Definition 2.5 A matrix $\mathbf{A}$ is called **Hermitian** if $\mathbf{A} = \mathbf{A}^H$, where $\mathbf{A}^H$ is the conjugate transpose of $\mathbf{A}$. Let a_{ij} be the (i, j)th element of $\mathbf{A}$, then $a_{ij} = \overline{a_{ji}}$, the complex conjugate of a_{ji}.

Definition 2.6 A matrix $\mathbf{U}$ is **unitary** if $\mathbf{U}^H \mathbf{U} = \mathbf{I}$.

1. If $\mathbf{A}$ is Hermitian, the eigenvalues of $\mathbf{A}$ are real-valued.
2. If $\mathbf{A}$ is Hermitian, there exists a unitary matrix $\mathbf{U}$ such that $\mathbf{A} = \mathbf{U}\mathbf{\Lambda}\mathbf{U}^H$, where $\mathbf{\Lambda}$ is a diagonal matrix of eigenvalues of $\mathbf{A}$.

2.1.7 Spectral Decomposition

Let $\mathbf{A}$ be a real symmetric $n \times n$ matrix. Then there exists an orthogonal matrix $\mathbf{P}$ such that

$$\mathbf{P}^T \mathbf{A} \mathbf{P} = \mathrm{diag}\{\lambda_1, \ldots, \lambda_n\} = \mathbf{\Delta}, \tag{2.2}$$

where $\lambda_1, \ldots, \lambda_n$ are eigenvalues of $\mathbf{A}$.

Since $\mathbf{P}$ is orthogonal, $\mathbf{P}^T = \mathbf{P}^{-1}$. Suppose that matrix $\mathbf{P}$ has columns $\mathbf{p}_1, \ldots, \mathbf{p}_n$. Then, from (2.2), we have

$$\mathbf{A}\mathbf{P} = \mathbf{P}\mathrm{diag}\{\lambda_1, \ldots, \lambda_n\}$$

and $\mathbf{A}\mathbf{p}_i = \lambda_i \mathbf{p}_i$. Therefore, (2.2) can also be written as

$$\mathbf{A} = \lambda_1 \mathbf{p}_1 \mathbf{p}_1^T + \cdots + \lambda_n \mathbf{p}_n \mathbf{p}_n^T = \mathbf{P}\mathbf{\Delta}\mathbf{P}^T$$

and is called **Spectral Decomposition** of $\mathbf{A}$.

If $\mathbf{A}$ is a Hermitian matrix, then all the eigenvalues of $\mathbf{A}$ are real and then similarly as above

$$\mathbf{A} = \lambda_1 \mathbf{p}_1 \mathbf{p}_1^H + \cdots + \lambda_n \mathbf{p}_n \mathbf{p}_n^H = \mathbf{P}\mathbf{\Delta}\mathbf{P}^H.$$

If $\lambda_i > 0$ for all i, then

$$\mathbf{A}^{-1} = \sum_{i=1}^{n} \frac{1}{\lambda_i} \mathbf{p}_i \mathbf{p}_i^T, \quad \text{or} \quad \mathbf{A}^{-1} = \sum_{i=1}^{n} \frac{1}{\lambda_i} \mathbf{p}_i \mathbf{p}_i^H.$$

2.1.8   Cholesky Decomposition

The Cholesky decomposition of a Hermitian positive-definite matrix $\mathbf{A}$ (i.e. $\mathbf{x}^H \mathbf{A} \mathbf{x} > 0$ for $\mathbf{x} \neq 0$) is a decomposition of the form

$$\mathbf{A} = \mathbf{L}\mathbf{L}^H,$$

where $\mathbf{L}$ is a lower triangular matrix with real and positive diagonal entries and $\mathbf{L}^H$ denotes the conjugate transpose of $\mathbf{L}$. Every Hermitian positive definite matrix has a unique Cholesky decomposition and hence every real symmetric positive definite matrix has a unique Cholesky decomposition. The Cholesky decomposition can also be expressed in terms of a unique upper triangular matrix of real and positive diagonal entries.

2.1.9   Singular Value Decomposition

If $\mathbf{A}$ is an $n \times m$ real or complex matrix of rank k, then there exists an $n \times n$ orthogonal matrix $\mathbf{U}$, an $m \times m$ orthogonal matrix $\mathbf{V}$, and an $n \times m$ matrix $\boldsymbol{\Sigma}$, such that

$$\mathbf{A} = \mathbf{U}\boldsymbol{\Sigma}\mathbf{V}, \tag{2.3}$$

where $\boldsymbol{\Sigma}$ is defined as

$$\boldsymbol{\Sigma} = \begin{bmatrix} \mathbf{S} & \mathbf{0} \\ \mathbf{0} & \mathbf{0} \end{bmatrix}, \quad \mathbf{S} = \mathrm{diag}\{\sigma_1^2, \ldots, \sigma_k^2\},$$

and $\sigma_1^2 \geq \cdots \geq \sigma_k^2 > 0$ are k non-zero eigenvalues of $\mathbf{A}^T\mathbf{A}$ or $\mathbf{A}^H\mathbf{A}$ depending on whether $\mathbf{A}$ is a real or a complex matrix.

2.1.10   Vector Differentiation

Let $\boldsymbol{\beta} = (\beta_1, \beta_2, \ldots, \beta_p)^T$ be a $p \times 1$ column vector, and $Q(\boldsymbol{\beta})$ be a real-valued function of $\boldsymbol{\beta}$. Then, the first and second order derivatives of $Q(\boldsymbol{\beta})$ are the following $p \times 1$ vector and $p \times p$ matrix, respectively,

$$\frac{dQ(\boldsymbol{\beta})}{d\boldsymbol{\beta}} = \left(\frac{\partial Q(\boldsymbol{\beta})}{\partial \beta_1}, \frac{\partial Q(\boldsymbol{\beta})}{\partial \beta_1}, \cdots, \frac{\partial Q(\boldsymbol{\beta})}{\partial \beta_p} \right)^T,$$

$$\frac{d^2 Q(\boldsymbol{\beta})}{d\boldsymbol{\beta} d\boldsymbol{\beta}^T} = \begin{pmatrix} \frac{\partial^2 Q(\boldsymbol{\beta})}{\partial \beta_1^2} & \frac{\partial^2 Q(\boldsymbol{\beta})}{\partial \beta_1 \partial \beta_2} & \cdots & \frac{\partial^2 Q(\boldsymbol{\beta})}{\partial \beta_1 \partial \beta_p} \\ \frac{\partial^2 Q(\boldsymbol{\beta})}{\partial \beta_1 \partial \beta_2} & \frac{\partial^2 Q(\boldsymbol{\beta})}{\partial \beta_2^2} & \cdots & \frac{\partial^2 Q(\boldsymbol{\beta})}{\partial \beta_2 \partial \beta_p} \\ \vdots & \vdots & \vdots & \vdots \\ \frac{\partial^2 Q(\boldsymbol{\beta})}{\partial \beta_1 \partial \beta_p} & \frac{\partial^2 Q(\boldsymbol{\beta})}{\partial \beta_2 \partial \beta_p} & \cdots & \frac{\partial^2 Q(\boldsymbol{\beta})}{\partial \beta_p^2} \end{pmatrix}.$$

1. For a row vector $\mathbf{c}$ of order p, $\dfrac{d(\boldsymbol{\beta}^T \mathbf{c})}{d\boldsymbol{\beta}} = \mathbf{c}$.

2. For a symmetric $p \times p$ matrix $\mathbf{C}$, $\dfrac{d(\boldsymbol{\beta}^T \mathbf{C}\boldsymbol{\beta})}{d\boldsymbol{\beta}} = 2\mathbf{C}\boldsymbol{\beta}$.

2.1.11   Matrix Inversion Formulas for Patterned Matrices

1. When both $\mathbf{A}$ and $\mathbf{D}$ are symmetric matrices such that all the inverses exist, then

$$\begin{pmatrix} \mathbf{A} & \mathbf{B} \\ \mathbf{B}^T & \mathbf{D} \end{pmatrix} = \begin{pmatrix} \mathbf{A}^{-1} + \mathbf{F}\mathbf{E}^{-1}\mathbf{E}^T & -\mathbf{F}\mathbf{E}^{-1} \\ -\mathbf{E}^{-1}\mathbf{F}^T & \mathbf{E}^{-1} \end{pmatrix}$$

$$= \begin{pmatrix} \mathbf{G}^{-1} & -\mathbf{G}^{-1}\mathbf{H} \\ -\mathbf{H}^T\mathbf{G}^{-1} & \mathbf{D}^{-1} + \mathbf{H}^T\mathbf{G}^{-1}\mathbf{H} \end{pmatrix},$$

where $\mathbf{E} = \mathbf{D} - \mathbf{B}^T\mathbf{A}^{-1}\mathbf{B}$, $\mathbf{F} = \mathbf{A}^{-1}\mathbf{B}$, $\mathbf{G} = \mathbf{A} - \mathbf{B}\mathbf{D}^{-1}\mathbf{B}^T$, and $\mathbf{H} = \mathbf{B}\mathbf{D}^{-1}$.

2. Let $\mathbf{A}$ be a non-singular matrix, and $\mathbf{u}$ and $\mathbf{v}$ are two column vectors. Then

$$(\mathbf{A} + \mathbf{u}\mathbf{v}^T)^{-1} = \mathbf{A}^{-1} - \frac{\mathbf{A}^{-1}\mathbf{u}\mathbf{v}^T\mathbf{A}^{-1}}{1 + \mathbf{v}^T\mathbf{A}^{-1}\mathbf{u}},$$

$$(\mathbf{A} - \mathbf{u}\mathbf{v}^T)^{-1} = \mathbf{A}^{-1} + \frac{\mathbf{A}^{-1}\mathbf{u}\mathbf{v}^T\mathbf{A}^{-1}}{1 - \mathbf{v}^T\mathbf{A}^{-1}\mathbf{u}}.$$

3. Let $\mathbf{A}$ and $\mathbf{B}$ be non-singular matrices of order $m \times m$ and $n \times n$, respectively, and $\mathbf{U}$ be $m \times n$ and $\mathbf{V}$ be $n \times m$ matrices. Then,

$$(\mathbf{A} + \mathbf{U}\mathbf{B}\mathbf{V})^{-1} = \mathbf{A}^{-1} - \mathbf{A}^{-1}\mathbf{U}\mathbf{B}(\mathbf{B} + \mathbf{B}\mathbf{V}\mathbf{A}^{-1}\mathbf{U}\mathbf{B})^{-1}\mathbf{B}\mathbf{V}\mathbf{A}^{-1}.$$

With $\mathbf{B} = \mathbf{I}_n$, $\mathbf{U} = \pm\mathbf{u}$, and $\mathbf{V} = \mathbf{v}^T$, the above identity reduces to expressions in 2.

4. Let $\mathbf{A}$ be an $n \times p$ matrix of rank p, and $\mathbf{a}_{(j)}$ be the jth column of $\mathbf{A}$, $j = 1, \ldots, p$. Then, (j, j) element of $(\mathbf{A}^T\mathbf{A})^{-1}$ is

$$(\mathbf{A}^T\mathbf{A})_{jj}^{-1} = [\mathbf{a}_{(j)}(\mathbf{I}_n - \mathbf{P}_j)\mathbf{a}_{(j)}]^{-1},$$

where $\mathbf{P}_j = \mathbf{A}^{(j)}(\mathbf{A}^{(j)^T}\mathbf{A}^{(j)})^{-1}\mathbf{A}^{(j)^T}$ and $\mathbf{A}^{(j)}$ is $\mathbf{A}$ with its jth column omitted.

2.2 Linear and Separable Regression

In this section, we briefly discuss some basic results in linear regression and some related topics, which we are going to use in this book. For more details, interested readers are referred to the textbook by Seber and Lee [19]. The multiple linear regression model with p $(n \gg p)$ covariates is given by

$$y_i = \beta_0 + \beta_1 x_{i1} + \cdots + \beta_{p-1} x_{i,p-1} + \varepsilon_i, \quad i = 1, \ldots, n, \tag{2.4}$$

where $y_1, \ldots, y_n$ are observed values of the response variable y; x_{ij} is the ith value of the jth covariate x_j; $\beta_j, j = 0, 1, \ldots, p-1$ are unknown parameters to be estimated; y_i is measured with ε_i error. In matrix notation,

$$\begin{pmatrix} y_1 \\ y_2 \\ \vdots \\ y_n \end{pmatrix} = \begin{pmatrix} 1 & x_{11} & x_{12} & \cdots & x_{1,p-1} \\ 1 & x_{21} & x_{22} & \cdots & x_{2,p-1} \\ \vdots & \vdots & \vdots & \vdots & \vdots \\ 1 & x_{n1} & x_{n2} & \cdots & x_{n,p-1} \end{pmatrix} \begin{pmatrix} \beta_0 \\ \beta_1 \\ \vdots \\ \beta_{p-1} \end{pmatrix} + \begin{pmatrix} \varepsilon_1 \\ \varepsilon_2 \\ \vdots \\ \varepsilon_n \end{pmatrix}$$

or

$$\mathbf{Y} = \mathbf{X}\boldsymbol{\beta} + \boldsymbol{\varepsilon}.$$

Here, $\mathbf{Y}$ is called the response vector, the $n \times p$ matrix $\mathbf{X}$ is the regression matrix, $\boldsymbol{\beta}$ is the parameter vector, and $\boldsymbol{\varepsilon}$ is the error vector. In general, the columns are chosen in a way that the columns are linearly independent. Therefore, $\mathbf{X}$ is a full rank matrix and rank$(\mathbf{X}) = p$.

Assume that $\mathbf{X}$ has rank p and $\varepsilon_1, \ldots, \varepsilon_n$ are uncorrelated random variables with mean zero and finite variance σ^2. So, $E(\boldsymbol{\varepsilon}) = \mathbf{0}$ and $\text{Var}(\boldsymbol{\varepsilon}) = \sigma^2 \mathbf{I}_n$. The least squares estimators of $\beta_0, \beta_1, \ldots, \beta_{p-1}$ minimizes

$$\sum_{i=1}^{n} \varepsilon_i^2 = \sum_{i=1}^{n} \left[y_i - \beta_0 - \beta_1 x_{i1} - \cdots - \beta_{p-1} x_{i,p-1} \right]^2$$

$$\Rightarrow \boldsymbol{\varepsilon}^T \boldsymbol{\varepsilon} = (\mathbf{Y} - \mathbf{X}\boldsymbol{\beta})^T (\mathbf{Y} - \mathbf{X}\boldsymbol{\beta})$$

$$= \mathbf{Y}^T \mathbf{Y} - 2\boldsymbol{\beta}^T \mathbf{X}^T \mathbf{Y} + \boldsymbol{\beta}^T \mathbf{X}^T \mathbf{X}\boldsymbol{\beta}.$$

Differentiating $\boldsymbol{\varepsilon}^T \boldsymbol{\varepsilon}$ with respect to $\boldsymbol{\beta}$ and using the fact $\boldsymbol{\beta}^T \mathbf{X}^T \mathbf{Y} = \mathbf{Y}^T \mathbf{X}\boldsymbol{\beta}$ in $\dfrac{\partial \boldsymbol{\varepsilon}^T \boldsymbol{\varepsilon}}{\partial \boldsymbol{\beta}} = \mathbf{0}$, we have

$$-2\mathbf{X}^T \mathbf{Y} + 2\mathbf{X}^T \mathbf{X}\boldsymbol{\beta} = \mathbf{0}$$

$$\Rightarrow \mathbf{X}^T \mathbf{X}\boldsymbol{\beta} = \mathbf{X}^T \mathbf{Y}.$$

The regression matrix $\mathbf{X}$ has rank p, hence $\mathbf{X}^T\mathbf{X}$ is positive definite and non-singular. Therefore, the above matrix equation has a unique solution and is

$$\widehat{\boldsymbol{\beta}} = (\mathbf{X}^T\mathbf{X})^{-1}\mathbf{X}^T\mathbf{Y}.$$

The following identity confirms that $\widehat{\boldsymbol{\beta}}$ is the minimum.

$$(\mathbf{Y} - \mathbf{X}\boldsymbol{\beta})^T(\mathbf{Y} - \mathbf{X}\boldsymbol{\beta}) = (\mathbf{Y} - \mathbf{X}\widehat{\boldsymbol{\beta}})^T(\mathbf{Y} - \mathbf{X}\widehat{\boldsymbol{\beta}}) + (\widehat{\boldsymbol{\beta}} - \boldsymbol{\beta})^T\mathbf{X}^T\mathbf{X}(\widehat{\boldsymbol{\beta}} - \boldsymbol{\beta}).$$

The following quantities are important in linear regression and its diagnostics.

- **Fitted values:** $\widehat{\mathbf{Y}} = \mathbf{X}\widehat{\boldsymbol{\beta}}$.
- **Residuals:** $\widehat{\boldsymbol{\varepsilon}} = \mathbf{Y} - \widehat{\mathbf{Y}} = \mathbf{Y} - \mathbf{X}\widehat{\boldsymbol{\beta}} = (\mathbf{I}_n - \mathbf{P}_{\mathbf{X}})\mathbf{Y}$ where $\mathbf{P}_{\mathbf{X}} = \mathbf{X}(\mathbf{X}^T\mathbf{X})^{-1}\mathbf{X}^T$ is the projection matrix on the column space of $\mathbf{X}$.
- **Residual sum of squares (RSS):**

$$\begin{aligned}
\widehat{\boldsymbol{\varepsilon}}^T\widehat{\boldsymbol{\varepsilon}} &= (\mathbf{Y} - \mathbf{X}\widehat{\boldsymbol{\beta}})^T(\mathbf{Y} - \mathbf{X}\widehat{\boldsymbol{\beta}}) \\
&= \mathbf{Y}^T\mathbf{Y} - \widehat{\boldsymbol{\beta}}^T\mathbf{X}^T\mathbf{Y} \\
&= \mathbf{Y}^T\mathbf{Y} - \widehat{\boldsymbol{\beta}}^T\mathbf{X}^T\mathbf{X}\widehat{\boldsymbol{\beta}}.
\end{aligned}$$

Properties of the least squares estimators:

Under assumptions $E[\boldsymbol{\varepsilon}] = \mathbf{0}$ and $\mathrm{Var}(\boldsymbol{\varepsilon}) = \sigma^2\mathbf{I}_n$, it can be shown that

$$E(\widehat{\boldsymbol{\beta}}) = \boldsymbol{\beta}, \quad \mathrm{Var}(\widehat{\boldsymbol{\beta}}) = \sigma^2(\mathbf{X}^T\mathbf{X})^{-1}.$$

Estimation of σ^2:

If $\mathbf{Y} = \mathbf{X}\boldsymbol{\beta} + \boldsymbol{\varepsilon}$, where $n \times p$ matrix $\mathbf{X}$ is a full rank matrix of rank p and $E[\boldsymbol{\varepsilon}] = \mathbf{0}$, $\mathrm{Var}(\boldsymbol{\varepsilon}) = \sigma^2\mathbf{I}_n$, then

$$s^2 = \frac{(\mathbf{Y} - \mathbf{X}\widehat{\boldsymbol{\beta}})^T(\mathbf{Y} - \mathbf{X}\widehat{\boldsymbol{\beta}})}{n - p} = \frac{\mathrm{RSS}}{n - p}$$

is an unbiased estimator of σ^2.

Distribution under normality:

Suppose $\mathbf{Y} \sim \mathcal{N}_n(\mathbf{X}\boldsymbol{\beta}, \sigma^2\mathbf{I}_n)$ with full rank matrix $\mathbf{X}$ of order $n \times p$. Then

1. $\widehat{\boldsymbol{\beta}} \sim \mathcal{N}_p(\boldsymbol{\beta}, \sigma^2(\mathbf{X}^T\mathbf{X})^{-1})$;
2. $\widehat{\boldsymbol{\beta}}$ is independent of s^2;
3. $\frac{(n-p)s^2}{\sigma^2} \sim \chi^2_{n-p}$;
4. $(\widehat{\boldsymbol{\beta}} - \boldsymbol{\beta})^T\mathbf{X}^T\mathbf{X}(\widehat{\boldsymbol{\beta}} - \boldsymbol{\beta}) \sim \chi^2_p$.

2.2.1 Separable Regression Technique

The separable regression technique, originally introduced by Richards [22], is applicable to nonlinear models of the following form where some parameters are linear.

$$y_t = \sum_{j=1}^{p} A_j f_j(\theta, t) + e_t, \quad t = 1, \ldots, n, \tag{2.5}$$

where $A_1, \ldots, A_p$ and θ are unknown parameters; θ can be vector-valued; $f_j(\theta, t)$s are known nonlinear functions of θ; and $\{e_t\}$ is the sequence of error random variables. Here, $A_1, \ldots, A_p$ are linear parameters whereas θ is a nonlinear parameter. In order to find the LSEs of $A_1, \ldots, A_p$ and θ, write model (2.5) in matrix notation

$$\begin{pmatrix} y_1 \\ y_2 \\ \vdots \\ y_n \end{pmatrix} = \begin{pmatrix} f_1(\theta, 1) & f_2(\theta, 1) & \cdots & f_p(\theta, 1) \\ f_1(\theta, 2) & f_2(\theta, 2) & \cdots & f_p(\theta, 2) \\ \vdots & \vdots & \vdots & \vdots \\ f_1(\theta, n) & f_2(\theta, n) & \cdots & f_p(\theta, n) \end{pmatrix} \begin{pmatrix} A_1 \\ A_2 \\ \vdots \\ A_p \end{pmatrix} + \begin{pmatrix} e_1 \\ e_2 \\ \vdots \\ e_n \end{pmatrix}$$

or

$$\mathbf{Y} = \mathbf{X}(\theta)v + \mathbf{e}.$$

Then LSEs of $v = (A_1, \ldots, A_p)^T$ and θ minimize

$$Q(v, \theta) = (\mathbf{Y} - \mathbf{X}(\theta)v)^T (\mathbf{Y} - \mathbf{X}(\theta)v) \tag{2.6}$$

with respect to v and θ. For a fixed θ, $Q(v, \theta)$ is minimized with respect to v at

$$\widehat{v}(\theta) = (\mathbf{X}^T(\theta)\mathbf{X}(\theta))^{-1}\mathbf{X}^T(\theta)\mathbf{Y}.$$

Replacing v by $\widehat{v}(\theta)$ in (2.6), we have

$$\begin{aligned} Q(\widehat{v}(\theta), \theta) &= (\mathbf{Y} - \mathbf{X}(\theta)(\mathbf{X}^T(\theta)\mathbf{X}(\theta))^{-1}\mathbf{X}^T(\theta)\mathbf{Y})^T (\mathbf{Y} - \mathbf{X}(\theta)(\mathbf{X}^T(\theta)\mathbf{X}(\theta))^{-1}\mathbf{X}^T(\theta)\mathbf{Y}) \\ &= \mathbf{Y}^T (\mathbf{I} - P_{\mathbf{X}(\theta)})^T (\mathbf{I} - P_{\mathbf{X}(\theta)})\mathbf{Y} \\ &= \mathbf{Y}^T (\mathbf{I} - P_{\mathbf{X}(\theta)})\mathbf{Y} \\ &= \mathbf{Y}^T \mathbf{Y} - \mathbf{Y}^T P_{\mathbf{X}(\theta)}\mathbf{Y}. \end{aligned}$$

Here, $P_{\mathbf{X}(\theta)} = \mathbf{X}(\theta)(\mathbf{X}^T(\theta)\mathbf{X}(\theta))^{-1}\mathbf{X}^T(\theta)$ is the projection matrix on the column space of $\mathbf{X}(\theta)$ and so is symmetric and idempotent.

Therefore, $Q(\widehat{v}(\theta), \theta)$ is a function of θ only. The LSEs are obtained in two steps. First minimize $Q(\widehat{v}(\theta), \theta)$ with respect to θ, denote the minimizer as $\widehat{\theta}$ and then estimate the linear parameters $A_1, \ldots, A_p$ using $\widehat{v}(\widehat{\theta})$.

Separable regression technique separates the linear parameters from the nonlinear parameters in the least squares estimation method. The implementation is done in two steps. This simplifies the estimation method in many complicated nonlinear models.

2.3 Nonlinear Regression Models: Results of Jennrich

Nonlinear regression plays an important role in Statistical Signal Processing. In this section, we present some of the basic theoretical results on nonlinear regression analysis established by Jennrich [8], see also Wu [25] or Kundu [11] in this respect. For different aspects of nonlinear regression models, one is referred to the textbooks by Bates and Watts [1] or Seber and Wild [23].

Consider a general nonlinear regression model of the following form.

$$y_t = f_t(\theta^0) + \varepsilon_t, \quad t = 1, 2, \ldots, n, \quad (2.7)$$

where $f_t(\theta)$ is a known real-valued continuous function on a compact subset Θ of a Euclidean space and $\{\varepsilon_t\}$ is a sequence of i.i.d. random variables with zero mean and variance $\sigma^2 > 0$. The values of θ^0 and σ^2 are unknown.

Any $\widehat{\theta} \in \Theta$ that minimizes

$$Q(\theta) = \frac{1}{n} \sum_{t=1}^{n} (y_t - f_t(\theta))^2$$

is called the least squares estimator of θ^0 based on n observations of $\{y_t\}_{t=1}^{n}$.

Definition 2.7 (*Tail Product and Tail Cross Product*) Let $\{x_t\}_{t=1}^{\infty}$ and $\{y_t\}_{t=1}^{\infty}$ be two sequences of real numbers and let

$$(x, y)_n = \frac{1}{n} \sum_{t=1}^{n} x_t y_t.$$

If $\lim_{n \to \infty} (x, y)_n$ exists, its limit (x, y) is called the tail product of x and y. Let g and h be two sequences of real-valued functions on Θ such that $(g(\alpha), h(\beta))_n$ converges uniformly to $(g(\alpha), h(\beta))$ for all $\alpha, \beta \in \Theta$. If $< g, h >$ denotes the function on $\Theta \times \Theta$ which takes (α, β) to $(g(\alpha), h(\beta))$, then this function is called the tail cross product of g and h. If, in addition, the components of g and h are continuous, then $< g, h >$, as a uniform limit of continuous functions, is also continuous.

Lemma 2.1 *If $f : A \times B \to I\!R$ is a continuous function, where $A \subseteq I\!R$ and $B \subseteq I\!R$ are closed, if B_1 is a bounded subset of B, then $\sup_{y \in B_1} f(x, y)$ is a continuous function of x.* ∎

Theorem 2.1 *Let $\mathscr{E}$ be a Euclidean space and $\Theta \subset \mathbb{R}$ be a compact set. Let f : $\mathscr{E} \times \Theta \to \mathbb{R}$ where f is bounded and continuous. If $F_1, F_2, \ldots$ are distribution functions on $\mathscr{E}$ which converge in distribution to F, then*

$$\int f(x, \theta)\,dF_n(x) \longrightarrow \int f(x, \theta)\,dF(x)$$

uniformly for all $\theta \in \Theta$.   ∎

Theorem 2.2 *Let $\mathscr{E}$ be a Euclidean space and $\Theta \subseteq \mathbb{R}^p$ where Θ is compact. Let $f : \mathscr{E} \times \Theta \to \mathbb{R}$ where for each x, f is a continuous function of θ, and for each θ, f is a measurable function of x. Assume that $|f(x, \theta)| \le g(x)$ for all x and θ, where g is bounded, continuous and $\int g(x)\,dF(x) < \infty$ where F is some distribution function on $\mathscr{E}$. If $x_1, x_2, \ldots$ is a random sample from F, then for almost every $x = (x_1, x_2, \ldots)$*

$$\frac{1}{n} \sum_{t=1}^{n} f(x_t, \theta) \longrightarrow \int f(x, \theta)\,dF(x)$$

uniformly, for all $\theta \in \Theta$.   ∎

2.3.1 Existence of Least Squares Estimators

Lemma 2.2 *Let $Q : \Theta \times Y \to \mathbb{R}$ where Θ is a countable subset of a Euclidean space $\mathbb{R}^p$, and Y is a measurable space. If for fixed $\theta \in \Theta$, Q is a measurable function of y and for each $y \in Y$, it is finite for each $\theta \in \Theta$, then there exists a measurable function $\widehat{\theta}$ from Y into Θ, such that for all $y \in Y$*

$$Q(\widehat{\theta}(y), y) = \inf_{\theta \in \Theta} Q(\theta, y).$$

■

Lemma 2.3 *Let $Q : \Theta \times Y \to \mathbb{R}$ where Θ is a compact subset of a Euclidean space $\mathbb{R}^p$, and Y is a measurable space. If for fixed $\theta \in \Theta$, Q is a measurable function of y and for each $y \in Y$, it is a continuous function of θ, then there exists a measurable function $\widehat{\theta}$ from Y into Θ, such that for all $y \in Y$*

$$Q(\widehat{\theta}(y), y) = \inf_{\theta \in \Theta} Q(\theta, y).$$

■

2.3.2 Consistency and Asymptotic Normality of the Least Squares Estimators

In this subsection, the sufficient conditions given by Jennrich [8] are stated under which the least squares estimators will be strongly consistent and asymptotically normal.

Assumption 2.1 The sequence of additive error $\{\varepsilon_t\}$, $t = 1, 2, \ldots$ is a sequence of i.i.d. real-valued random variables with $E(\varepsilon_t) = 0$ and $\text{Var}(\varepsilon_t) = \sigma^2$.

Assumption 2.2 The tail cross product of the function $f_t(\theta) = f(x_t, \theta)$ with itself exists, and

$$\lim_{n \to \infty} \frac{1}{n} \sum_{t=1}^{n} [f_t(\theta) - f_t(\theta^0)]^2 = Q(\theta)$$

has a unique minimum at $\theta = \theta^0$, where θ^0 is an interior point of Θ.
Note that under the assumption of the existence of the tail cross product of a sequence of functions $\{f_t(\theta)\}$, the above limit exists.

Lemma 2.4 *Let* $\{\varepsilon_t\}$, $t = 1, 2, \ldots$ *be a sequence of real-valued random variables that satisfies the conditions of Assumption 2.1 and suppose that* $\{x_t\}$ *is a sequence of real numbers whose tail product with itself exists. Then* $\lim_{n \to \infty} \frac{1}{n} \sum_{i=1}^{n} x_i \varepsilon_i = 0$ *for almost every sequence* $\{\varepsilon_t\}$. ∎

Theorem 2.3 *Let* $\{\varepsilon_t\}$, $t = 1, 2, \ldots$ *be a sequence of real-valued random variables that satisfies the conditions of Assumption 2.1 and* $\{g_t\}$, $t = 1, 2, \ldots$ *be a sequence of real-valued continuous functions on* Θ, *where* Θ *is a compact subset of* $\mathbb{R}^p$. *If the tail cross product of* $\{g_t\}$ *with itself exists, then for almost every sequence* $\{\varepsilon_t\}$,

$$\frac{1}{n} \sum_{t=1}^{n} g_t(\theta)\varepsilon_t \longrightarrow 0, \quad \forall \theta \in \Theta.$$

∎

Theorem 2.4 *Let* $\{\widehat{\theta}_n\}$ *be a sequence of least square estimators of model (2.7). Let* $\{\varepsilon_t\}$ *and* $\{f_t\}$ *satisfy Assumptions 2.1 and 2.2, respectively. Then,* $\widehat{\theta}_n$ *and* $\widehat{\sigma}_n^2 = Q(\widehat{\theta}_n)$ *are strongly consistent estimators of* θ^0 *and* σ^2, *respectively.* ∎

In order to establish the asymptotic normality of a sequence of least squares estimators, $\widehat{\theta}_n$, we need the derivatives,

$$f'_{ti}(\theta) = \frac{\partial}{\partial \theta_i} f_t(\theta), \quad f''_{tij}(\theta) = \frac{\partial^2}{\partial \theta_i \partial \theta_j} f_t(\theta).$$

Let $f' = (f'_{ti})$ and $f'' = (f''_{tij})$ and consider the following assumptions.

Assumption 2.3 The derivatives f'_{ti} and f''_{tij} exist for $i, j = 1, \ldots, p$ and are continuous on Θ. All possible tail cross products between f_t, f'_{ti}, and f''_{tij} exist. For each $\theta \in \Theta$, let

$$a_{nij}(\theta) = \frac{1}{n} \sum_{t=1}^{n} f'_{ti} f'_{tj}, \quad i, j = 1, 2, \ldots, p.$$

Consider the matrix $\mathbf{A}_n(\theta) = (a_{nij}(\theta))$, $i, j = 1, 2, \ldots, p$. Let $a_{ij}(\theta) = \lim_{n \to \infty} a_{nij}(\theta)$ and $\mathbf{A}(\theta) = \lim_{n \to \infty} \mathbf{A}_n(\theta)$.

Assumption 2.4 The matrix $\mathbf{A}(\theta^0)$ is non-singular.

Theorem 2.5 *Let* $\{\widehat{\theta}_n\}$ *be a sequence of least squares estimators. Under Assumptions 2.1–2.4,*

$$\sqrt{n}(\widehat{\theta}_n - \theta^0) \xrightarrow{d} \mathcal{N}_p(0, \mathbf{A}^{-1}(\theta^0)).$$

■

2.4 Numerical Algorithms

Different numerical algorithms have been used quite extensively in developing efficient estimation procedures for signal processing models. We have presented a few of these methods which have been used in this monograph. More details are available in books like Ortega and Rheinboldt [16] or Kundu and Basu [12].

2.4.1 Iterative Method

In order to find a root of the equation $f(x) = 0$, it is usually not possible to find a zero ξ explicitly within a finite number of steps and we need to use approximation methods. These methods are mostly iterative and have the following form. Starting with an initial value x_0, successive approximates x_i, $i = 1, 2, \ldots$ to ξ are computed as

$$x_{i+1} = h(x_i), \quad i = 1, 2, \ldots$$

where $h(\cdot)$ is an iteration function. The representation $x = h(x)$ from $f(x) = 0$ may not be a unique representation, and there may be many ways of doing this. For example, the equation

$$x^3 + x - 1 = 0$$

can be expressed as

$$x = 1 - x^3, \quad x = (1 - x)^{\frac{1}{3}}, \quad x = (1 + x^2)^{-1}, \ldots$$

If ξ is a fixed point of $h(h(\xi) = \xi)$, if all fixed points are also zeros of f, and if h is continuous in a neighborhood of each of its fixed points of h, then each limit point of the sequence $x_i, i = 1, 2, \ldots$ is a fixed point of h and hence a zero of f (Steor and Bulirsch [20]).

Next, the question is how to find a suitable iteration function. Systematically, iteration functions h can be found as follows: If ξ is a zero of a function $f : \mathbb{R} \rightarrow \mathbb{R}$, and f is sufficiently differentiable in a neighborhood $\mathcal{N}(\xi)$ of this point, then the Taylor series expansion of f about $x_0 \in \mathcal{N}(\xi)$ is

$$f(\xi) = f(x_0) + (\xi - x_0)f'(x_0) + \frac{1}{2!}(\xi - x_0)^2 f''(x_0) + \cdots +$$
$$\frac{1}{k!}(\xi - x_0)^k f^{(k)}(x_0 + \lambda(\xi - x_0)), \quad 0 < \lambda < 1.$$

If the higher order terms are ignored, we have equations which express the point ξ approximately in terms of a given nearby point x_0. For example,

$$f(\xi) = f(x_0) + (\xi - x_0)f'(x_0) = 0, \tag{2.8}$$

$$f(\xi) = f(x_0) + (\xi - x_0)f'(x_0) + \frac{1}{2!}(\xi - x_0)^2 f''(x_0) = 0, \tag{2.9}$$

which produce the approximations

$$\xi_{1*} = x_0 - \frac{f(x_0)}{f'(x_0)}, \quad \xi_{2*} = x_0 - \frac{f'(x_0) \pm \sqrt{(f'(x_0))^2 - 2f(x_0)f''(x_0)}}{f''(x_0)}.$$

In general, ξ_{1*} and ξ_{2*} must be corrected further by the scheme from which they were themselves derived. Therefore, we have the iteration methods

$$x_{i+1} = h_1(x_i), \quad h_1(x) = x - \frac{f(x)}{f'(x)} \tag{2.10}$$

$$x_{i+1} = h_2(x_i), \quad h_2(x) = x - \frac{f'(x) \pm \sqrt{(f'(x))^2 - 2f(x)f''(x)}}{f''(x)}. \tag{2.11}$$

The first one given in (2.10) is the classical **Newton–Raphson** method. The one given in (2.11) is an obvious extension.

Now one important question is whether the sequence of approximations x_0, $x_1, \ldots x_n$ always converges to some number ξ. The answer is no. But there are sufficient conditions available for the convergence of the sequence. Next, if it converges, whether ξ will be a root of $x = h(x)$. Consider the equation

$$x_{n+1} = h(x_n).$$

As n increases, the left side tends to the root ξ and if h is continuous, the right side tends to $h(\xi)$. Therefore, in the limit we have $\xi = h(\xi)$ which shows that ξ is a root of the equation $x = h(x)$. Now, how to choose h and the initial approximation x_0 so that the sequence $\{x_n\}$ converges to the root, we have the following theorem.

Theorem 2.6 (Sastry [18]) *Let $x = \xi$ be a root of $f(x) = 0$ and let $\mathcal{N}(\xi)$ be a neighborhood (interval) containing $x = \xi$. Let $h(x)$ and $h'(x)$ be continuous in $\mathcal{N}(\xi)$, where $h(x)$ is defined by $x = h(x)$ which is equivalent to $f(x) = 0$. Then if $|h'(x)| < 1$ for all x in $\mathcal{N}(\xi)$, the sequence of approximations $\{x_n\}$ converges to the root ξ, provided that the initial approximation x_0 is chosen in $\mathcal{N}(\xi)$.* ∎

Some of the drawbacks of Newton–Raphson method are as follows: If the initial estimate is not close enough to the root, the Newton–Raphson method may not converge or may converge to the wrong root. The successive estimates of the Newton–Raphson method may converge too slowly or may not converge at all.

2.4.2 Downhill Simplex Method

The Downhill Simplex method is an algorithm for multidimensional optimization, that is, finding the minimum of a function of one or more than one independent variables and is due to Nelder and Mead [15]. The method requires only function evaluations and does not require derivatives. A simplex is a geometrical figure consisting of n dimensions of $n + 1$ points (vertices) and all interconnecting line segments, polygonal faces, etc. In two-dimensional space, it is a triangle, in three-dimensional space, it is a tetrahedron. The basic idea is to move by reflections, expansions, and contractions starting from an initial starting value.

The algorithm is quite simple, but it can be implemented in many ways. The main difference in implementation is based on the methods of construction of the initial simplex and termination criteria to end the iteration process. Usually the initial simplex S is constructed by generating $n + 1$ vertices $x_1, x_2, \ldots, x_{n+1}$ around an initial starting point, say $x_s \in \mathbb{R}^n$. The most used choice is $x_1 = x_s$. The rest of the vertices are generated in either of the following two ways.

(a) S is right angled at x_1 and $x_j = x_0 + s_j e_j$, $j = 2, \ldots, n + 1$ where s_j is the step size in the direction of e_j, the unit vector in $\mathbb{R}^n$.
(b) S is a regular simplex with specified lengths of all edges.

Suppose we are trying to minimize the function $f(x)$, and we have $x_1, \ldots, x_{n+1}$ vertices. Then one iteration of simplex algorithm consists of the following steps.

(1) **Order** according to the values at the vertices $f(x_1) < f(x_2 < \cdots < f(x_{n+1})$. If the termination criterion is satisfied, then stop.
(2) Calculate the centroid $\bar{x} = \dfrac{1}{n} \displaystyle\sum_{i=1}^{n} x_i$ of all points except the worst point x_{n+1}.

(3) **Reflection**: Compute the reflected point

$$\mathbf{x}_r = \bar{\mathbf{x}} + \alpha(\bar{\mathbf{x}} - \mathbf{x}_{n+1}), \quad \alpha > 0.$$

If $f(\mathbf{x}_1) < f(\mathbf{x}_r) < f(\mathbf{x}_n)$, then find a new simplex by replacing $\mathbf{x}_{n+1}$, the worst point, with the reflected point $\mathbf{x}_r$ and go to step 1.

(4) **Expansion**: If $f(\mathbf{x}_r) < f(\mathbf{x}_1)$, then compute the expanded point

$$\mathbf{x}_e = \mathbf{x}_r + \beta(\mathbf{x}_r - \bar{\mathbf{x}}), \quad \beta > 1.$$

If $f(\mathbf{x}_e) < f(\mathbf{x}_r)$, then obtain a new simplex by replacing $\mathbf{x}_{n+1}$ with $\mathbf{x}_e$ and go to step 1; else obtain a new simplex by replacing $\mathbf{x}_{n+1}$ with $\mathbf{x}_r$ and go to step 1.

(5) **Contraction**: If $f(\mathbf{x}_r) < f(\mathbf{x}_n)$, compute the contracted point

$$\mathbf{x}_c = \bar{\mathbf{x}} + \gamma(\mathbf{x}_{n+1} - \bar{\mathbf{x}}), \quad 0 < \gamma \le 0.5.$$

If $f(\mathbf{x}_c) < f(\mathbf{x}_{n+1})$, then obtain a new simplex by replacing $\mathbf{x}_{n+1}$ with $\mathbf{x}_c$ and go to step 1.

(6) **Shrink**: Replace all points except $\mathbf{x}_1$ with

$$\mathbf{x}_i = \mathbf{x}_1 + \rho(\mathbf{x}_i - \mathbf{x}_1), \quad 0 < \rho < 1$$

and go to step 1. ∎

The standard values used in most implementations are $\alpha = 1$, $\beta = 2$, $\gamma = 0.5$, and $\rho = 0.5$.

2.4.3 Fisher Scoring Algorithm

Consider a model with log-likelihood function $l(\boldsymbol{\theta})$, where $\boldsymbol{\theta}$ is the p-variate parameter vector. Then Fisher's method of scoring for finding $\widehat{\boldsymbol{\theta}}$, the MLE of $\boldsymbol{\theta}$, is given by the iterative process

$$\boldsymbol{\theta}^{(k+1)} = \boldsymbol{\theta}^{(k)} - \left\{ E\left[\frac{\partial^2 l(\boldsymbol{\theta})}{\partial \boldsymbol{\theta} \partial \boldsymbol{\theta}^T} \right] \right\}_{\boldsymbol{\theta}^{(k)}}^{-1} \left(\frac{\partial l(\boldsymbol{\theta})}{\partial \boldsymbol{\theta}} \right)_{\boldsymbol{\theta}^{(k)}}$$

where $\boldsymbol{\theta}^{(k)}$ is the estimate at the kth step. This algorithm is similar to Newton–Raphson method for maximizing the likelihood $l(\boldsymbol{\theta})$ but with the Hessian matrix $\dfrac{\partial^2 l(\boldsymbol{\theta})}{\partial \boldsymbol{\theta} \partial \boldsymbol{\theta}^T}$ replaced by its expected value, that is, the information matrix. Because of the following relationship, the information matrix is more likely to be positive definite at a given iteration;

$$-E\left[\frac{\partial^2 l(\boldsymbol{\theta})}{\partial \boldsymbol{\theta} \partial \boldsymbol{\theta}^T}\right] = E\left[\frac{\partial l(\boldsymbol{\theta})}{\partial \boldsymbol{\theta}} \frac{\partial l(\boldsymbol{\theta})}{\partial \boldsymbol{\theta}^T}\right].$$

This is satisfied under fairly general conditions.

2.5 Miscellaneous

2.5.1 Some Useful Definitions

Let $\{a_n\}_{n=1}^{\infty}$ be a sequence of real numbers and $\{g_n\}_{n=1}^{\infty}$ be a sequence of positive real numbers. Let $\{X_n\}$ be a sequence of random variables.

Definition 2.8 a_n is of **smaller order** than g_n, if

$$\lim_{n \to \infty} g_n^{-1} a_n = 0$$

and write $a_n = o(g_n)$.

Definition 2.9 a_n is **at most of order** g_n, if there exists a positive real number M such that $g_n^{-1}|a_n| \le M$ for all n, and we write $a_n = O(g_n)$.

Definition 2.10 X_n is of **smaller order in probability** than g_n if

$$g_n^{-1} X_n \xrightarrow{P} 0,$$

and we write $X_n = o_p(g_n)$.

Definition 2.11 X_n is **at most of order** g_n **in probability**, if for every $\varepsilon > 0$, there exists a positive real number M_ε such that

$$P\left[|X_n| \ge M_\varepsilon g_n\right] \le \varepsilon,$$

for all n and we write $X_n = O_p(g_n)$.

2.5.2 Some Useful Theorems

In this section, we present some basic probability results which we have used in this monograph. For detailed proofs one is referred to Chung [3].

Theorem 2.7 (Borel–Cantelli Lemma) *Let* $\{A_n\}$, $n = 1, 2, \ldots$ *be a sequence of sets, and* $A = \overline{\lim}_{n \to \infty} A_n$. *Also let* P *be a probability set function. Then*

(a) $\displaystyle\sum_{n=1}^{\infty} P(A_n) < \infty \Rightarrow P(A) = 0,$

(b) $\displaystyle\sum_{n=1}^{\infty} P(A_n) = \infty,$ *and the events A_n are independent* $\Rightarrow P(A) = 1.$

∎

Theorem 2.8 (Lindeberg-Feller Central Limit Theorem) *Let $\{X_n\}$ be a sequence of independent random variables with distribution function F_n. Let $E(X_n) = \mu_n$ and $V(X_n) = \sigma_n^2 \neq 0$ exist. Define*

$$Y_n = \frac{\sum_{i=1}^{n}(X_i - \mu_i)}{B_n}, \qquad B_n = \left(\sum_{i=1}^{n}\sigma_i^2\right)^{1/2}.$$

Then the relations

$$\lim_{n\to\infty}\max_{1\leq i\leq n}\frac{\sigma_i}{B_n} \to 0 \quad and \quad Y_n \xrightarrow{d} \mathcal{N}(0,1)$$

hold if and only if, for every $\varepsilon > 0$,

$$\lim_{n\to\infty}\frac{1}{B_n^2}\sum_{i=1}^{n}\int_{|x-\mu_i|>\varepsilon B_n}(x-\mu_i)^2 dF_i(x) = 0.$$

∎

Theorem 2.9 (Central Limit Theorem for Finite Moving Average) *Let $\{X_t : t \in (0, \pm1, \pm2 \ldots)\}$ be defined by*

$$X_t = \mu + \sum_{j=0}^{m} b_j e_{t-j}$$

where $b_0 = 1$, $\displaystyle\sum_{j=0}^{m} b_j \neq 0$, and $\{e_t\}$ is a sequence of uncorrelated $(0, \sigma^2)$ random variables. Assume that

$$n^{1/2}\bar{e}_n \xrightarrow{d} \mathcal{N}(0, \sigma^2), \quad \bar{e}_n = \frac{1}{n}\sum_{j=1}^{n} e_j.$$

Then

$$n^{1/2}(\bar{x}_n - \mu) \xrightarrow{d} \mathcal{N}\left(0, \sigma^2\left[\sum_{j=0}^{m} b_j\right]^2\right),$$

where $\bar{x}_n = \dfrac{1}{n}\displaystyle\sum_{t=1}^{n} X_t.$

∎

Theorem 2.10 (Central Limit Theorem for Linear Process) *Let* $\{X_t\}$ *be a covariance stationary time series satisfying*

$$X_t = \sum_{j=0}^{\infty} \alpha_j e_{t-j}$$

where $\sum_{j=0}^{\infty} |\alpha_j| < \infty$ *and* $\sum_{j=0}^{\infty} \alpha_j \neq 0$, *and* $\{e_t\}$ *is a sequence of uncorrelated* $(0, \sigma^2)$ *random variables. Let* $\gamma_X(\cdot)$ *denote the autocovariance function of* $\{X_t\}$. *Assume that* $n^{1/2}\bar{e}_n$ *converges in distribution to* $\mathcal{N}(0, \sigma^2)$ *random variable. Then*

$$n^{-\frac{1}{2}} \sum_{t=1}^{n} X_t \xrightarrow{d} \mathcal{N}\left(0, \sum_{h=-\infty}^{\infty} \gamma_X(h)\right),$$

where

$$\sum_{h=-\infty}^{\infty} \gamma_X(h) = \left(\sum_{j=0}^{\infty} \alpha_j\right)^2 \sigma^2.$$

∎

Theorem 2.11 (Central Limit Theorem for Linear Function of Weakly Stationary Process; Fuller [5]) *Let* $\{X_t : t \in T = (0, \pm 1, \pm 2 \ldots)\}$ *be a time series defined by*

$$X_t = \sum_{j=0}^{\infty} \alpha_j e_{t-j}$$

where $\sum_{j=0}^{\infty} |\alpha_j| < \infty$ *and* $\{e_t\}$ *is a sequence of independent* $(0, \sigma^2)$ *random variables with distribution function* $F_t(e)$ *such that*

$$\lim_{\delta \to \infty} \sup_{t \in T} \int_{|e| > \delta} e^2 dF_t(e) = 0.$$

Let $\{C_t\}_{t=1}^{\infty}$ *be a sequence of fixed real numbers satisfying*

(a) $\lim_{n \to \infty} \sum_{t=1}^{n} C_t^2 = \infty$,

(b) $\lim_{n \to \infty} \dfrac{C_n^2}{\sum_{t=1}^{n} C_t^2} = 0$,

(c) $\lim_{n \to \infty} \dfrac{\sum_{t=1}^{n-h} C_t C_{t+|h|}}{\sum_{t=1}^{n} C_t^2} = g(h), \quad h = 0, \pm 1, \pm 2.$

Let $V = \sum_{h=-\infty}^{\infty} g(h)\gamma_X(h) \neq 0$ where $\gamma_X(\cdot)$ is the autocovariance function of $\{X_t\}$.
Then

$$\left[\sum_{t=1}^{n} C_t^2\right]^{-1/2} \sum_{t=1}^{n} C_t X_t \xrightarrow{d} \mathcal{N}(0, V).$$

∎

2.5.3   Stable Distribution

Definition 2.12 A random variable X has **stable distribution** if for any positive numbers A and B, there exists a positive number C and a real number D such that

$$X_1 + X_2 \overset{d}{=} CX + D \tag{2.12}$$

where X_1 and X_2 are independent copies of X.

A random variable X is called **strictly stable** if (2.12) holds for $D = 0$.

Definition 2.13 A random variable X is said to have a **stable distribution** if it has a domain of attraction, that is, there is a sequence of i.i.d. random variables $Z_1, Z_2, \ldots$ and sequences of positive numbers $\{b_n\}$ and real numbers $\{a_n\}$ such that

$$\frac{1}{b_n} \sum_{i=1}^{n} Z_i - a_n \xrightarrow{d} X.$$

Definition 2.14 A random variable X is said to have a **stable distribution** if there are parameters $0 < \alpha \leq 2, \sigma \geq 0, -1 \leq \beta \leq 1$, and μ real such that its characteristic function has the following form

$$E \exp[itX] = \begin{cases} \exp\left\{-\sigma^\alpha |t|^\alpha (1 - i\beta(sign(t))\tan(\frac{\pi\alpha}{2})) + i\mu t\right\} & \text{if } \alpha \neq 1, \\ \exp\{-\sigma |t|(1 + i\beta\frac{2}{\pi}(sign(t))ln|t|) + i\mu t\} & \text{if } \alpha = 1. \end{cases}$$

The parameter α is called the **index of stability** or the **characteristic exponent** and

$$sign(t) = \begin{cases} 1 & \text{if } t > 0 \\ 0 & \text{if } t = 0 \\ -1 & \text{if } t < 0. \end{cases}$$

The above three definitions of stable distribution are equivalent. The parameters σ (**scale parameter**), β (**skewness parameter**), and μ are unique (β is irrelevant when $\alpha = 2$) and is denoted by $X \sim S_\alpha(\sigma, \beta, \mu)$.

Definition 2.15 A symmetric (around 0) random variable X is said to have **symmetric α stable ($S\alpha S$) distribution,** with scale parameter σ, and stability index α, if the characteristic function of the random variable X is

$$Ee^{itX} = e^{-\sigma^\alpha |t|^\alpha}. \tag{2.13}$$

We denote the distribution of X as $S\alpha S(\sigma)$. Note that an $S\alpha S$ random variable with $\alpha = 2$ is $\mathcal{N}(0, 2\sigma^2)$ and with $\alpha = 1$, it is a Cauchy random variable, whose density function is

$$f_\sigma(x) = \frac{\sigma}{\pi(x^2 + \sigma^2)}.$$

The $S\alpha S$ distribution is a special case of the general Stable distribution with non-zero shift and skewness parameters. The random variables following $S\alpha S$ distribution with small characteristic exponent are highly impulsive. As α decreases, both the occurrence rate and strength of the outliers increase which results in very impulsive processes.

For detailed treatments of $S\alpha S$ distribution, the readers are referred to the book of Samorodnitsky and Taqqu [17].

2.5.3.1 Some Basic Properties of Stable Random Variable

1. Let X_1 and X_2 be independent random variables with $X_j \sim S_\alpha(\sigma_j, \beta_j, \mu_j)$, $j = 1, 2$, then $X_1 + X_2 \sim S_\alpha(\sigma, \beta, \mu)$ with

$$\sigma = (\sigma_1^\alpha + \sigma_2^\alpha)^{1/\alpha}, \qquad \beta = \frac{\beta_1 \sigma_1^\alpha + \beta_2 \sigma_2^\alpha}{\sigma_1^\alpha + \sigma_2^\alpha}, \qquad \mu = \mu_1 + \mu_2.$$

2. Let $X \sim S_\alpha(\sigma, \beta, \mu)$ and let a be a real constant. Then $X + a \sim S_\alpha(\sigma, \beta, \mu + a)$.
3. Let $X \sim S_\alpha(\sigma, \beta, \mu)$ and let a be a non-zero real constant. Then

$$aX \sim S_\alpha(|a|\sigma, sign(a)\beta, a\mu) \qquad \text{if } \alpha \neq 1$$
$$aX \sim S_\alpha(|a|\sigma, sign(a)\beta, a\mu - \tfrac{2}{\pi}a(ln|a|)\sigma\beta) \text{ if } \alpha = 1.$$

4. For any $0 < \alpha < 2$, $X \sim S_\alpha(\sigma, \beta, 0) \iff -X \sim S_\alpha(\sigma, -\beta, 0)$.
5. $X \sim S_\alpha(\sigma, \beta, \mu)$ is symmetric if and only if $\beta = 0$ and $\mu = 0$. It is symmetric about μ if and only if $\beta = 0$.

Proposition 2.1 *Let γ be uniform on $(-\pi/2, \pi/2)$ and let W be exponential with mean 1. Assume γ and W are independent. Then*

$$Y = \frac{\sin(\alpha\gamma)}{(\cos\gamma)^{1/\alpha}} \left(\frac{\cos((1-\alpha)\gamma)}{W} \right)^{\frac{1-\alpha}{\alpha}} \tag{2.14}$$

is distributed as $S_\alpha(1, 0, 0) = S_\alpha S(1)$. ∎

This proposition is useful in simulating α-stable random variables. In the Gaussian case, $\alpha = 2$ and (2.14) reduces to $W^{\frac{1}{2}} \sin(2\gamma)/\cos(\gamma)$ which is the Box–Muller method for generating $\mathcal{N}(0, 2)$ random variables (Box and Muller [2]).

Definition 2.16 Let $\mathbf{X} = (X_1, X_2, \ldots, X_d)$ be an α-**stable random vector** in $\mathbb{R}^d$, then

$$\Phi(\mathbf{t}) = \Phi(t_1, t_2, \ldots, t_d) = E \exp\{i(\mathbf{t}^T \mathbf{X})\} = E \exp\{i \sum_{k=1}^{d} t_k X_k\}$$

denote its characteristic function. $\Phi(\mathbf{t})$ is also called the joint characteristic function of the random variables $X_1, X_2, \ldots, X_d$.

Result 2.1 *Let $\mathbf{X}$ be a random vector in $\mathbb{R}^d$.*
(a) If all linear combinations of the random variables $X_1, \ldots, X_d$ are symmetric stable, then $\mathbf{X}$ is a symmetric stable random vector in $\mathbb{R}^d$.
(b) If all linear combination are stable with index of stability greater than or equal to one, then $\mathbf{X}$ is a stable vector in $\mathbb{R}^d$.

2.5.4 Some Number Theoretic Results and A Conjecture

We need some number theoretic results and a conjecture in establishing the asymptotic results related to chirp signal model. The following trigonometric results have been used quite extensively in establishing different results related to sinusoidal model. These results can be easily established, see, for example, Mangulis [14].

Result 2.2

$$\frac{1}{n} \sum_{t=1}^{n} \cos(\omega t) = o\left(\frac{1}{n}\right) \tag{2.15}$$

$$\frac{1}{n} \sum_{t=1}^{n} \sin(\omega t) = o\left(\frac{1}{n}\right) \tag{2.16}$$

$$\frac{1}{n} \sum_{t=1}^{n} \cos^2(\omega t) = \frac{1}{2} + o\left(\frac{1}{n}\right) \tag{2.17}$$

$$\frac{1}{n} \sum_{t=1}^{n} \sin^2(\omega t) = \frac{1}{2} + o\left(\frac{1}{n}\right) \tag{2.18}$$

$$\frac{1}{n^{k+1}} \sum_{t=1}^{n} t^k \cos^2(\omega t) = \frac{1}{2(k+1)} + o\left(\frac{1}{n}\right) \tag{2.19}$$

$$\frac{1}{n^{k+1}} \sum_{t=1}^{n} t^k \sin^2(\omega t) = \frac{1}{2(k+1)} + o\left(\frac{1}{n}\right) \tag{2.20}$$

$$\frac{1}{n^{k+1}} \sum_{t=1}^{n} t^k \cos(\omega t) \sin(\omega t) = o\left(\frac{1}{n}\right). \tag{2.21}$$

We need similar results in case of chirp models also.

Result 2.3 *Let* $\theta_1, \theta_2 \in (0, \pi)$ *and* $k = 0, 1, \ldots$ *Then except for countable number of points, the followings are true:*

$$\lim_{n \to \infty} \frac{1}{n} \sum_{t=1}^{n} \cos(\theta_1 t + \theta_2 t^2) = 0 \tag{2.22}$$

$$\lim_{n \to \infty} \frac{1}{n} \sum_{t=1}^{n} \sin(\theta_1 t + \theta_2 t^2) = 0 \tag{2.23}$$

$$\lim_{n \to \infty} \frac{1}{n^{k+1}} \sum_{t=1}^{n} t^k \cos^2(\theta_1 t + \theta_2 t^2) = \frac{1}{2(k+1)} \tag{2.24}$$

$$\lim_{n \to \infty} \frac{1}{n^{k+1}} \sum_{t=1}^{n} t^k \sin^2(\theta_1 t + \theta_2 t^2) = \frac{1}{2(k+1)} \tag{2.25}$$

$$\lim_{n \to \infty} \frac{1}{n^{k+1}} \sum_{t=1}^{n} t^k \cos(\theta_1 t + \theta_2 t^2) \sin(\theta_1 t + \theta_2 t^2) = 0. \tag{2.26}$$

Proof Using the number theoretic results of Vinogradov [24], the above results can be established, see Lahiri, Kundu, and Mitra [13] for details. ∎

The following conjecture was proposed by Vinogradov [24], and it still remains an open problem, although extensive simulation results suggest the validity of the conjecture. This conjecture also has been used to establish some of the crucial results related to the chirp model.

Conjecture 2.1 *Let* $\theta_1, \theta_2, \theta_1', \theta_2' \in (0, \pi)$ *and* $k = 0, 1, \ldots$ *Then except for countable number of points, the followings are true:*

$$\lim_{n \to \infty} \frac{1}{\sqrt{n} n^k} \sum_{t=1}^{n} t^k \cos(\theta_1 t + \theta_2 t^2) \sin(\theta_1' t + \theta_2' t^2) = 0. \tag{2.27}$$

In addition if $\theta_2 \neq \theta_2'$, *then*

$$\lim_{n \to \infty} \frac{1}{\sqrt{n} n^k} \sum_{t=1}^{n} t^k \cos(\theta_1 t + \theta_2 t^2) \cos(\theta_1' t + \theta_2' t^2) = 0$$

$$\lim_{n \to \infty} \frac{1}{\sqrt{n} n^k} \sum_{t=1}^{n} t^k \sin(\theta_1 t + \theta_2 t^2) \sin(\theta_1' t + \theta_2' t^2) = 0.$$

2.6 Prony's Equation and Exponential Models

2.6.1 Prony's Equation

Now we provide one important result which has been used quite extensively in the Statistical Signal Processing literature and it is known as Prony's equation. Prony, a Chemical engineer, proposed the following method more than two hundred years back in 1795, mainly to estimate the unknown parameters of the real exponential model. It is available in several numerical analysis text books, see, for example, Froberg [4] or Hildebrand [7]. Prony observed that for any arbitrary real non-zero constants $\alpha_1, \ldots, \alpha_M$ and for distinct constants $\beta_1, \ldots, \beta_M$, if

$$\mu(t) = \alpha_1 e^{\beta_1 t} + \cdots + \alpha_M e^{\beta_M t}; \quad t = 1, \ldots, n, \tag{2.28}$$

then there exists $(M + 1)$ constants $\{g_0, \ldots, g_M\}$, such that

$$\mathbf{Ag} = \mathbf{0}, \tag{2.29}$$

where

$$\mathbf{A} = \begin{bmatrix} \mu(1) & \cdots & \mu(M+1) \\ \vdots & \ddots & \vdots \\ \mu(n-M) & \cdots & \mu(n) \end{bmatrix}, \quad \mathbf{g} = \begin{bmatrix} g_0 \\ \vdots \\ g_M \end{bmatrix} \text{ and } \mathbf{0} = \begin{bmatrix} 0 \\ \vdots \\ 0 \end{bmatrix}. \tag{2.30}$$

Note that without loss of generality, we can always put restrictions on $g_0, \ldots, g_M$ such that $\sum_{j=0}^{M} g_j^2 = 1$ and $g_0 > 0$. The sets of linear equations (2.29) is known as Prony's equations. The roots of the following polynomial equation

$$p(x) = g_0 + g_1 x + \cdots + g_M x^M = 0 \tag{2.31}$$

are $e^{\beta_1}, \ldots, e^{\beta_M}$. Therefore, there is a one to one correspondence between $\{\beta_1, \ldots, \beta_M\}$ and $\{g_0, g_1, \ldots, g_M\}$, such that

$$\sum_{j=0}^{M} g_j^2 = 1, \quad g_0 > 0. \tag{2.32}$$

Moreover, $\{g_0, g_1, \ldots, g_M\}$ do not depend on $\{\alpha_1, \ldots, \alpha_M\}$. One natural question is, how to recover $\{\alpha_1, \ldots, \alpha_M\}$ and $\{\beta_1, \ldots, \beta_M\}$ from a given $\mu(1), \ldots, \mu(n)$. It can be done as follows. Note that the rank of the matrix $\mathbf{A}$ as defined in (2.30) is M. Therefore, there exists unique $\{g_0, g_1, \ldots, g_M\}$, such that (2.29) and (2.32) hold simultaneously. From that $\{g_0, g_1, \ldots, g_M\}$, using (2.31), $\{\beta_1, \ldots, \beta_M\}$ can be recovered. Now to recover $\{\alpha_1, \ldots, \alpha_M\}$, write (2.28) as

$$\boldsymbol{\mu} = \mathbf{X}\boldsymbol{\alpha}, \tag{2.33}$$

where $\boldsymbol{\mu} = (\mu(1), \ldots, \mu(n))^T$ and $\boldsymbol{\alpha} = (\alpha_1, \ldots, \alpha_M)^T$ are $n \times 1$ and $M \times 1$ vectors, respectively. The $n \times M$ matrix $\mathbf{X}$ is as follows:

$$\mathbf{X} = \begin{bmatrix} e^{\beta_1} & \cdots & e^{\beta_M} \\ \vdots & \ddots & \vdots \\ e^{n\beta_1} & \cdots & e^{n\beta_M} \end{bmatrix}. \tag{2.34}$$

Therefore $\boldsymbol{\alpha} = (\mathbf{X}^T\mathbf{X})^{-1}\mathbf{X}^T\boldsymbol{\mu}$. Note that $\mathbf{X}^T\mathbf{X}$ is a full rank matrix as $\beta_1, \ldots, \beta_M$ are distinct.

2.6.2 Undamped Exponential Model

Although Prony observed relation (2.29) for real exponential model, the same result is true for complex exponential model or popularly known as undamped exponential model. A complex exponential model can be expressed as

$$\mu(t) = A_1 e^{i\omega_1 t} + \cdots + A_M e^{i\omega_M t}; \quad t = 1, \ldots, n. \tag{2.35}$$

Here $A_1, \ldots, A_M$ are complex numbers, $0 < \omega_k < 2\pi$ for $k = 1, \ldots, M$, and $i = \sqrt{-1}$. In this case also there exists $\{g_0, \ldots, g_M\}$, such that they satisfy (2.29). Also the roots of the polynomial equation $p(z) = 0$, as given in (2.31), are $z_1 = e^{i\omega_1}, \ldots, z_M = e^{i\omega_M}$. Observe that

$$|z_1| = \cdots = |z_M| = 1, \quad \bar{z}_k = z_k^{-1}; \quad k = 1, \ldots, M. \tag{2.36}$$

Here $\bar{z}_k$ denotes the complex conjugate of z_k. Define the new polynomial

$$q(z) = z^{-M} \bar{p}(z) = \bar{g}_0 z^{-M} + \cdots + \bar{g}_M. \tag{2.37}$$

From (2.36), it is clear that $p(z)$ and $q(z)$ have the same roots. Therefore, we obtain

$$\frac{g_k}{g_M} = \frac{\bar{g}_{M-k}}{\bar{g}_0}; \quad k = 0, \ldots, M, \tag{2.38}$$

by comparing the coefficients of the two polynomials $p(z)$ and $q(z)$. If we denote

$$b_k = g_k \left(\frac{\bar{g}_0}{g_M} \right)^{-\frac{1}{2}}; \quad k = 0, \ldots, M, \tag{2.39}$$

then

$$b_k = \bar{b}_{M-k}; \quad k = 0, \ldots, M. \tag{2.40}$$

The condition (2.40) is the conjugate symmetric property and can be written as

$$\mathbf{b} = \mathbf{J}\bar{\mathbf{b}}, \tag{2.41}$$

where $\mathbf{b} = (b_0, \ldots, b_M)^T$ and $\mathbf{J}$ is an exchange matrix as defined in (2.1). From the above discussions, it is immediate that for $\mu(t)$ in (2.35), there exists a vector $\mathbf{g} = (g_0, \ldots, g_M)^T$, such that $\sum_{k=0}^{M} g_k^2 = 1$. This $\mathbf{g}$ satisfies (2.29), and also satisfies

$$\mathbf{g} = \mathbf{J}\bar{\mathbf{g}}. \tag{2.42}$$

For more details, see Kannan and Kundu [9].

2.6.3   Sum of Sinusoidal Model

Prony's method can also be applied to the sum of sinusoidal model. Suppose $\mu(t)$ can be written as follows:

$$\mu(t) = \sum_{k=1}^{p} [A_k \cos(\omega_k t) + B_k \sin(\omega_k t)]; \quad t = 1, \ldots, n. \tag{2.43}$$

Here $A_1, \ldots, A_p, B_1, \ldots, B_p$ are real numbers, the frequencies $\omega_1, \cdots, \omega_p$ are distinct and $0 < \omega_k < \pi$ for $k = 1, \ldots, p$. Then (2.43) can be written in the form:

$$\mu(t) = \sum_{k=1}^{p} C_k e^{i\omega_k t} + \sum_{k=1}^{p} D_k e^{-i\omega_k t}; \quad t = 1, \ldots, n, \tag{2.44}$$

where $C_k = (A_k - iB_k)/2$ and $D_k = (A_k + iB_k)/2$. Model (2.44) is in the same form as in (2.35). Therefore, there exists a vector $\mathbf{g} = (g_0, \ldots, g_{2p})$, such that $\sum_{j=0}^{2p} g_j^2 = 1$, which satisfies

$$\begin{bmatrix} \mu(1) & \cdots & \mu(2p+1) \\ \vdots & \ddots & \vdots \\ \mu(n-2p) & \cdots & \mu(n) \end{bmatrix} \begin{bmatrix} g_0 \\ \vdots \\ g_{2p} \end{bmatrix} = \begin{bmatrix} 0 \\ \vdots \\ 0 \end{bmatrix} \tag{2.45}$$

and

$$\mathbf{g} = \mathbf{J}\bar{\mathbf{g}}. \tag{2.46}$$

2.6.4 Linear Prediction

Observe that $\mu(t)$ as defined in (2.35) also satisfies the following backward linear prediction equation:

$$\begin{bmatrix} \mu(2) & \cdots & \mu(M+1) \\ \vdots & \ddots & \vdots \\ \mu(n-M+1) & \cdots & \mu(n) \end{bmatrix} \begin{bmatrix} d_1 \\ \vdots \\ d_M \end{bmatrix} = - \begin{bmatrix} \mu(1) \\ \vdots \\ \mu(n-M) \end{bmatrix}. \tag{2.47}$$

Clearly, $d_j = b_j/b_0$, for $j = 1, \ldots, M$. It is known as Mth order backward linear prediction equation. In this case, the following polynomial equation

$$p(z) = 1 + d_1 z + \cdots + d_M z^M = 0 \tag{2.48}$$

has roots at $e^{i\omega_1}, \ldots, e^{i\omega_M}$.

Along the same manner, it may be noted that $\mu(t)$ as defined in (2.35) also satisfies the following forward linear prediction equation:

$$\begin{bmatrix} \mu(M) & \cdots & \mu(1) \\ \vdots & \ddots & \vdots \\ \mu(n-1) & \cdots & \mu(n-M) \end{bmatrix} \begin{bmatrix} a_1 \\ \vdots \\ a_M \end{bmatrix} = - \begin{bmatrix} \mu(M+1) \\ \vdots \\ \mu(n) \end{bmatrix}, \tag{2.49}$$

where $a_1 = b_{M-1}/b_M, a_2 = b_{M-2}/b_M, \ldots, a_M = b_0/b_M$. It is known as Mth order forward linear prediction equation. Moreover, the following polynomial equation

$$p(z) = 1 + a_1 z + \cdots + a_M z^M = 0 \tag{2.50}$$

has roots at $e^{-i\omega_1}, \ldots, e^{-i\omega_M}$.

Consider the case when $\mu(t)$ has the form (2.43), then clearly both the polynomial equations (2.48) and (2.50) have roots at $e^{\pm i\omega_1}, \cdots, e^{\pm i\omega_p}$, hence the corresponding coefficients of both the polynomials must be equal. Therefore, in this case the forward backward linear prediction equation can be formed as follows:

$$
\begin{bmatrix}
\mu(2) & \cdots & \mu(M+1) \\
\vdots & \ddots & \vdots \\
\mu(n-M+1) & \cdots & \mu(n) \\
\mu(M) & \cdots & \mu(1) \\
\vdots & \ddots & \vdots \\
\mu(n-1) & \cdots & \mu(n-M)
\end{bmatrix}
\begin{bmatrix}
d_1 \\
\vdots \\
d_M
\end{bmatrix}
= -
\begin{bmatrix}
\mu(1) \\
\vdots \\
\mu(n-M) \\
\mu(M+1) \\
\vdots \\
\mu(n)
\end{bmatrix}.
\tag{2.51}
$$

In the signal processing literature, the linear prediction method has been used quite extensively for different purposes; see, for example, Kumaresan and Tufts [10].

2.6.5 Matrix Pencil

Suppose $\mathbf{A}$ and $\mathbf{B}$ are two $n \times n$, real or complex matrices. The collection of all the matrices $\mathscr{A}$ such that

$$
\mathscr{A} = \{\mathbf{C} : \mathbf{C} = \mathbf{A} - \lambda\mathbf{B}, \text{ where } \lambda \text{ is any complex number}\},
$$

is called a linear matrix pencil or matrix pencil, see Golub and van Loan [6], and it is denoted by $(\mathbf{A}, \mathbf{B})$. The set of all λ, such that $\det(\mathbf{A} - \lambda\mathbf{B}) = 0$, is called the eigenvalues of the matrix pencil $(\mathbf{A}, \mathbf{B})$. The eigenvalues of the matrix pencil $(\mathbf{A}, \mathbf{B})$ can be obtained by solving the general eigenvalue problem of the form:

$$
\mathbf{A}\mathbf{x} = \lambda\mathbf{B}\mathbf{x}.
$$

If $\mathbf{B}^{-1}$ exists, then

$$
\mathbf{A}\mathbf{x} = \lambda\mathbf{B}\mathbf{x} \quad \Leftrightarrow \quad \mathbf{B}^{-1}\mathbf{A}\mathbf{x} = \lambda\mathbf{x}.
$$

Efficient methods are available to compute the eigenvalues of $(\mathbf{A}, \mathbf{B})$ when $\mathbf{B}^{-1}$ does not exist, or it is nearly a singular matrix, see, for example, Golub and van Loan [6]. The matrix pencil has been used quite extensively in numerical linear algebra. Recently, extensive usage of the matrix pencil method can be found in the spectral estimation method also.

References

1. Bates, D. M., & Watts, D. G. (1988). *Nonlinear regression and its applications*. New York: Wiley.
2. Box, G. E. P., & Muller, M. E. (1958). A note on the generation of random normal deviates. *The Annals of Mathematical Statistics, 29*(2), 610–611.
3. Chung, K. L. (2001). *A course in probability theory*. San Diego: Academic Press.

4. Froberg, C. E. (1969). *Introduction to numerical analysis* (2nd ed.). Boston: Addision-Wesley Pub. Co.
5. Fuller, W. A. (1976). *Introduction to statistical time series.* New York: Wiley.
6. Golub, G., & van Loan, C. (1996). *Matrix computations* (3rd ed.). London: The Johns Hopkins University Press.
7. Hilderband, F. B. (1956). *An introduction to numerical analysis.* New York: McGraw-Hill.
8. Jennrich, R. I. (1969). Asymptotic properties of the non linear least squares estimators. *Annals of Mathematical Statistics, 40,* 633–643.
9. Kannan, N., & Kundu, D. (1994). On modified EVLP and ML methods for estimating super-imposed exponential signals. *Signal Processing, 39,* 223–233.
10. Kumaresan, R., & Tufts, D. W. (1982). Estimating the parameters of exponentially damped sinusoids and pole-zero modelling in noise. *IEEE Transactions on Acoustics, Speech and Signal Processing, 30,* 833–840.
11. Kundu, D. (1991). Assymptotic properties of the complex valued nonlinear regression model. *Communications in Statistics—Theory and Methods, 20,* 3793–3803.
12. Kundu, D., & Basu, A. (2004). *Statistical computing; existing methods and recent developments.* New Delhi: Narosa.
13. Lahiri, A., Kundu, D., & Mitra, A. (2012). Efficient algorithm for estimating the parameters of chirp signal. *Journal of Multivariate Analysis, 108,* 15–27.
14. Mangulis, V. (1965). *Handbook of series for scientists and engineers.* New York: Academic Press.
15. Nelder, J. A., & Mead, R. (1965). *Computer Journal, 7,* 308–313.
16. Ortega, J. M., & Rheinboldt, W. C. (1970). *Iterative solution of nonlinear equations in several variables.* New York: Academic Press.
17. Samorodnitsky, G., & Taqqu, M. S. (1994). *Stable non-Gaussian random processes; stochastic models with infinite variance.* New York: Chapman and Hall.
18. Sastry, S. S. (2005). *Introductory methods of numerical analysis* (4th ed.). New Delhi: Prentice-Hall of India Private Limited.
19. Seber, G. A. F., & Lee, A. J. (2018). *Linear regression analysis* (2nd ed.). New York: Wiley.
20. Stoer, J., & Bulirsch, R. (1992). *Introduction to numerical analysis* (2nd ed.). Berlin: Springer.
21. Rao, C. R. (1973). *Linear statistical inference and its applications* (2nd ed.). New York: Wiley.
22. Richards, F. S. G. (1961). A method of maximum likelihood estimation. *The Journal of the Royal Statistical Society, Series B,* 469–475.
23. Seber, A., & Wild, B. (1989). *Nonlinear regression.* New York: Wiley.
24. Vinogradov, I. M. (1954). The method of trigonometrical sums in the theory of numbers. Interscience, Translated from Russian, Revised and annoted by K.F. Roth and A. Devenport. Reprint of the 1954 translation (2004). Mineola, New York, U.S.A.: Dover publications Inc.
25. Wu, C. F. J. (1981). Asymptotic theory of non linear least squares estimation. *Annals of Statistics, 9,* 501–513.

Chapter 3
Estimation of Frequencies

In this section, we provide different estimation procedures of the frequencies of a periodic signal. We consider the following sum of sinusoidal model:

$$y(t) = \sum_{k=1}^{p} (A_k \cos(\omega_k t) + B_k \sin(\omega_k t)) + X(t); \quad t = 1, \ldots, n. \quad (3.1)$$

Here, A_k, B_k, ω_k for $k = 1, \ldots, p$ are unknown. In this chapter, p is assumed to be known; later in Chap. 5, we provide different estimation methods of p. The error component $X(t)$ has mean zero and finite variance, and it can have either one of the following forms.

Assumption 3.1 $\{X(t); t = 1, \ldots, n\}$ are i.i.d. random variables with $E(X(t)) = 0$ and $V(X(t)) = \sigma^2$.

Assumption 3.2 $\{X(t)\}$ is a stationary linear process with the following form:

$$X(t) = \sum_{j=0}^{\infty} a(j)e(t - j), \quad (3.2)$$

where $\{e(t); t = 1, 2, \ldots\}$ are i.i.d. random variables with $E(e(t)) = 0$, $V(e(t)) = \sigma^2$, and $\sum_{j=0}^{\infty} |a(j)| < \infty$.

All the available methods have been used under both these assumptions, although theoretical properties of these estimators obtained by different methods may not be available under both these assumptions.

In this chapter, we mainly discuss different estimation procedures; properties of these estimators are discussed in Chap. 4. Most of the methods deal with the estimation of frequencies. If the frequencies are known, model (3.1) can be treated as a linear regression model, therefore, the linear parameters A_k and B_k for $k = 1, \ldots, p$

S. Nandi and D. Kundu, *Statistical Signal Processing*,
https://doi.org/10.1007/978-981-15-6280-8_3

can be estimated using a simple least squares or generalized least squares method depending on the error structure. Similar methods can be adopted even when the frequencies are unknown.

3.1   ALSEs and PEs

In this section, first we assume that $p = 1$, later we provide the method for general p. Under Assumption 3.1 the most intuitive estimators are the LSEs, that is $\widehat{A}$, $\widehat{B}$, and $\widehat{\omega}$ which minimize

$$Q(A, B, \omega) = \sum_{t=1}^{n} (y(t) - A\cos(\omega t) - B\sin(\omega t))^2. \tag{3.3}$$

Finding the LSEs is a nonlinear optimization problem, and it is well-known to be a numerically difficult problem. The standard algorithm like Newton–Raphson, Gauss–Newton or their variants may be used to minimize (3.3). Often, it is observed that the standard algorithms may not converge even when the iterative process starts from a very good starting value. It is observed by Rice and Rosenblatt [23] that the least squares surface has several local minima near the true parameter value, and due to this reason most of the iterative procedures even when they converge, often converge to a local minimum rather than the global minimum.

Therefore, even if it is known that the LSEs are the most efficient estimators, finding the LSEs is a numerically challenging problem. Due to this reason, extensive work has been done in the statistical signal processing literature to find estimators which perform like the LSEs.

First, we provide the most popular method which has been used in practice to compute the frequency estimator. For $p = 1$, model (3.1) can be written as

$$\mathbf{Y} = \mathbf{Z}(\omega)\boldsymbol{\theta} + \mathbf{e}, \tag{3.4}$$

where

$$\mathbf{Y} = \begin{bmatrix} y(1) \\ \vdots \\ y(n) \end{bmatrix}, \quad \mathbf{Z}(\omega) = \begin{bmatrix} \cos(\omega) & \sin(\omega) \\ \vdots & \vdots \\ \cos(n\omega) & \sin(n\omega) \end{bmatrix}, \quad \boldsymbol{\theta} = \begin{bmatrix} A \\ B \end{bmatrix}, \quad \mathbf{e} = \begin{bmatrix} X(1) \\ \vdots \\ X(n) \end{bmatrix}. \tag{3.5}$$

Therefore, for a given ω, the LSEs of A and B can be obtained as

$$\widehat{\boldsymbol{\theta}}(\omega) = \begin{bmatrix} \widehat{A}(\omega) \\ \widehat{B}(\omega) \end{bmatrix} = \left(\mathbf{Z}^T(\omega)\mathbf{Z}(\omega)\right)^{-1} \mathbf{Z}^T(\omega)\mathbf{Y}. \tag{3.6}$$

Using Result 2.2, (3.6) can be written as

$$\begin{bmatrix} \widehat{A}(\omega) \\ \widehat{B}(\omega) \end{bmatrix} = \begin{bmatrix} 2\sum_{t=1}^{n} y(t)\cos(\omega t)/n \\ 2\sum_{t=1}^{n} y(t)\sin(\omega t)/n \end{bmatrix}. \tag{3.7}$$

Substituting $\widehat{A}(\omega)$ and $\widehat{B}(\omega)$ in (3.3), we obtain

$$\frac{1}{n}Q\left(\widehat{A}(\omega), \widehat{B}(\omega), \omega\right) = \frac{1}{n}\mathbf{Y}^T\mathbf{Y} - \frac{1}{n}\mathbf{Y}^T\mathbf{Z}(\omega)\mathbf{Z}^T(\omega)\mathbf{Y} + o(1). \tag{3.8}$$

Therefore, $\widehat{\omega}$ which minimizes $Q(\widehat{A}(\omega), \widehat{B}(\omega), \omega)/n$ is equivalent to $\widetilde{\omega}$, which maximizes

$$I(\omega) = \frac{1}{n}\mathbf{Y}^T\mathbf{Z}(\omega)\mathbf{Z}^T(\omega)\mathbf{Y} = \frac{1}{n}\left\{ \left(\sum_{t=1}^{n} y(t)\cos(\omega t) \right)^2 + \left(\sum_{t=1}^{n} y(t)\sin(\omega t) \right)^2 \right\}$$

in the sense $\widehat{\omega} - \widetilde{\omega} \overset{a.e.}{\to} 0$. The estimator of ω, which is obtained by maximizing $I(\omega)$ for $0 \leq \omega \leq \pi$, is known as the ALSE of ω.

The maximization of $I(\omega)$ can be performed by some standard algorithm like Newton–Raphson or Gauss–Newton method, although the computation of the ALSEs has the same type of problems as that of the LSEs. In practice, instead of maximizing $I(\omega)$ for $0 < \omega < \pi$, it is maximized at the Fourier frequencies, namely at the points $2\pi j/n, 0 \leq j < [n/2]$. Therefore, $\widetilde{\widetilde{\omega}} = 2\pi j_0/n$ is an estimator of ω, where

$$I(\omega_{j_0}) > I(\omega_k), \quad \text{for } k = 1, \ldots, [n/2], k \neq j_0.$$

It is also known as the PE of ω. Although it is not an efficient estimator, it is being used extensively as an initial guess of any iterative procedure to compute an efficient estimator of the frequency.

The method can be easily extended for the model when $p > 1$. The main idea is to remove the effect of the first component from the signal $\{y(t)\}$, and repeat the whole procedure. The details are explained in Sect. 3.12.

3.2 EVLP

The Equivariance Linear Prediction (EVLP) method was suggested by Bai et al. [2] for estimating the frequencies of model (3.1), under error Assumption 3.1. It mainly uses the fact that in the absence of $\{X(t); t = 1, \ldots, n\}$, $\{y(t); t = 1, \ldots, n\}$ satisfy (2.45). The idea behind the EVLP method is as follows: Consider an $(n - 2p) \times (p + 1)$ data matrix $\mathbf{Y}_D$ as

$$\mathbf{Y}_D = \begin{bmatrix} y(1) & \cdots & y(2p+1) \\ \vdots & \ddots & \vdots \\ y(n-2p) & \cdots & y(n) \end{bmatrix}. \tag{3.9}$$

If $\{X(t)\}$ is absent, $\mathrm{Rank}(\mathbf{Y}_D) = \mathrm{Rank}(\mathbf{Y}_D^T \mathbf{Y}_D/n) = 2p$. It implies that the symmetric matrix $(\mathbf{Y}_D^T \mathbf{Y}_D/n)$ has an eigenvalue zero with multiplicity one. Therefore, there exists an eigenvector $\mathbf{g} = (g_0, \ldots, g_{2p})$ that corresponds to the zero eigenvalue such that $\sum_{j=0}^{2p} g_j^2 = 1$, and the polynomial equation

$$p(x) = g_0 + g_1 x + \cdots + g_{2p} x^{2p} = 0 \tag{3.10}$$

has roots at $e^{\pm i\omega_1}, \ldots, e^{\pm i\omega_p}$.

Using this idea, Bai et al. [2] proposed that when $\{X(t)\}$ satisfies error Assumption 3.1, from the symmetric matrix $(\mathbf{Y}_D^T \mathbf{Y}_D/n)$ obtain the normalized eigenvector $\widehat{\mathbf{g}} = (\widehat{g}_0, \ldots, \widehat{g}_{2p})$, such that $\sum_{j=0}^{2p} \widehat{g}_j^2 = 1$, which corresponds to the minimum eigenvalue. Form the polynomial equation

$$\widehat{p}(x) = \widehat{g}_0 + \widehat{g}_1(x) + \cdots + \widehat{g}_{2p} x^{2p} = 0, \tag{3.11}$$

and obtain the estimates of $\omega_1, \ldots, \omega_p$ from these estimated roots.

It has been shown by Bai et al. [2] that as $n \to \infty$, the EVLP method provides consistent estimators of the unknown frequencies. It is interesting that although EVLP frequency estimators are consistent, the corresponding linear parameter estimators obtained by the least squares method as mentioned before are not consistent estimators. Moreover, it has been observed by extensive simulation studies by Bai et al. [3] that the performance of the EVLP estimators are not very satisfactory for small sample sizes.

3.3   MFBLP

In the signal processing literature, the forward linear prediction method or backward linear prediction, see Sect. 2.6.4, has been used to estimate the frequencies of the sinusoidal signals. It has been observed by Kumaresan [10] by using extensive simulation studies that the pth order linear prediction method does not work very well in estimating the frequencies in the presence of noise, particularly when the two frequencies are very close to each other.

Due to this reason, Kumaresan [10], see also Tufts and Kumaresan [27], used the extended order forward backward linear prediction method, and called it the

Modified Forward Backward Linear Prediction (MFBLP) method, and it can be described as follows. Choose an L, such that $n - 2p \geq L > 2p$, and set up the Lth order Backward and Forward Linear prediction equations as follows:

$$
\begin{bmatrix}
y(L) & \cdots & y(1) \\
\vdots & \ddots & \vdots \\
y(n-1) & \cdots & y(n-L) \\
y(2) & \cdots & y(L+1) \\
\vdots & \ddots & \vdots \\
y(n-L+1) & \cdots & y(n)
\end{bmatrix}
\begin{bmatrix}
b_1 \\
\vdots \\
b_L
\end{bmatrix}
= -
\begin{bmatrix}
y(L+1) \\
\vdots \\
y(n) \\
y(1) \\
\vdots \\
y(n-L)
\end{bmatrix}.
\tag{3.12}
$$

It has been shown by Kumaresan [10] that in the noiseless case $\{y(t)\}$ satisfies (3.12), and out of the L roots of the polynomial equation

$$
p(x) = x^L + b_1 x^{L-1} + \cdots + b_L = 0,
\tag{3.13}
$$

$2p$ roots are of the form $e^{\pm i\omega_k}$, and the rest of the $L - 2p$ roots are of the modulus not equal to one. Observe that (3.12) can be written as

$$
\mathbf{Ab} = -\mathbf{h}.
\tag{3.14}
$$

Tufts and Kumaresan [27] suggested using the truncated singular value decomposition solution of the vector $\mathbf{b}$ by setting the smaller singular values of the matrix $\mathbf{A}$ equal to zero. Therefore, if the singular value decomposition of $\mathbf{A}$ is as given in (2.3) where $\mathbf{v}_k$ and $\mathbf{u}_k$ for $k = 1, \ldots, 2p$ are the eigenvectors of $\mathbf{A}^T \mathbf{A}$ and $\mathbf{A}\mathbf{A}^T$, respectively, $\sigma_1^2 \geq \cdots \sigma_{2p}^2 > 0$ are the $2p$ nonzero eigenvalues of $\mathbf{A}^T\mathbf{A}$, then the solution $\widehat{\mathbf{b}}$ of the system of equations (3.14) becomes

$$
\widehat{\mathbf{b}} = -\sum_{k=1}^{2p} \frac{1}{\sigma_k} \left[\mathbf{u}_k^T \mathbf{h} \right] \mathbf{v}_k.
\tag{3.15}
$$

The effect of using the truncated singular value decomposition is to increase the signal-to-noise ratio in the noisy data, prior to obtaining the solution vector $\widehat{\mathbf{b}}$. Once $\widehat{\mathbf{b}}$ is obtained, get the L roots of the L-degree polynomial

$$
\widehat{p}(x) = x^L + \widehat{b}_1 x^{L-1} + \cdots + \widehat{b}_L,
\tag{3.16}
$$

and choose $2p$ roots which are closest to one in the absolute value. If the error variance is small, it is expected that out of those L roots, $2p$ roots will form p conjugate pairs, and from these papers the frequencies can be easily estimated.

It is observed by extensive simulation studies by Kumaresan [10] that the MFBLP performs very well if $L \approx 2n/3$, and the error variance is not too large. The main computation involved in this case is the computation of the singular value decomposition of a matrix $\mathbf{A}$ and then root findings of an L degree polynomial. Although the MFBLP performs very well for small sizes, it has been pointed out by Rao [22] that MFBLP estimators are not consistent.

3.4 NSD

The Noise Space Decomposition (NSD) method has been proposed by Kundu and Mitra [14], see also Kundu and Mitra [15] in this respect. The basic idea behind the NSD method can be described as follows. Consider the following $(n - L) \times (L + 1)$ matrix $\mathbf{A}$, where

$$
\mathbf{A} = \begin{bmatrix} \mu(1) & \cdots & \mu(L+1) \\ \vdots & \ddots & \vdots \\ \mu(n-L) & \cdots & \mu(n) \end{bmatrix},
\tag{3.17}
$$

for any integer L, such that $2p \leq L \leq n - 2p$, and $\mu(t)$ is the same as defined in (2.6.3). Let the spectral decomposition of $\mathbf{A}^T \mathbf{A}/n$ be

$$
\frac{1}{n}\mathbf{A}^T \mathbf{A} = \sum_{i=1}^{L+1} \sigma_i^2 \mathbf{u}_i \mathbf{u}_i^T,
\tag{3.18}
$$

where $\sigma_1^2 \geq \cdots \geq \sigma_{L+1}^2$ are the eigenvalues of $\mathbf{A}^T \mathbf{A}/n$ and $\mathbf{u}_1, \ldots, \mathbf{u}_{L+1}$ are the corresponding orthonormal eigenvectors. Since matrix $\mathbf{A}$ is of rank $2p$,

$$
\sigma_{2p+1}^2 = \cdots = \sigma_{L+1}^2 = 0,
$$

and the null space spanned by the columns of matrix $\mathbf{A}^T \mathbf{A}$ is of rank $L + 1 - 2p$. Using Prony's equations, one obtains

$$
\mathbf{AB} = \mathbf{0},
$$

where $\mathbf{B}$ is an $(L + 1) \times (L + 1 - 2p)$ matrix of rank $(L + 1 - 2p)$ as follows:

$$\mathbf{B} = \begin{bmatrix} g_0 & 0 & \cdots & 0 \\ g_1 & g_0 & \cdots & 0 \\ \vdots & \vdots & \ddots & \vdots \\ g_{2p} & g_{2p-1} & \cdots & 0 \\ 0 & g_{2p} & \cdots & g_0 \\ 0 & 0 & \ddots & g_1 \\ \vdots & \vdots & \cdots & \vdots \\ 0 & 0 & \cdots & g_{2p} \end{bmatrix},$$

(3.19)

and $g_0, \ldots, g_{2p}$ are the same as defined before. Moreover, the space spanned by the columns of $\mathbf{B}$ is the null space spanned by the columns of matrix $\mathbf{A}^T \mathbf{A}$.

Consider the following $(n - L) \times (L + 1)$ data matrix $\widetilde{\mathbf{A}}$ as follows:

$$\widetilde{\mathbf{A}} = \begin{bmatrix} y(1) & \cdots & y(L+1) \\ \vdots & \ddots & \vdots \\ y(n-L) & \cdots & y(n) \end{bmatrix}.$$

Let the spectral decomposition of $\widetilde{\mathbf{A}}^T \widetilde{\mathbf{A}}/n$ be

$$\frac{1}{n} \widetilde{\mathbf{A}}^T \widetilde{\mathbf{A}} = \sum_{i=1}^{L+1} \widetilde{\sigma}_i^2 \widetilde{\mathbf{u}}_i \widetilde{\mathbf{u}}_i^T,$$

where $\widetilde{\sigma}_1^2 > \cdots > \widetilde{\sigma}_{L+1}^2$ are ordered eigenvalues of $\widetilde{\mathbf{A}}^T \widetilde{\mathbf{A}}/n$, and $\widetilde{\mathbf{u}}_1, \ldots, \widetilde{\mathbf{u}}_{L+1}$ are orthonormal eigenvectors corresponding to $\widetilde{\sigma}_1^2, \ldots, \widetilde{\sigma}_{L+1}^2$, respectively. Construct $(L + 1) \times (L + 1 - 2p)$ matrix $\mathbf{C}$ as

$$\mathbf{C} = \begin{bmatrix} \widetilde{\mathbf{u}}_{2p+1} : \cdots : \widetilde{\mathbf{u}}_{L+1} \end{bmatrix}.$$

Partition matrix $\mathbf{C}$ as

$$\mathbf{C}^T = \begin{bmatrix} \mathbf{C}_{1k}^T : \mathbf{C}_{2k}^T : \mathbf{C}_{3k}^T \end{bmatrix},$$

for $k = 0, 1, \ldots, L - 2p$, where $\mathbf{C}_{1k}^T$, $\mathbf{C}_{2k}^T$, and $\mathbf{C}_{3k}^T$ are of orders $(L + 1 - 2p) \times k$, $(L + 1 - 2p) \times (2p + 1)$ and $(L + 1 - 2p) \times (L - k - 2p)$, respectively. Find an $(L + 1 - 2p)$ vector $\mathbf{x}_k$, such that

$$\begin{bmatrix} \mathbf{C}_{1k}^T \\ \mathbf{C}_{3k}^T \end{bmatrix} \mathbf{x}_k = \mathbf{0}.$$

Denote the vectors for $k = 0, 1, \cdots, L - 2p$

$$\mathbf{b}^k = \mathbf{C}_{2k}^T \mathbf{x}_k,$$

and consider the vector $\mathbf{b}$, the average of the vectors $\mathbf{b}^0, \ldots, \mathbf{b}^{L-2p}$, that is

$$\mathbf{b} = \frac{1}{L + 1 - 2p} \sum_{k=0}^{L-2p} \mathbf{b}^k = \left[\widehat{g}_0, \widehat{g}_1, \ldots, \widehat{g}_{2p} \right].$$

Construct the polynomial equation

$$\widehat{g}_0 + \widehat{g}_1 x + \cdots + \widehat{g}_{2p} x^{2p} = 0, \tag{3.20}$$

and obtain the estimates of the frequencies from complex conjugate roots of (3.20).

It has been shown by Kundu and Mitra [15] that the NSD estimators are strongly consistent, although asymptotic distribution of the NSD estimators has not yet been established. It is observed by extensive simulation studies that the performance of the NSD estimators are very good and it provides the best performance when $L \approx n/3$. The main computation of the NSD estimators involves computation of the singular value decomposition of an $(L + 1) \times (L + 1)$ matrix, and the root findings of a $2p$ degree polynomial equation.

3.5 ESPRIT

Estimation of Signal Parameters via Rotational Invariance Technique (ESPRIT) was first proposed by Roy [24] in his Ph.D. thesis, see also Roy and Kailath [25], which is based on the generalized eigenvalue-based method. The basic idea comes from Prony's system of homogeneous equations. For a given L, when $2p < L < n - 2p$, construct the two data matrices $\mathbf{A}$ and $\mathbf{B}$ both of the order $(n - L) \times L$ as given below:

$$\mathbf{A} = \begin{bmatrix} y(1) & \cdots & y(L) \\ \vdots & \ddots & \vdots \\ y(n - L) & \cdots & y(n - 1) \end{bmatrix}, \quad \mathbf{B} = \begin{bmatrix} y(2) & \cdots & y(L + 1) \\ \vdots & \ddots & \vdots \\ y(n - L + 1) & \cdots & y(n) \end{bmatrix}.$$

If $\mathbf{C}_1 = (\mathbf{A}^T \mathbf{A} - \sigma^2 \mathbf{I})$ and $\mathbf{C}_2 = \mathbf{B}^T \mathbf{A} - \sigma^2 \mathbf{K}$, where $\mathbf{I}$ is the identity matrix of order $L \times L$ and $\mathbf{K}$ is an $L \times L$ matrix with ones along the first lower diagonal off the major diagonal and zeros elsewhere, consider the matrix pencil $\mathbf{C}_1 - \gamma \mathbf{C}_2$.

It has been shown, see Pillai [18] for details, that out of the L eigenvalues of the matrix pencil $\mathbf{C}_1 - \gamma \mathbf{C}_2$, $2p$ nonzero eigenvalues are of the form $e^{\pm i \omega_k}, k = 1, \ldots, p$. Therefore, from those $2p$ nonzero eigenvalues the unknown frequencies $\omega_1, \ldots, \omega_p$ can be estimated. It is further observed that if $\sigma^2 = 0$, then $L - 2p$ eigenvalues of the matrix pencil $\mathbf{C}_1 - \gamma \mathbf{C}_2$ are zero.

The following problems are observed to implement the ESPRIT in practice. Note that both the matrices $\mathbf{C}_1$ and $\mathbf{C}_2$ involve σ^2, which is unknown. If σ^2 is very small it may be ignored, otherwise, it needs to be estimated, or some prior knowledge may be used. Another problem is to separate $2p$ nonzero eigenvalues from a total of L eigenvalues. Again if σ^2 is small it may not be much of a problem, but for large σ^2 separation of nonzero eigenvalues from the zero eigenvalues may not be a trivial issue. The major computational issue is to compute the eigenvalues of the matrix pencil $\mathbf{C}_1 - \gamma \mathbf{C}_2$, and the problem is quite ill-conditioned if σ^2 is small. In this case, the matrices $\mathbf{A}$ and $\mathbf{B}$ both are nearly singular matrices. To avoid both these issues the following method has been suggested.

3.6 TLS-ESPRIT

Total Least Squares ESPRIT (TLS-ESPRIT) has been proposed by Roy and Kailath [25] mainly to overcome some of the problems involved in implementing the ESPRIT algorithm in practice. Using the same notation as in Sect. 3.5, construct the following $2L \times 2L$ matrices $\mathbf{R}$ and $\mathbf{\Sigma}$ as follows:

$$\mathbf{R} = \begin{bmatrix} \mathbf{A}^T \\ \mathbf{B}^T \end{bmatrix} [\mathbf{A} \ \mathbf{B}] \quad \text{and} \quad \mathbf{\Sigma} = \begin{bmatrix} \mathbf{I} & \mathbf{K} \\ \mathbf{K}^T & \mathbf{I} \end{bmatrix}.$$

Let $\mathbf{e}_1, \ldots, \mathbf{e}_{2p}$ be $2p$ orthonormal eigenvectors of the matrix pencil $(\mathbf{R} - \gamma \mathbf{\Sigma})$ corresponding to the largest $2p$ eigenvalues. Now construct the two $L \times 2p$ matrices $\mathbf{E}_1$ and $\mathbf{E}_2$ from $\mathbf{e}_1, \ldots, \mathbf{e}_{2p}$ as

$$[\mathbf{e}_1 : \cdots : \mathbf{e}_{2p}] = \begin{bmatrix} \mathbf{E}_1 \\ \mathbf{E}_2 \end{bmatrix}$$

and then obtain the unique $4p \times 2p$ matrix $\mathbf{W}$ and the two $2p \times 2p$ matrices $\mathbf{W}_1$ and $\mathbf{W}_2$ as follows:

$$[\mathbf{E}_1 : \mathbf{E}_2]\mathbf{W} = \mathbf{0} \quad \text{and} \quad \mathbf{W} = \begin{bmatrix} \mathbf{W}_1 \\ \mathbf{W}_2 \end{bmatrix}.$$

Finally, obtain the $2p$ eigenvalues of $-\mathbf{W}_1\mathbf{W}_2^{-1}$. Again, it has been shown, see Pillai [18], that in the noiseless case, the above $2p$ eigenvalues are of the form $e^{\pm i\omega_k}$ for $k = 1, \ldots, p$. Therefore, the frequencies can be estimated from the eigenvalues of the matrix $-\mathbf{W}_1\mathbf{W}_2^{-1}$.

It is known that the performance of TLS-ESPRIT is very good, and it is much better than the ESPRIT method. The performance of both the methods depends on the values of L. In this case also the main computation involves the computation of the eigenvalues and eigenvectors of the $L \times L$ matrix pencil $(\mathbf{R} - \gamma \mathbf{\Sigma})$. Although the performance of the TLS-ESPRIT is very good, the consistency property of TLS-ESPRIT or ESPRIT has not yet been established.

3.7 Quinn's Method

Quinn [20] proposed his method in estimating the frequency of model (3.1) when $p = 1$. It can be easily extended for general p, see, for example, Kundu and Mitra [15]. Quinn's method can be applied in the presence of error Assumption 3.2. His method is based on the interpolation of the Fourier coefficients, and using the fact that the PE of the frequency when restricted over Fourier frequencies, has the convergence rate $O(1/n)$.

The method can be described as follows: Let

$$Z(j) = \sum_{t=1}^{n} y(t) e^{-i2\pi jt/n}; \quad j = 1, \ldots, n.$$

Algorithm 3.1

- Step 1: Let $\widetilde{\omega}$ be the maximizer of $|Z(j)|^2$, for $1 \leq j \leq n$.
- Step 2: Let $\widehat{\alpha}_1 = Re\{Z(\widetilde{\omega} - 1)/Z(\widetilde{\omega})\}, \widehat{\alpha}_2 = Re\{Z(\widetilde{\omega} + 1)/Z(\widetilde{\omega})\}, \widehat{\delta}_1 = \widehat{\alpha}_1/(1 - \widehat{\alpha}_1)$, and $\widehat{\delta}_2 = -\widehat{\alpha}_2/(1 - \widehat{\alpha}_2)$. If $\widehat{\delta}_1 > 0$ and $\widehat{\delta}_2 > 0$, put $\widehat{\delta} = \widehat{\delta}_2$, otherwise put $\widehat{\delta} = \widehat{\delta}_1$.
- Step 3: Estimate ω by $\widehat{\omega} = 2\pi(\widetilde{\omega} + \widehat{\delta})/n$.

Computationally, Quinn's method is very easy to implement. It is observed that Quinn's method produces consistent estimator of the frequency, and the asymptotic mean squared error of the frequency estimator is of the order $O(1/n^3)$. Although Quinn's method has been proposed for one component only, the method can be easily extended for the model when $p > 1$. The details are explained in Sect. 3.12.

3.8 IQML

The Iterative Quadratic Maximum Likelihood (IQML) method was proposed by Bresler and Macovski [4], and this is the first special purpose algorithm which has been used to compute the LSEs of the unknown parameters of model (3.1). It is well-known that in the presence of i.i.d. additive normal errors, the LSEs become the MLEs also. Rewrite model (3.1) as follows:

$$\mathbf{Y} = \mathbf{Z}(\omega)\theta + \mathbf{e}, \tag{3.21}$$

where

$$\mathbf{Y} = \begin{bmatrix} y(1) \\ \vdots \\ y(n) \end{bmatrix}, \mathbf{Z}(\boldsymbol{\omega}) = \begin{bmatrix} \cos(\omega_1) & \sin(\omega_1) & \cdots & \cos(\omega_p) & \sin(\omega_p) \\ \vdots & \vdots & \ddots & \vdots & \vdots \\ \cos(n\omega_1) & \sin(n\omega_1) & \cdots & \cos(n\omega_p) & \sin(n\omega_p) \end{bmatrix},$$

$$\boldsymbol{\theta}^T = [A_1, B_1, \ldots, A_p, B_p], \mathbf{e}^T = [X(1), \ldots, X(n)], \boldsymbol{\omega}^T = (\omega_1, \ldots, \omega_p).$$

Therefore, the LSEs of the unknown parameters can be obtained by minimizing

$$Q(\boldsymbol{\omega}, \boldsymbol{\theta}) = (\mathbf{Y} - \mathbf{Z}(\boldsymbol{\omega})\boldsymbol{\theta})^T (\mathbf{Y} - \mathbf{Z}(\boldsymbol{\omega})\boldsymbol{\theta}) \tag{3.22}$$

with respect to $\boldsymbol{\omega}$ and $\boldsymbol{\theta}$. Therefore, for a given $\boldsymbol{\omega}$, the LSE of $\boldsymbol{\theta}$ can be obtained as

$$\widehat{\boldsymbol{\theta}}(\boldsymbol{\omega}) = (\mathbf{Z}(\boldsymbol{\omega})^T \mathbf{Z}(\boldsymbol{\omega}))^{-1} \mathbf{Z}(\boldsymbol{\omega})^T \mathbf{Y}. \tag{3.23}$$

Substituting back (3.23) in (3.22), we obtain

$$R(\boldsymbol{\omega}) = Q(\widehat{\boldsymbol{\theta}}(\boldsymbol{\omega}), \boldsymbol{\omega}) = \mathbf{Y}^T (\mathbf{I} - \mathbf{P_Z})\mathbf{Y}. \tag{3.24}$$

Here, $\mathbf{P_Z} = \mathbf{Z}(\boldsymbol{\omega})(\mathbf{Z}(\boldsymbol{\omega})^T \mathbf{Z}(\boldsymbol{\omega}))^{-1} \mathbf{Z}(\boldsymbol{\omega})^T$ is the projection matrix on the space spanned by the columns of $\mathbf{Z}(\boldsymbol{\omega})$. Note that $\mathbf{I} - \mathbf{P_Z} = \mathbf{P_B}$, where the matrix $\mathbf{B} = \mathbf{B(g)}$ is the same as defined in (3.19), and $\mathbf{P_B} = \mathbf{B}(\mathbf{B}^T\mathbf{B})^{-1}\mathbf{B}^T$ is the projection matrix orthogonal to $\mathbf{P_Z}$. The IQML method mainly suggests how to minimize $\mathbf{Y}^T \mathbf{P_B}\mathbf{Y}$ with respect to the unknown vector $\mathbf{g} = (g_0, \ldots, g_{2p})^T$, which is equivalent to minimize (3.24) with respect to the unknown parameter vector $\boldsymbol{\omega}$. First, observe that

$$\mathbf{Y}^T \mathbf{P_B}\mathbf{Y} = \mathbf{g}^T \mathbf{Y}_D^T (\mathbf{B}^T\mathbf{B})^{-1}\mathbf{Y}_D \, \mathbf{g}, \tag{3.25}$$

where $\mathbf{Y}_D$ is an $(n - 2p) \times (2p + 1)$ matrix as defined in (3.9). The following algorithm has been suggested by Bresler and Macovski [4] to minimize $\mathbf{Y}^T\mathbf{P_B}\mathbf{Y} = \mathbf{g}^T\mathbf{Y}_D^T(\mathbf{B}^T\mathbf{B})^{-1}\mathbf{Y}_D \, \mathbf{g}$.

Algorithm 3.2

- Step 1: Suppose at the kth step, the value of the vector $\mathbf{g}$ is $\mathbf{g}_{(k)}$.
- Step 2: Compute matrix $\mathbf{C}_{(k)} = \mathbf{Y}_D^T(\mathbf{B}_{(k)}^T\mathbf{B}_{(k)})^{-1}\mathbf{Y}_D$; here $\mathbf{B}_{(k)}$ is obtained by replacing $\mathbf{g}$ with $\mathbf{g}_{(k)}$ in matrix $\mathbf{B}$ given in (3.19).
- Step 3: Solve the quadratic optimization problem

$$\min_{\mathbf{x}:\|\mathbf{x}\|=1} \mathbf{x}^T \mathbf{C}_{(k)}\mathbf{x},$$

and suppose the solution is $\mathbf{g}_{(k+1)}$.
- Step 4: Check the convergence whether $|\mathbf{g}_{(k+1)} - \mathbf{g}_{(k)}| < \varepsilon$, where ε is some pre-assigned value. If the convergence is met, go to step (5), otherwise go to step (1).

- Step 5: Obtain the estimate of ω from the estimate of $\mathbf{g}$.

Although no proof of convergence is available for the above algorithm, it works quite well in practice.

3.9   Modified Prony Algorithm

Modified Prony algorithm was proposed by Kundu [11], which also involves the minimization of

$$\Psi(\mathbf{g}) = \mathbf{Y}^T \mathbf{P_B} \mathbf{Y} \tag{3.26}$$

with respect to the vector $\mathbf{g}$, where the matrix $\mathbf{B}(\mathbf{g})$ and the projection matrix $\mathbf{P_B}$ is the same as defined in Sect. 3.8. Now observe that $\Psi(\mathbf{g})$ is invariant under scalar multiplication, that is,

$$\Psi(\mathbf{g}) = \Psi(c\mathbf{g}),$$

for any constant $c \in I\!R$. Therefore,

$$\min_{\mathbf{g}} \Psi(\mathbf{g}) = \min_{\mathbf{g};\ \mathbf{g}^T\mathbf{g}=1} \Psi(\mathbf{g}).$$

To minimize $\Psi(\mathbf{g})$ with respect to $\mathbf{g}$, differentiate $\Psi(\mathbf{g})$ with respect to different components of $\mathbf{g}$ and equating them to zero lead to solving the following nonlinear equation

$$\mathbf{C}(\mathbf{g})\mathbf{g} = \mathbf{0}; \quad \mathbf{g}^T\mathbf{g} = 1, \tag{3.27}$$

where $\mathbf{C} = \mathbf{C}(\mathbf{g})$ is a $(2p+1) \times (2p+1)$ symmetric matrix whose (i, j)th element, for $i, j = 1, \ldots, 2p+1$, is given by

$$c_{ij} = \mathbf{Y}^T \mathbf{B}_i (\mathbf{B}^T \mathbf{B})^{-1} \mathbf{B}_j^T \mathbf{Y} - \mathbf{Y}^T \mathbf{B} (\mathbf{B}^T \mathbf{B})^{-1} \mathbf{B}_j^T \mathbf{B}_i (\mathbf{B}^T \mathbf{B})^{-1} \mathbf{B}^T \mathbf{Y}.$$

Here, the elements of matrix $\mathbf{B}_j^T$, for $j = 1, \ldots, 2p+1$, are only zeros and ones such that

$$\mathbf{B}(\mathbf{g}) = \sum_{j=0}^{2p} g_j \mathbf{B}_j.$$

The problem (3.27) is a nonlinear eigenvalue problem and the following iterative scheme has been suggested by Kundu [11] to solve the set of nonlinear equations:

$$(\mathbf{C}(\mathbf{g}^{(k)}) - \lambda^{(k+1)}\mathbf{I})\mathbf{g}^{(k+1)} = \mathbf{0}; \quad \mathbf{g}^{(k+1)T}\mathbf{g}^{(k+1)} = 1. \tag{3.28}$$

Here, $\mathbf{g}^{(k)}$ denotes the kth iterate of the above iterative process, and $\lambda^{(k+1)}$ is the eigenvalue of $\mathbf{C}(\mathbf{g}^{(k)})$ which is closest to 0. The iterative process is stopped when

$|\lambda^{(k+1)}|$ is sufficiently small compared to $||\mathbf{C}||$, where $||\mathbf{C}||$ denotes a matrix norm of matrix $\mathbf{C}$. The proof of convergence of the modified Prony algorithm can be found in Kundu [12].

3.10 Constrained Maximum Likelihood Method

In the IQML or the modified Prony algorithm, the symmetric structure of vector $\mathbf{g}$ as derived in (2.42) has not been utilized. The constrained MLEs, proposed by Kannan and Kundu [9], utilized that symmetric structure of vector $\mathbf{g}$. The problem is same as to minimize $\Psi(\mathbf{g})$ as given in (3.26) with respect to $g_0, g_1, \ldots, g_{2p}$.

Again differentiating $\Psi(\mathbf{g})$ with respect to $g_0, \ldots, g_{2p}$ and equating them to zero lead to the matrix equation of the form

$$\mathbf{C(x)x = 0}. \tag{3.29}$$

Here, $\mathbf{C}$ is a $(p+1) \times (p+1)$ matrix and $\mathbf{x} = (x_0, \ldots, x_p)^T$ vector. The elements of matrix $\mathbf{C}$ say c_{ij} for $i, j = 0, \ldots, p$ are as follows:

$$c_{ij} = \mathbf{Y}^T \mathbf{U}_i (\mathbf{B}^T \mathbf{B})^{-1} \mathbf{U}_j^T \mathbf{Y} + \mathbf{Y}^T \mathbf{U}_j (\mathbf{B}^T \mathbf{B})^{-1} \mathbf{U}_i^T \mathbf{Y}$$
$$- \mathbf{Y}^T \mathbf{B} (\mathbf{B}^T \mathbf{B})^{-1} (\mathbf{U}_i^T \mathbf{U}_j + \mathbf{U}_j^T \mathbf{U}_i) (\mathbf{B}^T \mathbf{B})^{-1} \mathbf{B}^T \mathbf{Y}.$$

Here, matrix $\mathbf{B}$ is the same as defined in (3.19), with g_{p+k} being replaced by g_{p-k}, for $k = 1, \ldots, p$. $\mathbf{U}_1, \ldots, \mathbf{U}_p$ are $n \times (n - 2p)$ matrices with entries 0 and 1 only, such that

$$\mathbf{B} = \sum_{j=0}^{p} g_j \mathbf{U}_j.$$

Similar iterative scheme as the modified Prony algorithm has been used to solve for $\widehat{\mathbf{x}} = (\widehat{x}_0, \ldots, \widehat{x}_p)^T$, the solution of (3.29). Once $\widehat{\mathbf{x}}$ is obtained, $\widehat{\mathbf{g}}$ can be easily obtained as follows:

$$\widehat{\mathbf{g}} = (\widehat{x}_0, \ldots, \widehat{x}_{p-1}, \widehat{x}_p, \widehat{x}_{p-1}, \ldots, \widehat{x}_0)^T.$$

From $\widehat{\mathbf{g}}$, the estimates of $\omega_1, \ldots, \omega_p$ can be obtained along the same line as before. The proof of convergence of the algorithm has been established by Kannan and Kundu [9]. The performances of the constrained MLEs are very good as expected.

3.11 Expectation Maximization Algorithm

Expectation Maximization (EM) algorithm, developed by Dempster, Laird, and Rubin [6] is a general method for solving the maximum likelihood estimation problem when the data are incomplete. The details on EM algorithm can be found in a book by McLachlan and Krishnan [16]. Although this algorithm has been originally used for incomplete data, sometimes it can be used quite successfully even when the data are complete. It has been used quite effectively to estimate unknown parameters of model (3.1) by Feder and Weinstein [7] under the assumption that the errors are i.i.d. normal random variables.

For a better understanding, we briefly explain the EM algorithm here. Let $\mathbf{Y}$ denote the observed (may be incomplete) data with the probability density function $f_{\mathbf{Y}}(\mathbf{y}; \boldsymbol{\theta})$ indexed by the parameter vector $\boldsymbol{\theta} \in \boldsymbol{\Theta} \subset I\!\!R^k$, and let $\mathbf{X}$ denote the complete data vector related to $\mathbf{Y}$ by

$$H(\mathbf{X}) = \mathbf{Y},$$

where $H(\cdot)$ is a many to one non-invertible function. Therefore, the density function of $\mathbf{X}$, say $f_{\mathbf{X}}(\mathbf{x}; \boldsymbol{\theta})$ can be written as

$$f_{\mathbf{X}}(\mathbf{x}; \boldsymbol{\theta}) = f_{\mathbf{X}|\mathbf{Y}=\mathbf{y}}(\mathbf{x}; \boldsymbol{\theta}) f_{\mathbf{Y}}(\mathbf{y}; \boldsymbol{\theta}) \quad \forall H(\mathbf{x}) = \mathbf{y}. \tag{3.30}$$

Here $f_{\mathbf{X}|\mathbf{Y}=\mathbf{y}}(\mathbf{x}; \boldsymbol{\theta})$ is the conditional probability density function of $\mathbf{X}$ given $\mathbf{Y} = \mathbf{y}$. From (3.30) after taking the logarithm on both sides, we obtain

$$\ln f_{\mathbf{Y}}(\mathbf{y}; \boldsymbol{\theta}) = \ln f_{\mathbf{X}}(\mathbf{x}; \boldsymbol{\theta}) - \ln f_{\mathbf{X}|\mathbf{Y}=\mathbf{y}}(\mathbf{x}; \boldsymbol{\theta}). \tag{3.31}$$

Taking the conditional expectation given $\mathbf{Y} = \mathbf{y}$ at the parameter value $\boldsymbol{\theta}'$ on both sides of (3.31) gives

$$\ln f_{\mathbf{Y}}(\mathbf{y}; \boldsymbol{\theta}) = E\{\ln f_{\mathbf{X}}(\mathbf{x}; \boldsymbol{\theta})|\mathbf{Y} = \mathbf{y}, \boldsymbol{\theta}'\} - E\{\ln f_{\mathbf{X}|\mathbf{Y}=\mathbf{y}}(\mathbf{x}; \boldsymbol{\theta})|\mathbf{Y} = \mathbf{y}, \boldsymbol{\theta}'\}. \tag{3.32}$$

If we define $L(\boldsymbol{\theta}) = \ln f_{\mathbf{Y}}(\mathbf{y}; \boldsymbol{\theta}), U(\boldsymbol{\theta}, \boldsymbol{\theta}') = E\{\ln f_{\mathbf{X}}(\mathbf{x}; \boldsymbol{\theta})|\mathbf{Y} = \mathbf{y}, \boldsymbol{\theta}'\}$ and $V(\boldsymbol{\theta}, \boldsymbol{\theta}') = E\{\ln f_{\mathbf{X}|\mathbf{Y}=\mathbf{y}}(\mathbf{x}; \boldsymbol{\theta})|\mathbf{Y} = \mathbf{y}, \boldsymbol{\theta}'\}$, then (3.32) becomes

$$L(\boldsymbol{\theta}) = U(\boldsymbol{\theta}, \boldsymbol{\theta}') - V(\boldsymbol{\theta}, \boldsymbol{\theta}').$$

Here, $L(\boldsymbol{\theta})$ is the log-likelihood function of the observed data and that needs to be maximized to obtain the MLEs of $\boldsymbol{\theta}$. Due to Jensen's inequality, see, for example, Chung [5], $V(\boldsymbol{\theta}, \boldsymbol{\theta}') \leq V(\boldsymbol{\theta}', \boldsymbol{\theta}')$, therefore, if,

$$U(\boldsymbol{\theta}, \boldsymbol{\theta}') > U(\boldsymbol{\theta}', \boldsymbol{\theta}'),$$

then

$$L(\boldsymbol{\theta}) > L(\boldsymbol{\theta}'). \tag{3.33}$$

The relation (3.33) forms the basis of the EM algorithm. The algorithm starts with an initial guess and we denote it by $\widehat{\boldsymbol{\theta}}^{(m)}$, the current estimate of θ after m-iterations. Then $\widehat{\boldsymbol{\theta}}^{(m+1)}$ can be obtained as follows:

$$\text{E-Step : Compute } U(\boldsymbol{\theta}, \widehat{\boldsymbol{\theta}}^{(m)})$$

$$\text{M-Step : Compute } \widehat{\boldsymbol{\theta}}^{(m)} = \arg\max_{\theta} U(\boldsymbol{\theta}, \widehat{\boldsymbol{\theta}}^{(m)}).$$

Now we show how the EM algorithm can be used to estimate the unknown frequencies and amplitudes of model (3.1) when the errors are i.i.d. normal random variables with mean zero and variance σ^2. Under these assumptions, the log-likelihood function without the constant term takes the following form:

$$l(\boldsymbol{\omega}) = -n \ln \sigma - \sum_{t=1}^{n} \frac{1}{\sigma^2} \left(y(t) - \sum_{k=1}^{p} (A_k \cos(\omega_k t) + B_k \sin(\omega_k t)) \right)^2. \quad (3.34)$$

It is clear that if $\widehat{A}_k$, $\widehat{B}_k$, and $\widehat{\omega}_k$ are the MLEs of A_k, B_k, and ω_k, respectively, for $k = 1, \ldots, p$, then the MLE of σ^2 can be obtained as

$$\widehat{\sigma}^2 = \frac{1}{n} \sum_{t=1}^{n} \left(y(t) - \sum_{k=1}^{p} (\widehat{A}_k \cos(\widehat{\omega}_k t) + \widehat{B}_k \sin(\widehat{\omega}_k t)) \right)^2.$$

It is clear that the MLEs of A_k, B_k, and ω_k for $k = 1, \ldots, p$ can be obtained by minimizing

$$\frac{1}{n} \sum_{t=1}^{n} \left(y(t) - \sum_{k=1}^{p} (A_k \cos(\omega_k t) + B_k \sin(\omega_k t)) \right)^2, \quad (3.35)$$

with respect to the unknown parameters. The EM algorithm can be developed to compute the MLEs of A_k, B_k, and ω_k for $k = 1, \ldots, p$, in this case. In developing the EM algorithm, Feder and Weinstein [7] assumed that the noise variance σ^2 is known, and without loss of generality it can be taken as 1.

To implement the EM algorithm, rewrite the data vector $\mathbf{y}(t)$ as follows:

$$\mathbf{y}(t) = \left(y_1(t), \ldots, y_p(t) \right)^T, \quad (3.36)$$

where

$$y_k(t) = A_k \cos(\omega_k t) + B_k \sin(\omega_k t) + X_k(t).$$

Here, $X_k(t)$ for $k = 1, \ldots, p$ are obtained by arbitrarily decomposing the total noise $X(t)$ into p components, so that

$$\sum_{k=1}^{p} X_k(t) = X(t).$$

Therefore, if $\mathbf{H} = [1, \ldots, 1]$ is a $p \times 1$ vector, then model (3.1) can be written as

$$y(t) = \sum_{k=1}^{p} y_k(t) = \mathbf{H}\mathbf{y}(t).$$

If we choose $X_1(t), \ldots, X_p(t)$ to be independent normal random variables with mean zero and variance $\beta_1, \ldots, \beta_p$, respectively, then

$$\sum_{k=1}^{p} \beta_k = 1, \quad \beta_k > 0.$$

With the above notation, the EM algorithm takes the following form. If $\widehat{A}_k^{(m)}$, $\widehat{B}_k^{(m)}$, and $\widehat{\omega}_k^{(m)}$ denote the estimates of A_k, B_k, and ω_k, respectively after m-iterations, then
E-Step:

$$\widehat{y}_k^{(m)}(t) = \widehat{A}_k^{(m)} \cos(\widehat{\omega}_k^{(m)} t) + \widehat{B}_k^{(m)} \sin(\widehat{\omega}_k^{(m)} t)$$

$$+\beta_k \left[y(t) - \sum_{k=1}^{p} \left\{ \widehat{A}_k^{(m)} \cos(\widehat{\omega}_k^{(m)} t) + \widehat{B}_k^{(m)} \sin(\widehat{\omega}_k^{(m)} t) \right\} \right]. \quad (3.37)$$

M-Step:

$$(\widehat{A}_k^{(m+1)}, \widehat{B}_k^{(m+1)}, \widehat{\omega}_k^{(m+1)}) = \arg \min_{A,B,\omega} \sum_{t=1}^{n} \left(\widehat{y}_k^{(m)}(t) - A \cos(\omega t) - B \sin(\omega t) \right)^2.$$

$$(3.38)$$

It is interesting to observe that $\widehat{A}_k^{(m+1)}$, $\widehat{B}_k^{(m+1)}$, $\widehat{\omega}_k^{(m+1)}$ are the MLEs of A_k, B_k, ω_k, respectively, based on $\widehat{y}_k^{(m)}(t), t = 1, \ldots, n$. The most important feature of this algorithm is that it decomposes the complicated optimization problem into p separate simple 1-D optimization problems.

Feder and Weinstein [7] did not mention how to choose $\beta_1, \ldots, \beta_p$ and how the EM algorithm can be used when the error variance σ^2 is unknown. The choice of $\beta_1, \ldots, \beta_p$ plays an important role in the performance of the algorithm. One choice might be to take $\beta_1 = \cdots = \beta_p$, alternatively dynamical choice of $\beta_1, \ldots, \beta_p$ might provide better results.

We propose the following EM algorithm when σ^2 is unknown. Suppose $\widehat{A}_k^{(m)}$, $\widehat{B}_k^{(m)}$, $\widehat{\omega}_k^{(m)}$, and $\widehat{\sigma}^{2(m)}$ are the estimates of A_k, B_k, ω_k, and σ^2, respectively, at the m-step of the EM algorithm. They may be obtained from the periodogram estimates. Choose $\beta_k^{(m)}$ as

$$\beta_k^{(m)} = \frac{\widehat{\sigma}^{2(m)}}{p}; \quad k = 1, \dots, p.$$

In the E-step of (3.37) replace β_k by $\beta_k^{(m)}$, and in the M-step after computing (3.38) also obtain

$$\widehat{\sigma}^{2(m+1)} = \frac{1}{n} \sum_{t=1}^{n} \left(y(t) - \sum_{k=1}^{p} \left\{ \widehat{A}_k^{(m+1)} \cos(\widehat{\omega}_k^{(m+1)}t) + \widehat{B}_k^{(m+1)} \sin(\widehat{\omega}_k^{(m+1)}t) \right\} \right)^2 .$$

The iteration continues unless the convergence criterion is met. The proof of convergence or the properties of the estimators has not yet been established. Further work is needed along that direction.

3.12 Sequential Estimators

One major drawback of different estimators discussed so far is in their computational complexity. Sometimes to reduce the computational complexity, efficiency has been sacrificed. Prasad, Kundu, and Mitra [19] suggested sequential estimators, where computational complexity has been reduced and at the same time the efficiency of the estimators has not been sacrificed.

Prasad, Kundu, and Mitra [19] considered model (3.1) under error Assumptions 3.1 and 3.2. The sequential method is basically a modification of the approximate least squares method as described in Sect. 3.1. Using the same notation as in (3.5), the method can be described as follows.

Algorithm 3.3

- Step 1: Compute $\widehat{\omega}_1$, which can be obtained by minimizing $R_1(\omega)$, with respect to ω, where

$$R_1(\omega) = \mathbf{Y}^T (\mathbf{I} - \mathbf{P}_{\mathbf{Z}(\omega)}) \mathbf{Y}. \tag{3.39}$$

Here, $\mathbf{Z}(\omega)$ is the same as defined in Sect. 3.1, and $\mathbf{P}_{\mathbf{Z}(\omega)}$ is the projection matrix on the column space of $\mathbf{Z}(\omega)$.
- Step 2: Construct the following vector

$$\mathbf{Y}^{(1)} = \mathbf{Y} - \mathbf{Z}(\widehat{\omega}_1)\widehat{\alpha}_1, \tag{3.40}$$

where

$$\widehat{\alpha}_1 = \left[\mathbf{Z}^T(\widehat{\omega}_1)\mathbf{Z}(\widehat{\omega}_1) \right]^{-1} \mathbf{Z}^T(\widehat{\omega}_1)\mathbf{Y}.$$

- Step 3: Compute $\widehat{\omega}_2$, which can be obtained by minimizing $R_2(\omega)$, with respect to ω, where $R_2(\omega)$ is obtained by replacing $\mathbf{Y}$ with $\mathbf{Y}^{(1)}$ in (3.39).
- Step 4: The process continues up to p steps.

The main advantage of the proposed algorithm is that it significantly reduces the computational burden. The minimization of $R_k(\omega)$ for each k is a 1-D optimization problem, and it can be performed quite easily. It has been shown by the authors that the estimators obtained by this sequential procedure are strongly consistent and they have the same rate of convergence as the LSEs. Moreover, if the process continues even after p steps, it has been shown that the estimators of A and B converge to zero almost surely.

3.13   Quinn and Fernandes Method

Quinn and Fernandes [21] proposed the following method by exploiting the fact that there is a second-order filter which annihilates a sinusoid at a given frequency. First, consider model (3.1) with $p = 1$. It follows from Prony's equation that the model satisfies the following equation:

$$y(t) - 2\cos(\omega)y(t-1) + y(t-2) = X(t) - 2\cos(\omega)X(t-1) + X(t-2).$$
(3.41)

Therefore, $\{y(t)\}$ forms an autoregressive moving average, namely ARMA(2,2) process. It may be noted that the process does not have a stationary or invertible solution. As expected the above process does not depend on the linear parameters A and B, but only depends on the nonlinear frequency ω. It is clear from (3.41) that using the above ARMA(2,2) structure of $\{y(t)\}$, the estimate of ω is possible to obtain.

Rewrite (3.41) as follows:

$$y(t) - \beta y(t-1) + y(t-2) = X(t) - \alpha X(t-1) + X(t-2),$$
(3.42)

and the problem is to estimate the unknown parameter with the constraint $\alpha = \beta$, based on the observation $\{y(t)\}$ for $t = 1, \dots, n$. It is important to note that if the standard ARMA-based technique is used to estimate the unknown parameter, it can only produce estimator which has the asymptotic variance of the order $O(n^{-1})$. On the other hand, it is known that the LSE of ω has the asymptotic variance of the order $O(n^{-3})$. Therefore, some "non-standard" ARMA-based technique needs to be used to obtain an efficient frequency estimator.

If α is known, and $X(1), \dots, X(n)$ are i.i.d. normal random variables, then the MLE of β can be obtained by minimizing

$$Q(\beta) = \sum_{t=1}^{n} (\xi(t) - \beta\xi(t-1) + \xi(t-2))^2,$$
(3.43)

with respect to β, where $\xi(t) = 0$ for $t < 1$, and for $t \geq 1$,

$$\xi(t) = y(t) + \alpha\xi(t-1) - \xi(t-2).$$

The value of β which minimizes (3.43) can easily be obtained as

$$\alpha + \frac{\sum_{t=1}^{n} y(t)\xi(t-1)}{\sum_{t=1}^{n} \xi^2(t-1)}. \tag{3.44}$$

Therefore, one way can be to put the new value of α in (3.44) and then re-estimate β. This basic idea has been used by Quinn and Fernandes [21] with a proper acceleration factor, which ensures the convergence of the iterative procedure also. The algorithm is as follows.

Algorithm 3.4

- Step 1: Put $\alpha^{(1)} = 2\cos(\widehat{\omega}^{(1)})$, where $\widehat{\omega}^{(1)}$ is an initial estimator of ω.
- Step 2: For $j \geq 1$, compute

$$\xi(t) = y(t) + \alpha^{(j)}\xi(t-1) - \xi(t-2); \quad t = 1, \ldots, n.$$

- Step 3: Obtain

$$\beta^{(j)} = \alpha^{(j)} + \frac{\sum_{t=1}^{n} y(t)\xi(t-1)}{\sum_{t=1}^{n} \xi^{(}t-1)}.$$

- Step 4: If $|\alpha^{(j)} - \beta^{(j)}|$ is small then stop the iteration procedure, and obtain estimate of ω as $\widehat{\omega} = \cos^{-1}(\beta^{(j)}/2)$. Otherwise obtain $\alpha^{(j+1)} = \beta^{(j)}$, and go back to Step 2.

In the same paper, the authors extended the algorithm for general model (3.1) also based on the observation that a certain difference operator annihilates all the sinusoidal components. If $y(1), \ldots, y(n)$ are obtained from model (3.1), then from Prony's equations it again follows that there exists $\alpha_0, \ldots, \alpha_{2p}$, so that

$$\sum_{j=0}^{2p} \alpha_j y(t-j) = \sum_{j=0}^{2p} \alpha_j X(t-j), \tag{3.45}$$

where

$$\sum_{j=0}^{2p} \alpha_j z^j = \prod_{j=1}^{p} (1 - 2\cos(\omega_j)z + z^2). \tag{3.46}$$

It is clear from (3.46) that $\alpha_0 = \alpha_{2p} = 1$, and $\alpha_{2p-j} = \alpha_j$ for $j = 0, \ldots, p-1$. Therefore, in this case $y(1), \ldots, y(n)$ form an ARMA$(2p, 2p)$ process, and all the zeros of the corresponding auxiliary polynomial are on the unit circle. It can also be observed from the Prony's equations that no other polynomial of order less than $2p$ has this property.

Following exactly the same reasoning as before, Quinn and Fernandes [21] suggested the following algorithm for the multiple sinusoidal model.

Algorithm 3.5

- Step 1: If $\widetilde{\omega}_1, \ldots, \widetilde{\omega}_p$ are initial estimators of $\omega_1, \ldots, \omega_p$, compute $\alpha_1, \ldots, \alpha_p$ from

$$\sum_{j=0}^{2p} \alpha_j z^j = \prod_{j=1}^{p} (1 - 2z \cos(\widetilde{\omega}_j) + z^2).$$

- Step 2: Compute for $t = 1, \ldots, n$

$$\xi(t) = y(t) - \sum_{j=1}^{2p} \alpha_j \xi(t - j)$$

and for $j = 1, \ldots, p - 1$ compute the $p \times 1$ vector

$$\boldsymbol{\eta}(t - 1) = \left[\widetilde{\xi}(t - 1), \ldots, \widetilde{\xi}(t - p + 1), \, \xi(t - p) \right]^T,$$

where

$$\widetilde{\xi}(t - j) = \xi(t - j) + \xi(t - 2p + j),$$

and $\xi(t) = 0$, for $t < 1$.

- Step 3: Let $\boldsymbol{\alpha} = \left[\alpha_1, \ldots, \alpha_p \right]^T$ and compute $\boldsymbol{\beta} = \left[\beta_1, \ldots, \beta_p \right]^T$, where

$$\boldsymbol{\beta} = \boldsymbol{\alpha} - 2 \left\{ \sum_{t=1}^{n} \boldsymbol{\eta}(t-1)\boldsymbol{\eta}(t-1)^T \right\}^{-1} \sum_{t=1}^{n} y(t)\boldsymbol{\eta}(t-1).$$

- Step 4: If $\max\limits_{j} |\beta_j - \alpha_j|$ is small stop the iteration, and obtain estimates of $\omega_1, \ldots, \omega_p$. Otherwise set $\boldsymbol{\alpha} = \boldsymbol{\beta}$ and go to Step 2.

This algorithm works very well for small p. For large p the performance is not very satisfactory, as it involves a $2p \times 2p$ matrix inversion, which is quite ill-conditioned. Due to this reason, the elements of the vectors $\boldsymbol{\alpha}$ and $\boldsymbol{\beta}$ obtained by this algorithm can be quite large.

Quinn and Fernandes [21] suggested the following modified algorithm which works very well even for large p also.

Algorithm 3.6

- Step 1: If $\widetilde{\omega}_1, \ldots, \widetilde{\omega}_p$ are initial estimators of $\omega_1, \ldots, \omega_p$, respectively, compute $\theta_k = 2 \cos(\widetilde{\omega}_k)$, for $k = 1, \ldots, p$.
- Step 2: Compute for $t = 1, \ldots, n, j = 1, \ldots, p, \zeta_j(t)$, where $\zeta_j(-1) = \zeta_j(-2) = 0$, and they satisfy

$$\zeta_j(t) - \theta_j \zeta_j(t - 1) + \zeta_j(t - 2) = y(t).$$

- Step 3: Compute $\boldsymbol{\theta} = [\theta_1, \ldots, \theta_p]^T$, $\boldsymbol{\zeta}(t) = [\zeta_1(t), \ldots, \zeta_p(t)]^T$, and

$$\boldsymbol{\psi} = \boldsymbol{\theta} - 2 \left\{ \sum_{t=1}^{n} \boldsymbol{\zeta}(t-1)\boldsymbol{\zeta}(t-1)^T \right\}^{-1} \sum_{t=1}^{n} y(t)\boldsymbol{\zeta}(t-1).$$

- Step 4: If $|\boldsymbol{\psi} - \boldsymbol{\theta}|$ is small stop the iteration, and obtain the estimates of $\omega_1, \ldots, \omega_p$. Otherwise set $\boldsymbol{\theta} = \boldsymbol{\psi}$ and go to Step 2.

Comments: Note that in both the above algorithms which have been proposed by Quinn and Fernandes [21] for $p > 1$, involve the computation of a $2p \times 2p$ matrix inversion. Therefore, if p is very large, then it becomes a computationally challenging problem.

3.14 Amplified Harmonics Method

The estimation of frequencies of model (3.1) based on amplified harmonics was proposed by Truong-Van [26]. The main idea of Truong-Van [26] is to construct a process which enhances the amplitude of a particular frequency quasi-linearly with t, whereas the amplitudes of the other frequencies remain constant in time. The method is as follows. For each frequency ω_k, and any estimate $\omega_k^{(0)}$ near ω_k, define the process $\xi(t)$ as the solution of the following linear equation:

$$\xi(t) - 2\alpha_k^{(0)}\xi(t-1) + \xi(t-2) = y(t); \quad t \geq 1, \tag{3.47}$$

with the initial conditions $\xi(0) = \xi(-1) = 0$, and $\alpha_k^{(0)} = \cos(\omega_k^{(0)})$. Using the results of Ahtola and Tiao [1], it can be shown that

$$\xi(t) = \sum_{j=0}^{t-1} v(j; \omega_k^{(0)}) y(t-k); \quad t \geq 1, \tag{3.48}$$

where

$$v(j; \omega) = \frac{\sin(\omega(j+1))}{\sin(\omega)}.$$

Note that the process $\{\xi(t)\}$ depends on $\omega_k^{(0)}$, but we do not make it explicit. If it is needed, we denote it by $\{\xi(t; \omega_k^{(0)})\}$. It has been shown by the author that the process $\{\xi(t)\}$ acts like an amplifier of the frequency ω_k. It has been further shown that such an amplifier exists and an estimator $\widehat{\omega}_k$ of ω_k was proposed by observing the fact that $y(t)$ and $\xi(t-1; \widehat{\omega}_k)$ are orthogonal to each other, that is,

$$\sum_{t=2}^{n} \xi(t-1; \widehat{\omega}_k) y(t) = 0. \tag{3.49}$$

Truong-Van [26] proposed two different algorithms mainly to solve (3.49). The first algorithm (Algorithm 3.7) has been proposed when the initial guess values are very close to the true values and the algorithm is based on Newton's method to solve (3.49). The second algorithm (Algorithm 3.8) has been proposed when the initial guess values are not very close to the true values, and this algorithm also tries to find a solution of the nonlinear equation (3.49) using least squares approach.

Algorithm 3.7

- Step 1: Find an initial estimator $\omega_k^{(0)}$ of ω_k.
- Step 2: Compute

$$\omega_k^{(1)} = \omega_k^{(0)} - F(\omega_k^{(0)})(F'(\omega_k^{(0)}))^{-1},$$

where

$$F(\omega) = \sum_{t=2}^{n} \xi(t-1; \omega)y(t), \quad \text{and} \quad F'(\omega) = \sum_{t=2}^{n} \frac{d}{d\omega}\xi(t-1; \omega)y(t).$$

- Step 3: If $\omega_k^{(0)}$ and $\omega_k^{(1)}$ are close to each other stop the iteration, otherwise replace $\omega_k^{(0)}$ by $\omega_k^{(1)}$ and continue the process.

Algorithm 3.8

- Step 1: Find an initial estimator $\omega_k^{(0)}$ of ω_k, and compute $\alpha_k^{(0)} = \cos(\omega_k^{(0)})$.
- Step 2: Compute

$$\alpha_k^{(1)} = \alpha_k^{(0)} + \left(2 \sum_{t=2}^{n} \xi^2(t-1; \omega_k^{(0)})\right)^{-1} F(\omega_k^{(0)}).$$

- Step 3: If $\alpha_k^{(0)}$ and $\alpha_k^{(1)}$ are close to each other stop the iteration, otherwise replace $\alpha_k^{(0)}$ by $\alpha_k^{(1)}$ and continue the process. From the estimate of α_k, the estimate of ω_k can be easily obtained.

3.15   Weighted Least Squares Estimators

Weighted Least Squares Estimators (WLSEs) are proposed by Irizarry [8]. The main idea is to produce asymptotically unbiased estimators, which may have lower variances than the LSEs depending on the weight function. Irizarry [8] considered model (3.1), and WLSEs of the unknown parameters can be obtained by minimizing

$$S(\omega, \theta) = \sum_{t=1}^{n} w\left(\frac{t}{n}\right) \left(y(t) - \sum_{k=1}^{p} \{A_k \cos(\omega_k t) + B_k \sin(\omega_k t)\}\right)^2, \qquad (3.50)$$

with respect to $\boldsymbol{\omega} = (\omega_1, \ldots, \omega_p), \boldsymbol{\theta} = (A_1, \ldots, A_p, B_1, \ldots, B_p)$. The weight function $w(s)$ is non-negative, bounded, of bounded variation, has support $[0, 1]$. Moreover, it is such that $W_0 > 0$ and $W_1^2 - W_0 W_2 \neq 0$, where

$$W_n = \int_0^1 s^n w(s) ds.$$

It is assumed that the weight function is known a priori. In this case, it can be seen along the same line as the LSEs that if we denote $\widehat{\omega}_1, \ldots, \widehat{\omega}_k, \widehat{A}_1, \ldots, \widehat{A}_k, \widehat{B}_1, \ldots, \widehat{B}_k$ as the WLSEs of $\omega_1, \ldots, \omega_k, A_1, \ldots, A_k, B_1, \ldots, B_k$, respectively, then they can be obtained as follows. First obtain $\widehat{\omega}_1, \ldots, \widehat{\omega}_k$, which maximize $Q(\boldsymbol{\omega})$, with respect to $\omega_1, \ldots, \omega_k$, where

$$Q(\boldsymbol{\omega}) = \sum_{k=1}^p \left| \frac{1}{n} \sum_{t=1}^n w\left(\frac{t}{n}\right) y(t) e^{it\omega_k} \right|^2, \tag{3.51}$$

and then $\widehat{A}_k$ and $\widehat{B}_k$ are obtained as

$$\widehat{A}_k = \frac{2 \sum_{t=1}^n w\left(\frac{t}{n}\right) y(t) \cos(\widehat{\omega}_k t)}{\sum_{t=1}^n w\left(\frac{t}{n}\right)} \quad \text{and} \quad \widehat{B}_k = \frac{2 \sum_{t=1}^n w\left(\frac{t}{n}\right) y(t) \sin(\widehat{\omega}_k t)}{\sum_{t=1}^n w\left(\frac{t}{n}\right)},$$

for $k = 1, \ldots, p$. Irizarry [8] proved that WLSEs are strongly consistent estimators of the corresponding parameters, and they are asymptotically normally distributed under a fairly general set of assumptions on the weight function and on the error random variables. The explicit expression of the variance covariance matrix is also provided, which as expected depends on the weight function. It appears that with the proper choice of the weight function, the asymptotic variances of the WLSEs can be made smaller than the corresponding asymptotic variances of the LSEs, although it has not been explored. Moreover, it has not been indicated how to maximize $Q(\boldsymbol{\omega})$ as defined in (3.51), with respect to the unknown parameter vector $\boldsymbol{\omega}$. It is a multidimensional optimization problem, and if p is large, it is a difficult problem to solve. It might be possible to use the sequential estimation procedure as suggested by Prasad, Kundu, and Mitra [19], see Sect. 3.12, in this case also. More work is needed in that direction.

3.16 Nandi and Kundu Algorithm

Nandi and Kundu [17] proposed a computationally efficient algorithm for estimating the parameters of sinusoidal signals in the presence of additive stationary noise, that is, under the error Assumption 3.2. The key features of the proposed algorithm are (i) the estimators are strongly consistent and they are asymptotically equivalent to the LSEs, (ii) the algorithm converges in three steps starting from the initial frequency estimators as the PEs over Fourier frequencies, and (iii) the algorithm does not use

the whole sample at each step. In the first two steps, it uses only some fractions of
the whole sample, and only in the third step it uses the whole sample. For notational
simplicity, we describe the algorithm when $p = 1$, and note that for general p, the
sequential procedure of Prasad, Kundu, and Mitra [19] can be easily used.

If at the jth stage the estimator of ω is denoted by $\omega^{(j)}$, then $\omega^{(j+1)}$ is calculated
as

$$\omega^{(j+1)} = \omega^{(j)} + \frac{12}{n_j} \operatorname{Im} \left[\frac{P(j)}{Q(j)} \right], \tag{3.52}$$

where

$$P(j) = \sum_{t=1}^{n_j} y(t) \left(t - \frac{n_j}{2} \right) e^{-i\omega^{(j)}t}, \tag{3.53}$$

$$Q(j) = \sum_{t=1}^{n_j} y(t) e^{-i\omega^{(j)}t}, \tag{3.54}$$

and n_j denotes the sample size used at the jth iteration. Suppose $\omega^{(0)}$ denotes the
periodogram estimator of ω, then the algorithm takes the following form.

Algorithm 3.9

- Step 1: Compute $\omega^{(1)}$ from $\omega^{(0)}$ using (3.52) with $n_1 = n^{0.8}$.
- Step 2: Compute $\omega^{(2)}$ from $\omega^{(1)}$ using (3.52) with $n_2 = n^{0.9}$.
- Step 3: Compute $\omega^{(3)}$ from $\omega^{(2)}$ using (3.52) with $n_3 = n$.

It should be mentioned that the fraction 0.8 or 0.9 which has been used in Step 1 or
Step 2, respectively, is not unique, and several other choices are also possible, see
Nandi and Kundu [17] for details. Moreover, it has been shown by the authors that
asymptotic properties of $\omega^{(3)}$ are the same as the corresponding LSE.

3.17   Superefficient Estimator

Kundu et al. [13] proposed a modified Newton–Raphson method to obtain superefficient estimators of the frequencies of model (3.1) in the presence of stationary noise
$\{X(t)\}$. It is observed that if the algorithm starts with an initial estimator with a convergence rate $O_p(1/n)$, and uses the Newton–Raphson algorithm with proper step
factor modification, then it produces superefficient frequency estimator, in the sense
its asymptotic variance is lower than the asymptotic variance of the corresponding
LSE. It is indeed a very counter-intuitive result because it is well-known that the
usual Newton–Raphson method cannot be used to compute the LSE, whereas with
proper step-factor modification, it can produce superefficient frequency estimator.

If we denote

$$S(\omega) = \mathbf{Y}^T \mathbf{Z}(\mathbf{Z}^T \mathbf{Z})^{-1}\mathbf{Z}^T \mathbf{Y}, \tag{3.55}$$

where $\mathbf{Y}$ and $\mathbf{Z}$ are same as defined in (3.5), then the LSE of ω can be obtained by maximizing $S(\omega)$ with respect to ω. The maximization of $S(\omega)$ using Newton–Raphson algorithm can be performed as follows:

$$\omega^{(j+1)} = \omega^{(j)} - \frac{S'(\omega^{(j)})}{S''(\omega^{(j)})}. \tag{3.56}$$

Here $\omega^{(j)}$ is the same as defined before, that is, the estimate of ω at the jth stage, moreover, $S'(\omega^{(j)})$ and $S''(\omega^{(j)})$ denote the first derivative and second derivative, respectively, of $S(\omega)$ evaluated at $\omega^{(j)}$. The standard Newton–Raphson algorithm is modified with a smaller correction factor as follows:

$$\omega^{(j+1)} = \omega^{(j)} - \frac{1}{4} \times \frac{S'(\omega^{(j)})}{S''(\omega^{(j)})}. \tag{3.57}$$

Suppose $\omega^{(0)}$ denotes the PE of ω, then the algorithm can be described as follows.

Algorithm 3.10

- Step 1: Take $n_1 = n^{6/7}$, and calculate

$$\omega^{(1)} = \omega^{(0)} - \frac{1}{4} \times \frac{S'_{n_1}(\omega^{(0)})}{S''_{n_1}(\omega^{(0)})},$$

where $S'_{n_1}(\omega^{(0)})$ and $S''_{n_1}(\omega^{(0)})$ are same as $S'(\omega^{(0)})$ and $S''(\omega^{(0)})$, respectively, computed using a subsample of size n_1.
- Step 2: With $n_j = n$, repeat

$$\omega^{(j+1)} = \omega^{(j)} - \frac{1}{4} \times \frac{S'_{n_j}(\omega^{(j)})}{S''_{n_j}(\omega^{(j)})}, \quad j = 1, 2, \ldots,$$

until a suitable stopping criterion is satisfied.

It is observed that any n_1 consecutive data points can be used at Step 1 to start the algorithm, and it is observed in the simulation study that the choice of the initial subsamples does not have any visible effect on the final estimator. Moreover, the factor $6/7$ in the exponent at Step 1 is not unique, and there are several other ways the algorithm can be initiated. It is observed in extensive simulation studies by Kundu et al. [13] that the iteration converges very quickly, and it produces frequency estimators which have lower variances than the corresponding LSE.

3.18 Conclusions

In this section, we discussed different estimation procedures for estimating the frequencies of the sum of sinusoidal model. It should be mentioned that although we have discussed seventeen different methods, the list is nowhere near complete. The main aim is to provide an idea of how the same problem due to its complicated nature has been attempted by different methods to get some satisfactory answers. Moreover, it is also observed that none of these methods work uniformly well for all values of the model parameters. It is observed that finding the efficient estimators is a numerically challenging problem. Due to this reason several suboptimal solutions have been suggested in the literature which do not have the same rate of convergence as the efficient estimators. The detailed theoretical properties of the different estimators are provided in the next chapter.

References

1. Ahtola, J., & Tiao, G. C. (1987). Distributions of least squares estimators of autoregressive parameters for a process with complex roots on the unit circle. *Journal of Time Series Analysis*, *8*, 1–14.
2. Bai, Z. D., Chen, X. R., Krishnaiah, P. R., & Zhao, L. C. (1987). Asymptotic properties of EVLP estimators for superimposed exponential signals in noise. Tech. Rep. 87-19, CMA, U. Pittsburgh.
3. Bai, Z. D., Rao, C. R., Chow, M., & Kundu, D. (2003). An efficient algorithm for estimating the parameters of superimposed exponential signals. *Journal of Statistical Planning and Inference*, *110*, 23–34.
4. Bresler, Y., & Macovski, A. (1986). Exact maximum likelihood parameter estimation of superimposed exponential signals in noise. *IEEE Transactions on Acoustics, Speech and Signal Processing*, *34*, 1081–1089.
5. Chung, K. L. (1974). *A course in probability theory*. New York: Academic Press.
6. Dempster, A. P., Laird, N. M., & Rubin, D. B. (1977). Maximum likelihood from incomplete data via EM algorithm. *Journal of the Royal Statistical Society Series B*, *39*, 1–38.
7. Feder, M., & Weinstein, E. (1988). Parameter estimation of superimposed signals using the EM algorithm. *IEEE Transactions on Acoustics, Speech and Signal Processing*, *36*, 477–489.
8. Irizarry, R. A. (2002). Weighted estimation of harmonic components in a musical sound signal. *Journal of Time Series Analysis*, *23*, 29–48.
9. Kannan, N., & Kundu, D. (1994). On modified EVLP and ML methods for estimating superimposed exponential signals. *Signal Processing*, *39*, 223–233.
10. Kumaresan, R. (1982). *Estimating the parameters of exponential signals*, Ph.D. thesis, U. Rhode Island.
11. Kundu, D. (1993). Estimating the parameters of undamped exponential signals. *Technometrics*, *35*, 215–218.
12. Kundu, D. (1994). Estimating the parameters of complex valued exponential signals. *Computational Statistics and Data Analysis*, *18*, 525–534.
13. Kundu, D., Bai, Z. D., Nandi, S., & Bai, L. (2011). Super efficient frequency estimation. *Journal of Statistical Planning and Inference*, *141*, 2576–2588.
14. Kundu, D., & Mitra, A. (1995). Consistent method of estimating the superimposed exponential signals. *Scandinavian Journal of Statistics*, *22*, 73–82.

15. Kundu, D., & Mitra, A. (1997). Consistent methods of estimating sinusoidal frequencies; a non iterative approach. *Journal of Statistical Computation and Simulation, 58*, 171–194.
16. McLachlan, G. J., & Krishnan, T. (2008). *The EM algorithm and extensions.* Hoboken, New Jersey: Wiley.
17. Nandi, S., & Kundu, D. (2006). A fast and efficient algorithm for estimating the parameters of sum of sinusoidal model. *Sankhya, 68*, 283–306.
18. Pillai, S. U. (1989). *Array processing.* New York: Springer.
19. Prasad, A., Kundu, D., & Mitra, A. (2008). Sequential estimation of the sum of sinusoidal model parameters. *Journal of Statistical Planning and Inference, 138*, 1297–1313.
20. Quinn, B. G. (1994). Estimating frequency by interpolation using Fourier coefficients. *IEEE Transactions on Signal Processing, 42*, 1264–1268.
21. Quinn, B. G., & Fernandes, J. M. (1991). A fast efficient technique for the estimation of frequency. *Biometrika, 78*, 489–497.
22. Rao, C. R. (1988). Some results in signal detection. In S. S. Gupta & J. O. Berger (Eds.), *Decision theory and related topics, IV, 2* (pp. 319–332). New York: Springer.
23. Rice, J. A., & Rosenblatt, M. (1988). On frequency estimation. *Biometrika, 75*, 477–484.
24. Roy, R. H. (1987). *ESPRIT - Estimation of Signal Parameters via Rotational Invariance Technique*, Ph.D. thesis, Stanford University.
25. Roy, R. H., & Kailath, T. (1989). ESPRIT-estimation of signal parameters via rotational invariance technique. *IEEE Transactions on Acoustics, Speech and Signal Processing, 43*, 984–995.
26. Truong-Van, B. (1990). A new approach to frequency analysis with amplified harmonics. *Journal of the Royal Statistical Society Series B, 52*, 203–221.
27. Tufts, D. W., & Kumaresan, R. (1982). Estimation of frequencies of multiple sinusoids: making linear prediction perform like maximum likelihood. *Proceedings of the IEEE, 70*, 975–989.

Chapter 4
Asymptotic Properties

4.1 Introduction

In this chapter, we discuss the asymptotic properties of some of the estimators
described in Chap. 3. Asymptotic results or results based on large samples deal with
properties of estimators under the assumption that the sample size increases indefi-
nitely. The statistical models, observed in the signal processing literature, are mostly
quite complicated nonlinear models. Even a single-component sinusoidal model is
highly nonlinear in its frequency parameter. Due to that, many statistical concepts for
small samples cannot be applied in case of sinusoidal models. The added problem is
that under different error assumptions, this model is mean nonstationary. Therefore,
it is not possible to obtain any finite sample property of the LSE or any other estimators, discussed in the previous chapter. All the results have to be asymptotic. The
most intuitive estimator is the LSE and the most popular one is the ALSE. These
two estimators are asymptotically equivalent, and we discuss their equivalence in
Sect. 4.4.3. The sinusoidal model is a nonlinear regression model, but it does not
satisfy, see Kundu [8], the standard sufficient conditions of Jennrich [7] or Wu [25]
for the LSE to be consistent. Jennrich [7] first proved the existence of the LSE in a
nonlinear regression model of the form $y(t) = f_t(\theta^0) + \varepsilon(t), t = 1, \ldots$. The almost
sure convergence of the LSE of the unknown parameter θ was shown under the
following assumption: Define

$$F_n(\theta_1, \theta_2) = \frac{1}{n} \sum_{t=1}^{n} (f_t(\theta_1) - f_t(\theta_2))^2,$$

then $F_n(\theta_1, \theta_2)$ converges uniformly to a continuous function $F(\theta_1, \theta_2)$ and

$$F(\theta_1, \theta_2) \neq 0 \quad \text{if and only if} \quad \theta_1 = \theta_2.$$

Consider a single-component sinusoidal model and assume that $A^0 = 1$ and $B^0 = 0$
in (4.1). Suppose the model satisfies Assumption 3.1 and ω^0 is an interior point of

© Springer Nature Singapore Pte Ltd. 2020
S. Nandi and D. Kundu, *Statistical Signal Processing*,
https://doi.org/10.1007/978-981-15-6280-8_4

$[0, \pi]$. In this simple situation, $F_n(\theta_1, \theta_2)$ does not converge uniformly to a continuous function. Wu [25] gave some sufficient conditions under which the LSE of θ^0 is strongly consistent when the growth rate requirement of $F_n(\theta_1, \theta_2)$ is replaced by a Lipschitz-type condition on the sequence $\{f_t\}$. In addition, the sinusoidal model does not satisfy Wu's Lipschitz-type condition also.

Whittle [24] first obtained some of the theoretical results. Recent results are by Hannan [4, 5], Walker [23], Rice and Rosenblatt [20], Quinn and Fernandes [19], Kundu [8, 10], Quinn [18], Kundu and Mitra [13], Irizarry [6], Nandi, Iyer, and Kundu [15], Prasad, Kundu, and Mitra [17], and Kundu et al. [11]. Walker [23] considered the sinusoidal model with one component and obtained the asymptotic properties of the PEs under the assumption that the errors are i.i.d. with mean zero and finite variance. The result has been extended by Hannan [4] when the errors are from a weakly stationary process and by Hannan [5] when the errors are from a strictly stationary random process with continuous spectrum. Some of the computational issues have been discussed in Rice and Rosenblatt [20]. The estimation procedure, proposed by Quinn and Fernandes [19], is based on fitting an ARMA(2,2) model and the estimator is strongly consistent and efficient. Kundu and Mitra [13] considered the model when errors are i.i.d. and proved directly the consistency and asymptotic normality of the LSEs. The result has been generalized by Kundu [10] when the errors are from a stationary linear process. The weighted LSEs are proposed by Irizarry [6] and extended the asymptotic results of the LSEs to the weighted one. Nandi, Iyer, and Kundu [15] proved the strong consistency of the LSEs when the errors are i.i.d. with mean zero, but may not have finite variance. They also obtained the asymptotic distribution of the LSEs when the error distribution is symmetric stable. It is well-known that the Newton–Raphson method does not work well in case of sinusoidal frequency model. Kundu et al. [11] proposed a modification of the Newton–Raphson method with a smaller step factor such that the resulting estimators have the same rate of convergence as the LSEs. Additionally, the asymptotic variances of the proposed estimators are less than those of the LSEs. Therefore, the estimators are named as the super efficient estimators.

4.2 Sinusoidal Model with One Component

We first discuss the asymptotic results of the estimators of the parameters of the sinusoidal model with one component. This is just to keep the mathematical expression simple. We talk about the model with p components at the end of the chapter. The model is now

$$y(t) = A^0 \cos(\omega^0 t) + B^0 \sin(\omega^0 t) + X(t), \quad t = 1, \dots, n. \qquad (4.1)$$

In this section, we explicitly write A^0, B^0, and ω^0 as the true values of the unknown parameters A, B, and ω, respectively. Write $\theta = (A, B, \omega)$ and θ^0 be the true value of θ and let $\widehat{\theta}$ and $\widetilde{\theta}$ be the LSE and ALSE of θ^0, respectively; $\{X(t)\}$ is a sequence of error random variables. To ensure the presence of the frequency component and to make sure that $\{y(t)\}$ is not pure noise, assume that A^0 and B^0 are not simultaneously equal to zero. For technical reasons, take $\omega^0 \in (0, \pi)$. At this moment, we do not explicitly mention the complete error structure. We assume in this chapter that the number of signal components (distinct frequency), p, is known in advance. The problem of estimation of p is considered in Chap. 5.

In the following, we discuss the consistency and asymptotic distribution of the LSE and ALSE of θ^0 under different error assumptions. Apart from Assumptions 3.1 and 3.2, the following two assumptions regarding the structure of the error process are required.

Assumption 4.1 $\{X(t)\}$ is a sequence of i.i.d. random variables with mean zero and $E|X(t)|^{1+\delta} < \infty$ for some $0 < \delta \leq 1$.

Assumption 4.2 $\{X(t)\}$ is a sequence of i.i.d. random variables distributed as $S\alpha S(\sigma)$.

Note 4.1

(a) Assumption 3.1 is a special case of Assumption 4.1 as both are the same with $\delta = 1$.
(b) If $\{X(t)\}$ satisfies Assumption 4.2, then Assumption 4.1 is also true with $1 + \delta < \alpha \leq 2$. Therefore, from now on, we take $1 + \delta < \alpha \leq 2$.
(c) Assumption 3.2 is a standard assumption for a stationary linear process; any finite dimensional stationary AR, MA or ARMA process can be represented as a linear process with absolute summable coefficients. A process is called a stationary linear process if it satisfies Assumption 3.2.

4.3   Strong Consistency of LSE and ALSE of θ^0

We recall that $Q(\theta)$ is the residual sum of squares, defined in (3.3) and $I(\omega)$ is the periodogram function, defined in (1.6).

Theorem 4.1 *If $\{X(t)\}$ satisfies either Assumptions 3.1, 4.1, or 3.2, then the LSE $\widehat{\theta}$ and the ALSE $\widetilde{\theta}$ are both strongly consistent estimators of θ^0, that is,*

$$\widehat{\theta} \xrightarrow{a.s.} \theta^0 \quad and \quad \widetilde{\theta} \xrightarrow{a.s.} \theta^0. \tag{4.2}$$

The following lemmas are required to prove Theorem 4.1.

Lemma 4.1 *Let $S_{C_1,K} = \left\{\theta; \theta = (A, B, \theta), |\theta - \theta^0| \geq 3C_1, |A| \leq K, |B| \leq K\right\}$. If for any $C_1 > 0$ and for some $K < \infty$,*

$$\liminf_{n \to \infty} \inf_{\theta \in S_{C_1,K}} \frac{1}{n}\left[Q(\theta) - Q(\theta^0)\right] > 0 \quad a.s., \tag{4.3}$$

then $\widehat{\theta}$ is a strongly consistent estimator of θ^0.

Lemma 4.2 *If* $\{X(t)\}$ *satisfies either Assumptions 3.1, 4.1, or 3.2, then*

$$\sup_{\omega \in (0,\pi)} \left| \frac{1}{n} \sum_{t=1}^{n} X(t) \cos(\omega t) \right| \to 0 \quad a.s. \quad as \ n \to \infty. \tag{4.4}$$

Corollary 4.1

$$\sup_{\omega \in (0,\pi)} \left| \frac{1}{n^{k+1}} \sum_{t=1}^{n} t^k X(t) \cos(\omega t) \right| \to 0 \quad a.s., \quad for \ k = 1, 2 \ldots$$

The result is true for sine functions also.

Lemma 4.3 *Write* $S_{C_2} = \{\omega : |\omega - \omega^0| > C_2\}$, *for any* $C_2 > 0$. *If for some* $C_2 > 0$,

$$\overline{\lim} \sup_{\omega \in S_{C_2}} \frac{1}{n} \left[I(\omega) - I(\omega^0) \right] < 0 \quad a.s.,$$

then $\widetilde{\omega}$, *the ALSE of* ω^0, *converges to* ω^0 *a.s. as* $n \to \infty$.

Lemma 4.4 *Suppose* $\widetilde{\omega}$ *is the ALSE of* ω^0. *Then* $n(\widetilde{\omega} - \omega^0) \to 0$ *a.s. as* $n \to \infty$.

Lemma 4.1 provides a sufficient condition for $\widehat{\boldsymbol{\theta}}$ to be strongly consistent whereas Lemma 4.3 gives a similar condition for $\widetilde{\omega}$, the ALSE of ω^0. Lemma 4.2 is used to verify conditions given in Lemmas 4.1 and 4.3. Lemma 4.4 is required to prove the strong consistency of the ALSEs of the amplitudes, $\widetilde{A}$ and $\widetilde{B}$.

The consistency results of the LSEs and the ALSEs are stated in Theorem 4.1 in concise form, and it is proved in two steps. First is the proof of the strong consistency of $\widehat{\boldsymbol{\theta}}$, the LSE of $\boldsymbol{\theta}^0$, and next is the proof of the strong consistency of $\widetilde{\boldsymbol{\theta}}$, the ALSE of $\boldsymbol{\theta}^0$. The proofs of Lemmas 4.1–4.4, required to prove Theorem 4.1, are given in Appendix A. We prove Lemma 4.1. The proof of Lemma 4.3 is similar to Lemma 4.1 and so it is omitted. Lemma 4.2 is proved separately under Assumptions 4.1 and 3.2.

4.3.1 Proof of the Strong Consistency of $\widehat{\boldsymbol{\theta}}$, the LSE of $\boldsymbol{\theta}^0$

In this proof, we denote $\widehat{\boldsymbol{\theta}}$ by $\widehat{\boldsymbol{\theta}}_n$ to write explicitly that $\widehat{\boldsymbol{\theta}}$ depends on n. If $\widehat{\boldsymbol{\theta}}_n$ is not consistent for $\boldsymbol{\theta}^0$, then either Case I or Case II occurs

Case I: For all subsequences $\{n_k\}$ of $\{n\}$, $|\widehat{A}_{n_k}| + |\widehat{B}_{n_k}| \to \infty$. This implies $\left[Q(\widehat{\boldsymbol{\theta}}_{n_k}) - Q(\boldsymbol{\theta}^0) \right] / n_k \to \infty$. At the same time, $\widehat{\boldsymbol{\theta}}_{n_k}$ is the LSE of $\boldsymbol{\theta}^0$ at $n = n_k$, therefore, $Q(\widehat{\boldsymbol{\theta}}_{n_k}) - Q(\boldsymbol{\theta}^0) < 0$. This leads to a contradiction.

Case II: For at least one subsequence $\{n_k\}$ of $\{n\}$, $\widehat{\theta}_{n_k} \in S_{C_1,K}$ for some $C_1 > 0$ and $0 < K < \infty$. Write $\left[Q(\theta) - Q(\theta^0)\right]/n = f_1(\theta) + f_2(\theta)$, where

$$f_1(\theta) = \frac{1}{n}\sum_{t=1}^{n}\left[A^0\cos(\omega^0 t) - A\cos(\omega t) + B^0\sin(\omega^0 t) - B\sin(\omega t)\right]^2,$$

$$f_2(\theta) = \frac{1}{n}\sum_{t=1}^{n}X(t)\left[A^0\cos(\omega^0 t) - A\cos(\omega t) + B^0\sin(\omega^0 t) - B\sin(\omega t)\right].$$

Define sets $S_{C_1,K}^{j} = \{\theta : |\theta_j - \theta_j^0| > C_1, |A| \le K, |B| \le K\}$ for $j = 1, 2, 3$, where θ_j is the jth element of θ, that is, $\theta_1 = A$, $\theta_2 = B$ and $\theta_3 = \omega$, and θ_j^0 is the true value of θ_j. Then $S_{C_1,K} \subset \cup_{j=1}^{3}S_{C_1,K}^{j} = S$, say, and

$$\liminf_{n\to\infty}\ \inf_{\theta\in S_{C_1,K}}\frac{1}{n}\left[Q(\theta) - Q(\theta^0)\right] \ge \liminf_{n\to\infty}\inf_{\theta\in S}\frac{1}{n}\left[Q(\theta) - Q(\theta^0)\right].$$

Using Lemma 4.2, $\lim_{n\to\infty} f_2(\theta) = 0$ a.s. Then

$$\liminf_{n\to\infty}\ \inf_{\theta\in S_{C_1,K}^{j}}\frac{1}{n}\left[Q(\theta) - Q(\theta^0)\right]$$

$$= \liminf_{n\to\infty}\ \inf_{\theta\in S_{C_1,K}^{j}}f_1(\theta) > 0 \quad \text{a.s. for } j = 1, \ldots, 3,$$

$$\Rightarrow \liminf_{n\to\infty}\ \inf_{\theta\in S_{C_1,K}}\frac{1}{ns}\left[Q(\theta) - Q(\theta^0)\right] > 0 \ \ a.s.$$

Therefore, for $j = 1$,

$$\liminf_{n\to\infty}\ \inf_{\theta\in S_{C_1,K}^{j}}\ f_1(\theta)$$

$$= \liminf_{n\to\infty}\ \inf_{|A-A^0|>C_1}\frac{1}{n}\sum_{t=1}^{n}\left[\left\{A^0\cos(\omega^0 t) - A\cos(\omega t)\right\}^2 + \right.$$

$$\left\{B^0\sin(\omega^0 t) - B\sin(\omega t)\right\}^2 +$$

$$\left. 2\left\{A^0\cos(\omega^0 t) - A\cos(\omega t)\right\}\left\{B^0\sin(\omega^0 t) - B\sin(\omega t)\right\}\right]$$

$$= \liminf_{n\to\infty}\ \inf_{|A-A^0|>C_1}\frac{1}{n}\sum_{i=1}^{n}\left[\left\{A^0\cos(\omega^0 t) - A\cos(\omega t)\right\}^2 > \frac{1}{2}C_1^2 > 0 \text{ a.s.}\right.$$

We have used Result (2.2) here. Similarly, the inequality holds for $j = 2, 3$. Therefore, using Lemma 4.1, we say that $\widehat{\theta}$ is a strongly consistent estimator of θ^0. ∎

4.3.2 Proof of Strong Consistency of $\widetilde{\theta}$, the ALSE of θ^0

We first prove the consistency of $\widetilde{\omega}$, the ALSE of ω^0, and then provide the proof of the linear parameter estimators. Consider

$$
\frac{1}{n}\left[I(\omega) - I(\omega^0)\right] = \frac{1}{n^2}\left[\left|\sum_{t=1}^{n} y(t)e^{-i\omega t}\right|^2 - \left|\sum_{t=1}^{n} y(t)e^{-i\omega^0 t}\right|^2\right]
$$

$$
= \frac{1}{n^2}\left[\left\{\sum_{t=1}^{n} y(t)\cos(\omega t)\right\}^2 + \left\{\sum_{t=1}^{n} y(t)\sin(\omega t)\right\}^2\right.
$$

$$
\left. - \left\{\sum_{t=1}^{n} y(t)\cos(\omega^0 t)\right\}^2 - \left\{\sum_{t=1}^{n} y(t)\sin(\omega^0 t)\right\}^2\right]
$$

$$
= \left\{\sum_{t=1}^{n}\left(A^0\cos(\omega^0 t) + B^0\sin(\omega^0 t) + X(t)\right)\cos(\omega t)\right\}^2
$$

$$
+ \left\{\sum_{t=1}^{n}\left(A^0\cos(\omega^0 t) + B^0\sin(\omega^0 t) + X(t)\right)\sin(\omega t)\right\}^2
$$

$$
- \left\{\sum_{t=1}^{n}\left(A^0\cos(\omega^0 t) + B^0\sin(\omega^0 t) + X(t)\right)\cos(\omega^0 t)\right\}^2
$$

$$
- \left\{\sum_{t=1}^{n}\left(A^0\cos(\omega^0 t) + B^0\sin(\omega^0 t) + X(t)\right)\sin(\omega^0 t)\right\}^2.
$$

Using Lemma 4.2, the terms of the form $\overline{\lim}\sup_{\omega\in S_{C_2}}(1/n)\sum_{t=1}^{n} X(t)\cos(\omega t) = 0$ a.s. and using trigonometric identities (2.15)–(2.21), we have

$$
\overline{\lim}\sup_{\omega\in S_{C_2}}\frac{1}{n}\left[I(\omega) - I(\omega^0)\right]
$$

$$
= -\lim_{n\to\infty}\left\{\frac{1}{n}\sum_{t=1}^{n} A^0\cos^2(\omega^0 t)\right\}^2 - \lim_{n\to\infty}\left\{\frac{1}{n}\sum_{t=1}^{n} B^0\sin^2(\omega^0 t)\right\}^2
$$

$$
= -\frac{1}{4}(A^{0^2} + B^{0^2}) < 0 \quad a.s.
$$

Therefore using Lemma 4.3, $\widetilde{\omega} \to \omega^0$ a.s.

We need Lemma 4.4 to prove that $\widetilde{A}$ and $\widetilde{B}$ are strongly consistent. Observe that

$$
\widetilde{A} = \frac{2}{n}\sum_{t=1}^{n} y(t)\cos(\widetilde{\omega}t) = \frac{2}{n}\sum_{t=1}^{n}\left(A^0\cos(\omega^0 t) + B^0\sin(\omega^0 t) + X(t)\right)\cos(\widetilde{\omega}t).
$$

Using Lemma 4.2, $(2/n) \sum_{t=1}^{n} X(t) \cos(\widetilde{\omega} t) \to 0$. Expand $\cos(\widetilde{\omega} t)$ by Taylor series around ω^0.

$$\widetilde{A} = \frac{2}{n} \sum_{t=1}^{n} \left(A^0 \cos(\omega^0 t) + B^0 \sin(\omega^0 t) \right) \left[\cos(\omega^0 t) - t(\widetilde{\omega} - \omega^0) \sin(\omega t) \right] \to A^0 \quad a.s.$$

using Lemma 4.4 and trigonometric results (2.15)–(2.21).   ∎

Remark 4.1 The proof of the consistency of LSE and ALSE of θ^0 for model (4.1) are extensively discussed. The consistency of the LSE of the parameter vector of the multiple sinusoidal model with $p > 1$ follows similarly as the consistency of $\widehat{\theta}$ and $\widetilde{\theta}$ of this section.

4.4 Asymptotic Distribution of LSE and ALSE of θ^0

This section discusses the asymptotic distribution of LSE and ALSE of θ^0 under different error assumptions. We discuss the asymptotic distribution of $\widehat{\theta}$, the LSE of θ^0 under Assumption 3.2. Then we consider the case under Assumption 4.2. Under Assumption 3.2, the asymptotic distribution of $\widehat{\theta}$, as well as $\widetilde{\theta}$, is multivariate normal whereas under Assumption 4.2, it is distributed as multivariate symmetric stable. First we discuss asymptotic distribution of $\widehat{\theta}$ in both the cases. Then the asymptotic distribution of $\widetilde{\theta}$ is shown to be the same as that of $\widehat{\theta}$ by proving the asymptotic equivalence of $\widehat{\theta}$ and $\widetilde{\theta}$ in either of the cases.

4.4.1 Asymptotic Distribution of $\widehat{\theta}$ Under Assumption 3.2

Assume that $\{X(t)\}$ satisfies Assumption 3.2. Let $Q'(\theta)$ be a 1×3 vector of the first derivative and $Q''(\theta)$, a 3×3 matrix of second derivatives of $Q(\theta)$, that is,

$$Q'(\theta) = \left(\frac{\partial Q(\theta)}{\partial A}, \frac{\partial Q(\theta)}{\partial B}, \frac{\partial Q(\theta)}{\partial \omega} \right), \tag{4.5}$$

$$Q''(\theta) = \begin{pmatrix} \frac{\partial^2 Q(\theta)}{\partial A^2} & \frac{\partial^2 Q(\theta)}{\partial A \partial B} & \frac{\partial^2 Q(\theta)}{\partial A \partial \omega} \\ \frac{\partial^2 Q(\theta)}{\partial B \partial A} & \frac{\partial^2 Q(\theta)}{\partial B^2} & \frac{\partial^2 Q(\theta)}{\partial B \partial \omega} \\ \frac{\partial^2 Q(\theta)}{\partial \omega \partial A} & \frac{\partial^2 Q(\theta)}{\partial \omega \partial B} & \frac{\partial^2 Q(\theta)}{\partial \omega^2} \end{pmatrix}. \tag{4.6}$$

The elements of $Q'(\theta)$ and $Q''(\theta)$ are

$$\frac{\partial Q(\theta)}{\partial A} = -2 \sum_{t=1}^{n} X(t) \cos(\omega t), \qquad \frac{\partial Q(\theta)}{\partial B} = -2 \sum_{t=1}^{n} X(t) \sin(\omega t),$$

$$\frac{\partial Q(\theta)}{\partial \omega} = 2 \sum_{t=1}^{n} t X(t) \left[A \sin(\omega t) - B \cos(\omega t) \right],$$

$$\frac{\partial^2 Q(\theta)}{\partial A^2} = 2 \sum_{t=1}^{n} \cos^2(\omega t), \quad \frac{\partial^2 Q(\theta)}{\partial B^2} = 2 \sum_{t=1}^{n} \sin^2(\omega t),$$

$$\frac{\partial^2 Q(\theta)}{\partial A \partial B} = 2 \sum_{t=1}^{n} \cos(\omega t) \sin(\omega t),$$

$$\frac{\partial^2 Q(\theta)}{\partial A \partial \omega} = -2 \sum_{t=1}^{n} t \cos(\omega t) \left[A \sin(\omega t) - B \cos(\omega t) \right] + 2 \sum_{t=1}^{n} t X(t) \sin(\omega t),$$

$$\frac{\partial^2 Q(\theta)}{\partial B \partial \omega} = -2 \sum_{t=1}^{n} t \sin(\omega t) \left[A \sin(\omega t) - B \cos(\omega t) \right] - 2 \sum_{t=1}^{n} t X(t) \cos(\omega t),$$

$$\frac{\partial^2 Q(\theta)}{\partial \omega^2} = 2 \sum_{t=1}^{n} t^2 \left[A \sin(\omega t) - B \cos(\omega t) \right]^2 + 2 \sum_{t=1}^{n} t^2 X(t) \left[A \cos(\omega t) + B \sin(\omega t) \right].$$

Consider a 3×3 diagonal matrix $\mathbf{D} = \text{diag} \left\{ n^{-1/2}, n^{-1/2}, n^{-3/2} \right\}$. Expanding $Q'(\widehat{\theta})$ around θ^0 using Taylor series expansion

$$Q'(\widehat{\theta}) - Q'(\theta^0) = (\widehat{\theta} - \theta^0) Q''(\bar{\theta}), \qquad (4.7)$$

where $\bar{\theta}$ is a point on the line joining $\widehat{\theta}$ and θ^0. As $\widehat{\theta}$ is the LSE of θ^0, $Q'(\widehat{\theta}) = 0$. Also $\widehat{\theta} \xrightarrow{a.s.} \theta^0$ using Theorem 4.1. Because $Q(\theta)$ is a continuous function of θ, we have

$$\lim_{n \to \infty} \mathbf{D} Q''(\bar{\theta}) \mathbf{D} = \lim_{n \to \infty} \mathbf{D} Q''(\theta^0) \mathbf{D} = \begin{pmatrix} 1 & 0 & \frac{1}{2} B^0 \\ 0 & 1 & -\frac{1}{2} A^0 \\ \frac{1}{2} B^0 & -\frac{1}{2} A^0 & \frac{1}{3} (A^{0^2} + B^{0^2}) \end{pmatrix} = \boldsymbol{\Sigma}, \quad \text{say.}$$

$$(4.8)$$

Therefore, (4.7) can be written as

$$(\widehat{\theta} - \theta^0) \mathbf{D}^{-1} = - \left[Q'(\theta^0) \mathbf{D} \right] \left[\mathbf{D} Q''(\bar{\theta}) \mathbf{D} \right]^{-1}, \qquad (4.9)$$

since $\mathbf{D} Q''(\bar{\theta}) \mathbf{D}$ is an invertible matrix a.e. for large n. Using a central limit theorem of Stochastic processes, Fuller [3], it follows that $Q'(\theta^0) \mathbf{D}$ tends to a 3-variate normal distribution with mean vector zero and variance covariance matrix equal to $2\sigma^2 c(\omega^0) \boldsymbol{\Sigma}$, where

$$c(\omega^0) = \left| \sum_{j=0}^{\infty} a(j) e^{-ij\omega^0} \right|^2 = \left[\sum_{j=0}^{\infty} a(j) \cos(j\omega^0) \right]^2 + \left[\sum_{j=0}^{\infty} a(j) \sin(j\omega^0) \right]^2. \quad (4.10)$$

Therefore, $(\widehat{\theta} - \theta^0)\mathbf{D}^{-1} \xrightarrow{d} \mathcal{N}_3(\mathbf{0}, 2\sigma^2 c(\omega^0) \boldsymbol{\Sigma}^{-1})$ and we can state the asymptotic distribution in the following theorem.

Theorem 4.2 *Under Assumption 3.2, the limiting distribution of $(n^{\frac{1}{2}}(\widehat{A} - A^0),$ $n^{\frac{1}{2}}(\widehat{B} - B^0), n^{\frac{3}{2}}(\widehat{\omega} - \omega^0))$ as $n \to \infty$ is a 3-variate normal distribution with mean vector zero and dispersion matrix $2\,\sigma^2 c(\omega^0)\boldsymbol{\Sigma}^{-1}$, where $c(\omega^0)$ is defined in (4.10) and $\boldsymbol{\Sigma}^{-1}$ has the following form:*

$$\boldsymbol{\Sigma}^{-1} = \frac{1}{A^{0^2} + B^{0^2}} \begin{bmatrix} A^{0^2} + 4B^{0^2} & -3A^0 B^0 & -3B^0 \\ -3A^0 B^0 & 4A^{0^2} + B^{0^2} & 3A^0 \\ -3B^0 & 3A^0 & 6 \end{bmatrix}. \quad (4.11)$$

Remark 4.2

1. The diagonal entries of matrix $\mathbf{D}$ correspond to the rates of convergence of $\widehat{A}$, $\widehat{B}$, and $\widehat{\omega}$, respectively. Therefore, $\widehat{A} - A^0 = O_p(n^{-1/2})$, $\widehat{B} - B^0 = O_p(n^{-1/2})$, and $\widehat{\omega} - \omega^0 = O_p(n^{-3/2})$.
2. Instead of Assumption 3.2, if $\{X(t)\}$ only satisfies Assumption 3.1, then $a(j) = 0$, for all $j \neq 0$ and $a(0) = 1$ in the derivation discussed above and so $c(\omega^0) = 1$. Therefore, in such situation $(\widehat{\theta} - \theta^0)\mathbf{D}^{-1} \xrightarrow{d} \mathcal{N}_3(\mathbf{0}, 2\sigma^2 \boldsymbol{\Sigma}^{-1})$.
3. Observe that $(\sigma^2/2\pi)c(\omega) = f(\omega)$, where $f(\omega)$ is the spectral density function of the error process $\{X(t)\}$ under Assumption 3.2.

4.4.2 Asymptotic Distribution of $\widehat{\theta}$ Under Assumption 4.2

In this section, the asymptotic distribution of $\widehat{\theta}$ under Assumption 4.2 is developed, that is, $\{X(t)\}$ is a sequence of i.i.d. symmetric stable random variables with stability index α and scale parameter σ (see Nandi, Iyer, and Kundu [15]).

Define two diagonal matrices of order 3×3 as follows:

$$\mathbf{D}_1 = \text{diag}\left\{ n^{-\frac{1}{\alpha}}, n^{-\frac{1}{\alpha}}, n^{-\frac{1+\alpha}{\alpha}} \right\}, \quad \mathbf{D}_2 = \text{diag}\left\{ n^{-\frac{\alpha-1}{\alpha}}, n^{-\frac{\alpha-1}{\alpha}}, n^{-\frac{2\alpha-1}{\alpha}} \right\}. \quad (4.12)$$

Note that $\mathbf{D}_1 \mathbf{D}_2 = \mathbf{D}^2$. Also if $\alpha = 2$, it corresponds to normal distribution and in that case $\mathbf{D}_1 = \mathbf{D}_2 = \mathbf{D}$. Using the same argument as in (4.8), we have

$$\lim_{n \to \infty} \mathbf{D}_2 Q''(\bar{\theta})\mathbf{D}_1 = \lim_{n \to \infty} \mathbf{D}_2 Q''(\theta^0)\mathbf{D}_1 = \boldsymbol{\Sigma}. \quad (4.13)$$

Similarly as in (4.9), we write

$$(\widehat{\theta} - \theta^0)\mathbf{D}_2^{-1} = -\left[Q'(\theta^0)\mathbf{D}_1\right]\left[\mathbf{D}_2 Q''(\bar{\theta})\mathbf{D}_1\right]^{-1}. \tag{4.14}$$

To find the distribution of $Q'(\theta^0)\mathbf{D}_1$, write

$$Q'(\theta^0)\mathbf{D}_1 = \left[-\frac{2}{n^{\frac{1}{\alpha}}}\sum_{t=1}^{n}X(t)\cos(\omega^0 t), -\frac{2}{n^{\frac{1}{\alpha}}}\sum_{t=1}^{n}X(t)\sin(\omega^0 t),\right.$$
$$\left.\frac{2}{n^{\frac{1+\alpha}{\alpha}}}\sum_{t=1}^{n}tX(t)\left[A^0\sin(\omega^0 t) - B^0\cos(\omega^0 t)\right]\right]$$
$$= \left(Z_n^1, Z_n^2, Z_n^3\right), \quad \text{(say)}. \tag{4.15}$$

Then the joint characteristic function of (Z_n^1, Z_n^2, Z_n^3) is

$$\phi_n(\mathbf{t}) = E\exp\{i(t_1 Z_n^1 + t_2 Z_n^2 + t_3 Z_n^3)\} = E\exp\left\{i\frac{2}{n^{1/\alpha}}\sum_{j=1}^{n}X(j)K_{\mathbf{t}}(j)\right\},$$
$$\tag{4.16}$$

where $\mathbf{t} = (t_1, t_2, t_3)$ and

$$K_{\mathbf{t}}(j) = -t_1\cos(\omega^0 j) - t_2\sin(\omega^0 j) + \frac{jt_3}{n}\left\{A^0\sin(\omega^0 j) - B^0\cos(\omega^0 j)\right\}. \tag{4.17}$$

Since $\{X(t)\}$ is a sequence of i.i.d. random variables

$$\phi_n(\mathbf{t}) = \prod_{j=1}^{n}\exp\left\{-2^{\alpha}\sigma^{\alpha}\frac{1}{n}|K_{\mathbf{t}}(j)|^{\alpha}\right\} = \exp\left\{-2^{\alpha}\sigma^{\alpha}\frac{1}{n}\sum_{j=1}^{n}|K_{\mathbf{t}}(j)|^{\alpha}\right\}. \tag{4.18}$$

Nandi, Iyer, and Kundu [15] argued that $(1/n)\sum_{j=1}^{n}|K_{\mathbf{t}}(j)|^{\alpha}$ converges, based on extensive numerical experiments. Assuming that it converges, it is proved in Nandi, Iyer, and Kundu [15] that it converges to a non-zero limit for $\mathbf{t} \neq \mathbf{0}$. The proof is given in Appendix B. Suppose

$$\lim_{n\to\infty}\frac{1}{n}\sum_{j=1}^{n}|K_{\mathbf{t}}(j)|^{\alpha} = \tau_{\mathbf{t}}(A^0, B^0, \omega^0, \alpha). \tag{4.19}$$

Therefore, the limiting characteristic function is

$$\lim_{n\to\infty}\phi_n(\mathbf{t}) = e^{-2^{\alpha}\sigma^{\alpha}\tau_{\mathbf{t}}(A^0, B^0, \omega^0, \alpha)}, \tag{4.20}$$

which indicates that even if $n \to \infty$, any linear combination of Z_n^1, Z_n^2, and Z_n^3, follows a $S\alpha S$ distribution. Using Theorem 2.1.5 of Samorodnitsky and Taqqu [21], it follows that

$$\lim_{n \to \infty} [Q'(\theta^0)\mathbf{D_1}] [\mathbf{D_2} Q''(\bar{\theta})\mathbf{D_1}]^{-1} \tag{4.21}$$

converges to a symmetric α stable random vector in $I\!R^3$ with the characteristic function $\phi(\mathbf{t}) = \exp\{-2^\alpha \sigma^\alpha \tau_{\mathbf{u}}(A^0, B^0, \omega^0, \alpha)\}$ where $\tau_{\mathbf{u}}$ is defined through (4.19) replacing $\mathbf{t}$ by $\mathbf{u}$. The vector $\mathbf{u}$ is defined as a function of $\mathbf{t}$ as $\mathbf{u} = (u_1, u_2, u_3)$ with

$$u_1(t_1, t_2, t_3, A^0, B^0) = \left[(A^{0^2} + 4B^{0^2})t_1 - 3A^0 B^0 t_2 - 6B^0 t_3 \right] \frac{1}{A^{0^2} + B^{0^2}},$$

$$u_2(t_1, t_2, t_3, A^0, B^0) = \left[-3A^0 B^0 t_1 + (4A^{0^2} + B^{0^2})t_2 + 6A^0 t_3 \right] \frac{1}{A^{0^2} + B^{0^2}},$$

$$u_3(t_1, t_2, t_3, A^0, B^0) = \left[-6B^0 t_1 + 6A^0 t_2 + 12t_3 \right] \frac{1}{A^{0^2} + B^{0^2}}.$$

Therefore, we have the following theorem.

Theorem 4.3 *Under Assumption 4.2, $(\widehat{\theta} - \theta^0)\mathbf{D_2}^{-1} = \left(n^{\frac{\alpha-1}{\alpha}}(\widehat{A} - A^0), n^{\frac{\alpha-1}{\alpha}}(\widehat{B} - B^0), n^{\frac{2\alpha-1}{\alpha}}(\omega^0 - \omega) \right)$ converges to a multivariate symmetric stable distribution in $I\!R^3$ having characteristic function equal to $\phi(\mathbf{t})$.*

4.4.3 Asymptotic Equivalence of $\widehat{\theta}$ and $\widetilde{\theta}$

In this section, it is shown that the asymptotic distribution of $\widetilde{\theta}$, the ALSE of θ^0, is equivalent to that of $\widehat{\theta}$ for large n. We have presented the asymptotic distribution of $\widehat{\theta}$ in two cases : (i) $\{X(t)\}$ is a sequence of i.i.d. symmetric stable random variables with stability index $1 < \alpha < 2$ and scale parameter σ, and (ii) $\{X(t)\}$ is a stationary linear process such that it can be expressed as (3.2) with absolute summable coefficients. In both the cases, the asymptotic distribution of $\widetilde{\theta}$ is the same as the LSE, $\widehat{\theta}$, and is stated in the following theorem.

Theorem 4.4 *Under Assumption 4.2 the asymptotic distribution of $(\widetilde{\theta} - \theta^0)\mathbf{D_2}^{-1}$ is the same as that of $(\widehat{\theta} - \theta^0)\mathbf{D_2}^{-1}$. Similarly, under Assumption 3.2 the asymptotic distribution of $(\widetilde{\theta} - \theta^0)\mathbf{D}^{-1}$ is the same as that of $(\widehat{\theta} - \theta^0)\mathbf{D}^{-1}$.*

Proof of Theorem 4.4 Observe that under Assumption 4.2

$$\frac{1}{n} Q(\theta)$$

$$= \frac{1}{n} \sum_{t=1}^{n} y(t)^2 - \frac{2}{n} \sum_{t=1}^{n} y(t) \{A \cos(\omega t) + B \sin(\omega t)\} + \frac{1}{n} \sum_{t=1}^{n} (A \cos(\omega t) + B \sin(\omega t))^2$$

$$= \frac{1}{n} \sum_{t=1}^{n} y(t)^2 - \frac{2}{n} \sum_{t=1}^{n} y(t) \{A \cos(\omega t) + B \sin(\omega t)\} + \frac{1}{2} \left(A^2 + B^2 \right) + O\left(\frac{1}{n}\right)$$

$$= C - \frac{1}{n} J(\theta) + O\left(\frac{1}{n}\right), \tag{4.22}$$

where

$$C = \frac{1}{n} \sum_{t=1}^{n} y(t)^2 \text{ and } \frac{1}{n} J(\theta) = \frac{2}{n} \sum_{t=1}^{n} y(t) \{A \cos(\omega t) + B \sin(\omega t)\} - \frac{1}{2} \left(A^2 + B^2 \right).$$

Write $\frac{1}{n} J'(\theta) = \left(\frac{1}{n} \frac{\partial J(\theta)}{\partial A}, \frac{1}{n} \frac{\partial J(\theta)}{\partial B}, \frac{1}{n} \frac{\partial J(\theta)}{\partial \omega} \right)$, then at θ^0

$$\frac{1}{n} \frac{\partial J(\theta^0)}{\partial A} = \frac{2}{n} \sum_{t=1}^{n} y(t) \cos(\omega^0 t) - A^0$$

$$= \frac{2}{n} \sum_{t=1}^{n} \{A^0 \cos(\omega^0 t) + B^0 \sin(\omega^0 t) + X(t)\} \cos(\omega^0 t) - A^0$$

$$= \frac{2}{n} \sum_{t=1}^{n} X(t) \cos(\omega^0 t) + \frac{2A^0}{n} \sum_{t=1}^{n} \cos^2(\omega^0 t)$$

$$+ \frac{2B^0}{n} \sum_{t=1}^{n} \sin(\omega^0 t) \cos(\omega^0 t) - A^0$$

$$= \frac{2}{n} \sum_{t=1}^{n} X(t) \cos(\omega^0 t) + A^0 + O\left(\frac{1}{n}\right) - A^0,$$

$$= \frac{2}{n} \sum_{t=1}^{n} X(t) \cos(\omega^0 t) + O\left(\frac{1}{n}\right).$$

Similarly $\left(\frac{1}{n}\right) \frac{\partial J(\theta^0)}{\partial B} = \left(\frac{2}{n}\right) \sum_{t=1}^{n} X(t) \sin(\omega^0 t) + O\left(\frac{1}{n}\right)$, and

$$\frac{1}{n} \frac{\partial J(\theta^0)}{\partial \omega} = \frac{2}{n} \sum_{t=1}^{n} t X(t) \{-A^0 \sin(\omega^0 t) + B^0 \cos(\omega^0 t)\} + O(1). \tag{4.23}$$

Comparing $\frac{1}{n}Q'(\theta^0)$ and $\frac{1}{n}J'(\theta^0)$, we have

$$\frac{1}{n}Q'(\theta^0) = -\frac{1}{n}J'(\theta^0) + \begin{bmatrix} O\left(\frac{1}{n}\right) \\ O\left(\frac{1}{n}\right) \\ O(1) \end{bmatrix}^T \Rightarrow Q'(\theta^0) = -J'(\theta^0) + \begin{bmatrix} O(1) \\ O(1) \\ O(n) \end{bmatrix}^T . \quad (4.24)$$

Note that $\widetilde{A} = \widetilde{A}(\omega)$ and $\widetilde{B} = \widetilde{B}(\omega)$, therefore, at $(\widetilde{A}, \widetilde{B}, \omega)$

$$J(\widetilde{A}, \widetilde{B}, \omega)$$

$$= 2\sum_{t=1}^{n} y(t)\left[\left\{\frac{2}{n}\sum_{k=1}^{n} y(k)\cos(\omega k)\right\}\cos(\omega t) + \left\{\frac{2}{n}\sum_{k=1}^{n} y(k)\sin(\omega k)\right\}\sin(\omega t)\right]$$

$$- \frac{n}{2}\left[\left\{\frac{2}{n}\sum_{t=1}^{n} y(t)\cos(\omega t)\right\}^2 + \left\{\frac{2}{n}\sum_{t=1}^{n} y(t)\sin(\omega t)\right\}^2\right]$$

$$= \frac{2}{n}\left\{\sum_{t=1}^{n} y(t)\cos(\omega t)\right\}^2 + \frac{2}{n}\left\{\sum_{t=1}^{n} y(t)\sin(\omega t)\right\}^2 = \frac{2}{n}\left|\sum_{t=1}^{n} y(t)e^{-i\omega t}\right|^2 = I(\omega).$$

Hence, the estimator of θ^0, which maximizes $J(\theta)$, is equivalent to $\widetilde{\theta}$, the ALSE of θ^0. Thus, for the ALSE $\widetilde{\theta}$, in terms of $J(\theta)$, is

$$(\widetilde{\theta} - \theta^0) = -J'(\theta^0)\left[J''(\bar{\theta})\right]^{-1}$$

$$\Rightarrow (\widetilde{\theta} - \theta^0)\mathbf{D}_2^{-1} = -\left[J'(\theta^0)\mathbf{D}_1\right]\left[\mathbf{D}_2 J''(\bar{\theta})\mathbf{D}_1\right]^{-1}$$

$$= -\left[\left(-Q'(\theta^0) + \begin{bmatrix} O(1) \\ O(1) \\ O(n) \end{bmatrix}^T\right)\mathbf{D}_1\right]\left[\mathbf{D}_2 J''(\bar{\theta})\mathbf{D}_1\right]^{-1}. \quad (4.25)$$

The matrices $\mathbf{D}_1$ and $\mathbf{D}_2$ are the same as defined in (4.12). One can show similarly as (4.8) and (4.13) that

$$\lim_{n\to\infty}\left[\mathbf{D}_2 J''(\bar{\theta})\mathbf{D}_1\right] = \lim_{n\to\infty}\left[\mathbf{D}_2 J''(\theta^0)\mathbf{D}_1\right] = -\mathbf{\Sigma} = -\lim_{n\to\infty}\left[\mathbf{D}_2 Q''(\theta^0)\mathbf{D}_1\right]. \quad (4.26)$$

Using (4.14) and (4.26) in (4.25), we have

$$(\widetilde{\theta} - \theta^0)\mathbf{D}_2^{-1} = -\left[Q'(\theta^0)\mathbf{D}_1\right]\left[\mathbf{D}_2 Q''(\bar{\theta})\mathbf{D}_1\right]^{-1} + \begin{pmatrix} O(1) \\ O(1) \\ O(n) \end{pmatrix}^T \mathbf{D}_1\left[\mathbf{D}_2 J''(\bar{\theta})\mathbf{D}_1\right]^{-1}$$

$$= (\widehat{\theta} - \theta^0)\mathbf{D}_2^{-1} + \begin{pmatrix} O(1) \\ O(1) \\ O(n) \end{pmatrix}^T \mathbf{D}_1\left[\mathbf{D}_2 J''(\bar{\theta})\mathbf{D}_1\right]^{-1}.$$

Since $\mathbf{D_2} J''(\theta^0) \mathbf{D_1}$ is an invertible matrix a.e. for large n and $\lim_{n\to\infty} \begin{pmatrix} O(1) \\ O(1) \\ O(n) \end{pmatrix}^T \mathbf{D_1} =$

$\mathbf{0}$, it follows that LSE, $\widehat{\theta}$, and ALSE, $\widetilde{\theta}$, of θ^0 of model (4.1) are asymptotically equivalent in distribution. Therefore, asymptotic distribution of $\widetilde{\theta}$ is the same as that of $\widehat{\theta}$.

Under Assumption 3.2, instead of Assumption 4.2, (4.25) follows similarly by replacing $\mathbf{D_1} = \mathbf{D_2} = \mathbf{D}$. This is the case corresponding to $\alpha = 2$, so that the second moment is finite. Similarly (4.26) and equivalence follows.   ∎

4.5 Super Efficient Frequency Estimator

In this section, we discuss the theoretical results behind the super efficient algorithm proposed by Kundu et al. [11]. This method modifies the widely used Newton–Raphson iterative method. In the previous section, we have seen that the least squares method estimates the frequency with convergence rate $O_p(n^{-3/2})$ and once the frequency is estimated with $O_p(n^{-3/2})$, the linear parameters can be estimated efficiently with rate of convergence $O_p(n^{-1/2})$. The modified Newton–Raphson method estimates the frequency with the same rate of convergence as the LSE and the asymptotic variance is smaller than that of the LSE.

The super efficient frequency estimator of ω^0 maximizes $S(\omega)$ where $S(\omega)$ is defined in (3.55) in Sect. 3.17. Suppose $\widehat{\omega}$ maximizes $S(\omega)$, then the estimators of A and B are obtained using the separable regression technique as

$$(\widehat{A}\ \widehat{B})^T = (\mathbf{Z}(\widehat{\omega})^T \mathbf{Z}(\widehat{\omega}))^{-1} \mathbf{Z}(\widehat{\omega})^T \mathbf{Y}, \tag{4.27}$$

where $\mathbf{Z}(\omega)$ is defined in (3.5).

The motivation behind using a correction factor, one-fourth of the standard Newton–Raphson correction factor, is based on the following limiting result. As assumed before, ω^0 is the true value of ω.

Theorem 4.5 *Assume that $\widetilde{\omega}$ is an estimate of ω^0 such that $\widetilde{\omega} - \omega^0 = O_p(n^{-1-\delta})$, $\delta \in \left(0, \frac{1}{2}\right]$. Suppose $\widetilde{\omega}$ is updated as $\widehat{\omega}$, using $\widehat{\omega} = \widetilde{\omega} - \frac{1}{4} \times \frac{S'(\widetilde{\omega})}{S''(\widetilde{\omega})}$, then*

(a) $\widehat{\omega} - \omega^0 = O_p(n^{-1-3\delta})$ if $\delta \leq \frac{1}{6}$,

(b) $n^{\frac{3}{2}}(\widehat{\omega} - \omega^0) \xrightarrow{d} \mathcal{N}\left(0, \frac{6\sigma^2 c(\omega^0)}{A^{0^2} + B^{0^2}}\right)$ if $\delta > \frac{1}{6}$,

where $c(\omega^0)$ is the same as defined in the asymptotic distribution of LSEs.

Proof of Theorem 4.5 Write

$$\mathbf{D} = \text{diag}\{1, 2, \ldots, n\}, \quad \mathbf{E} = \begin{bmatrix} 0 & 1 \\ -1 & 0 \end{bmatrix}, \tag{4.28}$$

$$\dot{\mathbf{Z}} = \frac{d}{d\omega}\mathbf{Z} = \mathbf{DZE}, \quad \ddot{\mathbf{Z}} = \frac{d^2}{d\omega^2}\mathbf{Z} = -\mathbf{D}^2\mathbf{Z}. \tag{4.29}$$

In this proof, we use $\mathbf{Z}(\omega) \equiv \mathbf{Z}$. Note that $\mathbf{EE} = -\mathbf{I}$, $\mathbf{EE}^T = \mathbf{I} = \mathbf{E}^T\mathbf{E}$ and

$$\frac{d}{d\omega}(\mathbf{Z}^T\mathbf{Z})^{-1} = -(\mathbf{Z}^T\mathbf{Z})^{-1}[\dot{\mathbf{Z}}^T\mathbf{Z} + \mathbf{Z}^T\dot{\mathbf{Z}}](\mathbf{Z}^T\mathbf{Z})^{-1}. \tag{4.30}$$

Compute the first- and second-order derivatives of $S(\omega)$ as

$$\frac{1}{2} S'(\omega) = \mathbf{Y}^T\dot{\mathbf{Z}}(\mathbf{Z}^T\mathbf{Z})^{-1}\mathbf{Z}^T\mathbf{Y} - \mathbf{Y}^T\mathbf{Z}(\mathbf{Z}^T\mathbf{Z})^{-1}\dot{\mathbf{Z}}^T\mathbf{Z}(\mathbf{Z}^T\mathbf{Z})^{-1}\mathbf{Z}^T\mathbf{Y},$$

$$\frac{1}{2} S''(\omega) = \mathbf{Y}^T\ddot{\mathbf{Z}}(\mathbf{Z}^T\mathbf{Z})^{-1}\mathbf{Z}^T\mathbf{Y} - \mathbf{Y}^T\dot{\mathbf{Z}}(\mathbf{Z}^T\mathbf{Z})^{-1}(\dot{\mathbf{Z}}^T\mathbf{Z} + \mathbf{Z}^T\dot{\mathbf{Z}})(\mathbf{Z}^T\mathbf{Z})^{-1}\mathbf{Z}^T\mathbf{Y}$$

$$+ \mathbf{Y}^T\dot{\mathbf{Z}}(\mathbf{Z}^T\mathbf{Z})^{-1}\dot{\mathbf{Z}}^T\mathbf{Y} - \mathbf{Y}^T\dot{\mathbf{Z}}(\mathbf{Z}^T\mathbf{Z})^{-1}\dot{\mathbf{Z}}^T\mathbf{Z}(\mathbf{Z}^T\mathbf{Z})^{-1}\mathbf{Z}^T\mathbf{Y}$$

$$+ \mathbf{Y}^T\mathbf{Z}(\mathbf{Z}^T\mathbf{Z})^{-1}(\dot{\mathbf{Z}}^T\mathbf{Z} + \mathbf{Z}^T\dot{\mathbf{Z}})(\mathbf{Z}^T\mathbf{Z})^{-1}\dot{\mathbf{Z}}^T\mathbf{Z}(\mathbf{Z}^T\mathbf{Z})^{-1}\mathbf{Z}^T\mathbf{Y}$$

$$- \mathbf{Y}^T\mathbf{Z}(\mathbf{Z}^T\mathbf{Z})^{-1}(\ddot{\mathbf{Z}}^T\mathbf{Z})(\mathbf{Z}^T\mathbf{Z})^{-1}\mathbf{Z}^T\mathbf{Y}$$

$$- \mathbf{Y}^T\mathbf{Z}(\mathbf{Z}^T\mathbf{Z})^{-1}(\dot{\mathbf{Z}}^T\dot{\mathbf{Z}})(\mathbf{Z}^T\mathbf{Z})^{-1}\mathbf{Z}^T\mathbf{Y}$$

$$+ \mathbf{Y}^T\mathbf{Z}(\mathbf{Z}^T\mathbf{Z})^{-1}\dot{\mathbf{Z}}^T\mathbf{Z}(\mathbf{Z}^T\mathbf{Z})^{-1}(\dot{\mathbf{Z}}^T\mathbf{Z} + \mathbf{Z}^T\dot{\mathbf{Z}})(\mathbf{Z}^T\mathbf{Z})^{-1}\mathbf{Z}^T\mathbf{Y}$$

$$- \mathbf{Y}^T\mathbf{Z}(\mathbf{Z}^T\mathbf{Z})^{-1}\dot{\mathbf{Z}}^T\mathbf{Z}(\mathbf{Z}^T\mathbf{Z})^{-1}\dot{\mathbf{Z}}^T\mathbf{Y}.$$

Assume $\tilde{\omega} - \omega^0 = O_p(n^{-1-\delta})$. So, for large n,

$$\left(\frac{1}{n}\mathbf{Z}^T\mathbf{Z}\right) \equiv \left(\frac{1}{n}\mathbf{Z}(\tilde{\omega})^T\mathbf{Z}(\tilde{\omega})\right)^{-1} = 2\,\mathbf{I}_2 + O_p\left(\frac{1}{n}\right)$$

and

$$\frac{1}{2n^3} S''(\tilde{\omega}) = \frac{2}{n^4}\mathbf{Y}^T\ddot{\mathbf{Z}}\mathbf{Z}^T\mathbf{Y} - \frac{4}{n^5}\mathbf{Y}^T\dot{\mathbf{Z}}(\dot{\mathbf{Z}}^T\mathbf{Z} + \mathbf{Z}^T\dot{\mathbf{Z}})\mathbf{Z}^T\mathbf{Y} + \frac{2}{n^4}\mathbf{Y}^T\dot{\mathbf{Z}}\dot{\mathbf{Z}}^T\mathbf{Y}$$

$$- \frac{4}{n^5}\mathbf{Y}^T\dot{\mathbf{Z}}\dot{\mathbf{Z}}^T\mathbf{Z}\mathbf{Z}^T\mathbf{Y} + \frac{8}{n^6}\mathbf{Y}^T\mathbf{Z}(\dot{\mathbf{Z}}^T\mathbf{Z} + \mathbf{Z}^T\dot{\mathbf{Z}})\dot{\mathbf{Z}}^T\mathbf{Z}\mathbf{Z}^T\mathbf{Y}$$

$$- \frac{4}{n^5}\mathbf{Y}^T\mathbf{Z}\ddot{\mathbf{Z}}^T\mathbf{Z}\mathbf{Z}^T\mathbf{Y} - \frac{4}{n^5}\mathbf{Y}^T\mathbf{Z}\dot{\mathbf{Z}}^T\dot{\mathbf{Z}}\mathbf{Z}^T\mathbf{Y}$$

$$+ \frac{8}{n^6}\mathbf{Y}^T\mathbf{Z}\dot{\mathbf{Z}}^T\mathbf{Z}(\dot{\mathbf{Z}}^T\mathbf{Z} + \mathbf{Z}^T\dot{\mathbf{Z}})\mathbf{Z}^T\mathbf{Y} - \frac{4}{n^5}\mathbf{Y}^T\mathbf{Z}\dot{\mathbf{Z}}^T\mathbf{Z}\dot{\mathbf{Z}}^T\mathbf{Y} + O_p\left(\frac{1}{n}\right).$$

Substituting $\dot{\mathbf{Z}}$ and $\ddot{\mathbf{Z}}$ in terms of $\mathbf{D}$ and $\mathbf{Z}$, we obtain

$$\frac{1}{2n^3}\,S''(\tilde{\omega}) = -\frac{2}{n^4}\mathbf{Y}^T\mathbf{D}^2\mathbf{Z}\mathbf{Z}^T\mathbf{Y} - \frac{4}{n^5}\mathbf{Y}^T\mathbf{DZE}(\mathbf{E}^T\mathbf{Z}^T\mathbf{DZ} + \mathbf{Z}^T\mathbf{DZE})\mathbf{Z}^T\mathbf{Y}$$

$$+ \frac{2}{n^4}\mathbf{Y}^T\mathbf{DZEE}^T\mathbf{Z}^T\mathbf{DY} - \frac{4}{n^5}\mathbf{Y}^T\mathbf{DZEE}^T\mathbf{Z}^T\mathbf{DZZ}^T\mathbf{Y}$$

$$+ \frac{8}{n^6}\mathbf{Y}^T\mathbf{Z}(\mathbf{E}^T\mathbf{Z}^T\mathbf{DZ} + \mathbf{Z}^T\mathbf{DZE})\mathbf{E}^T\mathbf{Z}^T\mathbf{DZZ}^T\mathbf{Y}$$

$$+ \frac{4}{n^5}\mathbf{Y}^T\mathbf{ZZ}^T\mathbf{D}^2\mathbf{ZZ}^T\mathbf{Y} - \frac{4}{n^5}\mathbf{Y}^T\mathbf{ZE}^T\mathbf{Z}^T\mathbf{D}^2\mathbf{ZEZ}^T\mathbf{Y}$$

$$+ \frac{8}{n^6}\mathbf{Y}^T\mathbf{ZE}^T\mathbf{Z}^T\mathbf{DZ}(\mathbf{E}^T\mathbf{Z}^T\mathbf{DZ} + \mathbf{Z}^T\mathbf{DZE})\mathbf{Z}^T\mathbf{Y}$$

$$- \frac{4}{n^5}\mathbf{Y}^T\mathbf{ZE}^T\mathbf{Z}^T\mathbf{DZE}^T\mathbf{Z}^T\mathbf{DY} + O_p\left(\frac{1}{n}\right).$$

Using results (2.15)–(2.21), one can see that

$$\frac{1}{n^2}\mathbf{Y}^T\mathbf{DZ} = \frac{1}{4}(A\ \ B) + O_p\left(\frac{1}{n}\right), \quad \frac{1}{n^3}\mathbf{Y}^T\mathbf{D}^2\mathbf{Z} = \frac{1}{6}(A\ \ B) + O_p\left(\frac{1}{n}\right), \quad (4.31)$$

$$\frac{1}{n^3}\mathbf{Z}^T\mathbf{D}^2\mathbf{Z} = \frac{1}{6}\mathbf{I}_2 + O_p\left(\frac{1}{n}\right), \quad \frac{1}{n}\mathbf{Z}^T\mathbf{Y} = \frac{1}{2}(A\ \ B)^T + O_p\left(\frac{1}{n}\right), \quad (4.32)$$

$$\frac{1}{n^2}\mathbf{Z}^T\mathbf{DZ} = \frac{1}{4}\mathbf{I}_2 + O_p\left(\frac{1}{n}\right). \quad (4.33)$$

Therefore,

$$\frac{1}{2n^3}\,S''(\tilde{\omega}) = (A^2 + B^2)\left[-\frac{1}{6} - 0 + \frac{1}{8} - \frac{1}{8} + 0 + \frac{1}{6} - \frac{1}{6} + 0 + \frac{1}{8}\right] + O_p\left(\frac{1}{n}\right)$$

$$= -\frac{1}{24}(A^2 + B^2) + O_p\left(\frac{1}{n}\right).$$

Write $S'(\omega)/2n^3 = I_1 + I_2$ and simplify I_1 and I_2 separately for large n.

$$I_1 = \frac{1}{n^3}\mathbf{Y}^T\dot{\mathbf{Z}}(\mathbf{Z}^T\mathbf{Z})^{-1}\mathbf{Z}^T\mathbf{Y} = \frac{2}{n^4}\mathbf{Y}^T\mathbf{DZEZ}^T\mathbf{Y},$$

$$I_2 = \frac{1}{n^3}\mathbf{Y}^T\mathbf{Z}(\mathbf{Z}^T\mathbf{Z})^{-1}\dot{\mathbf{Z}}^T\mathbf{Z}(\mathbf{Z}^T\mathbf{Z})^{-1}\mathbf{Z}^T\mathbf{Y} = \frac{1}{n^3}\mathbf{Y}^T\mathbf{Z}(\mathbf{Z}^T\mathbf{Z})^{-1}\mathbf{E}^T\mathbf{Z}^T\mathbf{DZ}(\mathbf{Z}^T\mathbf{Z})^{-1}\mathbf{Z}^T\mathbf{Y}$$

$$= \frac{1}{n^3}\mathbf{Y}^T\mathbf{Z}\left(2\mathbf{I} + O_p\left(\frac{1}{n}\right)\right)\mathbf{E}^T\left(\frac{1}{4}\mathbf{I} + O_p\left(\frac{1}{n}\right)\right)\left(2\mathbf{I} + O_p\left(\frac{1}{n}\right)\right)\mathbf{Z}^T\mathbf{Y}$$

$$= \frac{1}{n^3}\mathbf{Y}^T\mathbf{ZE}^T\mathbf{Z}^T\mathbf{Y} + O_p\left(\frac{1}{n}\right) = O_p\left(\frac{1}{n}\right),$$

and $(n^4 I_1)/2 = \mathbf{Y}^T\mathbf{DZEZ}^T\mathbf{Y}$ at $\tilde{\omega}$ for large n is simplified as

$$\mathbf{Y}^T \mathbf{DZEZ}^T \mathbf{Y} = \mathbf{Y}^T \mathbf{DZ}(\widetilde{\omega}) \mathbf{EZ}^T(\widetilde{\omega}) \mathbf{Y}$$

$$= \left(\sum_{t=1}^{n} y(t) t \cos(\widetilde{\omega}t) \right) \left(\sum_{t=1}^{n} y(t) \sin(\widetilde{\omega}t) \right) - \left(\sum_{t=1}^{n} y(t) t \sin(\widetilde{\omega}t) \right) \left(\sum_{t=1}^{n} y(t) \cos(\widetilde{\omega}t) \right).$$

Observe that $\sum_{t=1}^{n} y(t) e^{-i\omega t} = \sum_{t=1}^{n} y(t) \cos(\omega t) - i \sum_{t=1}^{n} y(t) \sin(\omega)$. Then along the same line as Bai et al. [2] (see also Nandi and Kundu [16])

$$\sum_{t=1}^{n} y(t) e^{-i\widetilde{\omega}t} = \sum_{t=1}^{n} \left[A^0 \cos(\omega^0 t) + B^0 \sin(\omega^0 t) + X(t) \right] e^{-i\widetilde{\omega}t}$$

$$= \sum_{t=1}^{n} \left[\frac{A^0}{2} \left(e^{i\omega^0 t} + e^{-i\omega^0 t} \right) + \frac{B^0}{2i} \left(e^{i\omega^0 t} - e^{-i\omega^0 t} \right) + X(t) \right] e^{-i\widetilde{\omega}t}$$

$$= \left(\frac{A^0}{2} + \frac{B^0}{2i} \right) \sum_{t=1}^{n} e^{i(\omega^0 - \widetilde{\omega})t} + \left(\frac{A^0}{2} - \frac{B^0}{2i} \right) \sum_{t=1}^{n} e^{-i(\omega^0 + \widetilde{\omega})t}$$

$$+ \sum_{t=1}^{n} X(t) e^{-i\widetilde{\omega}t}.$$

If $\widetilde{\omega} - \omega^0 = O_p(n^{-1-\delta})$, then it can be shown that $\sum_{t=1}^{n} e^{-i(\omega^0 + \widetilde{\omega})t} = O_p(1)$ and

$$\sum_{t=1}^{n} e^{i(\omega^0 - \widetilde{\omega})t} = n + i(\omega^0 - \widetilde{\omega}) \sum_{t=1}^{n} e^{i(\omega^0 - \omega^*)t} = n + O_p(n^{-1-\delta}) O_p(n^2)$$

$$= n + O_p(n^{1-\delta}),$$

where ω^* is a point between ω^0 and $\widetilde{\omega}$. Choose L_1 large enough, such that $L_1 \delta > 1$. Therefore, using Taylor series approximation of $e^{-i\widetilde{\omega}t}$ around ω^0 up to L_1th order terms is

$$\sum_{t=1}^{n} X(t) e^{-i\widetilde{\omega}t}$$

$$= \sum_{k=0}^{\infty} a(k) \sum_{t=1}^{n} e(t-k) e^{-i\widetilde{\omega}t}$$

$$= \sum_{k=0}^{\infty} a(k) \sum_{t=1}^{n} e(t-k) e^{-i\omega^0 t} + \sum_{k=0}^{\infty} a(k) \sum_{l=1}^{L_1-1} \frac{(-i(\widetilde{\omega} - \omega^0))^l}{l!} \sum_{t=1}^{n} e(t-k) t^l e^{-i\omega^0 t}$$

$$+ \sum_{k=0}^{\infty} a(k) \frac{\theta_1 (n(\widetilde{\omega} - \omega^0))^{L_1}}{L_1!} \sum_{t=1}^{n} |e(t-k)|,$$

with $|\theta_1| < 1$. Since $\{a(k)\}$ is absolutely summable, $\sum_{k=0}^{\infty} |a(k)| < \infty$,

$$\sum_{t=1}^{n} X(t)e^{-i\widetilde{\omega}t} = O_p(n^{\frac{1}{2}}) + \sum_{l=1}^{L_1-1} \frac{O_p(n^{-(1+\delta)l})}{l!} O_p(n^{l+\frac{1}{2}}) + O_p\left((n.n^{-1-\delta})^{L_1}.n\right)$$

$$= O_p(n^{\frac{1}{2}}) + O_p(n^{\frac{1}{2}+\delta-L_1\delta}) + O_p(n^{1-L_1\delta}) = O_p(n^{\frac{1}{2}}).$$

Therefore,

$$\sum_{t=1}^{n} y(t)e^{-i\widetilde{\omega}t} = \left(\frac{A^0}{2} + \frac{B^0}{2i}\right)\left(n + O_p(n^{1-\delta})\right) + O_p(1) + O_p(n^{\frac{1}{2}})$$

$$= \frac{n}{2}\left[(A^0 - iB^0) + O_p(n^{-\delta})\right] \quad \text{as} \quad \delta \in \left(0, \frac{1}{2}\right],$$

and

$$\sum_{t=1}^{n} y(t)\cos(\widetilde{\omega}t) = \frac{n}{2}\left(A^0 + O_p(n^{-\delta})\right), \quad \sum_{t=1}^{n} y(t)\sin(\widetilde{\omega}t) = \frac{n}{2}\left(B^0 + O_p(n^{-\delta})\right).$$

Similarly as above, observe that

$$\sum_{t=1}^{n} y(t)te^{-i\widetilde{\omega}t} = \sum_{t=1}^{n}\left(A^0\cos(\omega^0 t) + B^0\sin(\omega^0 t) + X(t)\right)te^{-i\widetilde{\omega}t}$$

$$= \frac{1}{2}(A^0 - iB^0)\sum_{t=1}^{n} t\, e^{i(\omega^0 - \widetilde{\omega})t}$$

$$+ \frac{1}{2}(A^0 + iB^0)\sum_{t=1}^{n} t\, e^{-i(\omega^0 + \widetilde{\omega})t} + \sum_{t=1}^{n} X(t)te^{-i\widetilde{\omega}t},$$

and following Bai et al. [2]

$$\sum_{t=1}^{n} t\, e^{-i(\omega^0 + \widetilde{\omega})t} = O_p(n),$$

$$\sum_{t=1}^{n} t\, e^{i(\omega^0 - \widetilde{\omega})t} = \sum_{t=1}^{n} t + i(\omega^0 - \widetilde{\omega})\sum_{t=1}^{n} t^2 - \frac{1}{2}(\omega^0 - \widetilde{\omega})^2 \sum_{t=1}^{n} t^3$$

$$- \frac{1}{6}i(\omega^0 - \widetilde{\omega})^3 \sum_{t=1}^{n} t^4 + \frac{1}{24}(\omega^0 - \widetilde{\omega})^4 \sum_{t=1}^{n} t^5 e^{i(\omega^0 - \omega^*)t}.$$

Again using $\widetilde{\omega} - \omega^0 = O_p(n^{-1-\delta})$, we have

$$\frac{1}{24}(\omega^0 - \widetilde{\omega})^4 \sum_{t=1}^{n} t^5 e^{i(\omega^0 - \omega^*)t} = O_p(n^{2-4\delta}).$$ (4.34)

Choose L_2 large enough such that $L_2\delta > 1$ and using Taylor series expansion of $e^{-i\widetilde{\omega}t}$, we have

$$\sum_{t=1}^{n} X(t)te^{-i\widetilde{\omega}t}$$

$$= \sum_{k=0}^{\infty} a(k) \sum_{t=1}^{n} e(t-k)te^{-i\widetilde{\omega}t}$$

$$= \sum_{k=0}^{\infty} a(k) \sum_{t=1}^{n} e(t-k)te^{-i\omega^0 t} + \sum_{k=0}^{\infty} a(k) \sum_{l=1}^{L_2-1} \frac{(-i(\widetilde{\omega} - \omega^0))^l}{l!} \sum_{t=1}^{n} e(t-k)t^{l+1}e^{-i\omega^0 t}$$

$$+ \sum_{k=0}^{\infty} a(k) \frac{\theta_2(n(\widetilde{\omega} - \omega^0))^{L_2}}{L_2!} \sum_{t=1}^{n} t|e(t-k)| \text{ (as before } |\theta_2| < 1)$$

$$= \sum_{k=0}^{\infty} a(k) \sum_{t=1}^{n} e(t-k)te^{-i\omega^0 t} + \sum_{l=1}^{L_2-1} O_p(n^{-(1+\delta)l})O_p(n^{l+\frac{3}{2}}) + \sum_{k=0}^{\infty} a(k)O_p(n^{\frac{5}{2}-L_2\delta})$$

$$= \sum_{k=0}^{\infty} a(k) \sum_{t=1}^{n} e(t-k)te^{-i\omega^0 t} + O_p(n^{\frac{5}{2}-L_2\delta}).$$

Therefore,

$$\sum_{t=1}^{n} y(t)t\cos(\widetilde{\omega}t)$$

$$= \frac{1}{2}\left[A^0\left(\sum_{t=1}^{n} t - \frac{1}{2}(\omega^0 - \widetilde{\omega})^2 \sum_{t=1}^{n} t^3\right) + B^0\left(\sum_{t=1}^{n}(\omega^0 - \widetilde{\omega})t^2 - \frac{1}{6}(\omega^0 - \widetilde{\omega})^3 \sum_{t=1}^{n} t^4\right)\right]$$

$$+ \sum_{k=0}^{\infty} a(k) \sum_{t=1}^{n} e(t-k)t\cos(\omega^0 t) + O_p(n^{\frac{5}{2}-L_2\delta}) + O_p(n) + O_p(n^{2-4\delta}).$$

Similarly,

$$\sum_{t=1}^{n} y(t)t\sin(\widetilde{\omega}t)$$

$$= \frac{1}{2}\left[B^0\left(\sum_{t=1}^{n} t - \frac{1}{2}(\omega^0 - \widetilde{\omega})^2 \sum_{t=1}^{n} t^3\right) - A^0\left(\sum_{t=1}^{n}(\omega^0 - \widetilde{\omega})t^2 - \frac{1}{6}(\omega^0 - \widetilde{\omega})^3 \sum_{t=1}^{n} t^4\right)\right]$$

$$+ \sum_{k=0}^{\infty} a(k) \sum_{t=1}^{n} e(t-k)t\sin(\omega^0 t) + O_p(n^{\frac{5}{2}-L_2\delta}) + O_p(n) + O_p(n^{2-4\delta}).$$

Hence,

$$\widehat{\omega} = \widetilde{\omega} - \frac{1}{4}\frac{S'(\widetilde{\omega})}{S''(\widetilde{\omega})}$$

$$= \widetilde{\omega} - \frac{1}{4}\frac{\frac{1}{2n^3}S'(\widetilde{\omega})}{-\frac{1}{24}(A^{02}+B^{02})+O_p\left(\frac{1}{n}\right)}$$

$$= \widetilde{\omega} - \frac{1}{4}\frac{\frac{2}{n^4}\mathbf{Y}^T\mathbf{DZEZ}^T\mathbf{Y}}{-\frac{1}{24}(A^{02}+B^{02})+O_p\left(\frac{1}{n}\right)}$$

$$= \widetilde{\omega} + 12\frac{\frac{1}{n^4}\mathbf{Y}^T\mathbf{DZEZ}^T\mathbf{Y}}{(A^{02}+B^{02})+O_p\left(\frac{1}{n}\right)}$$

$$= \widetilde{\omega} + 12\frac{\frac{1}{4n^3}(A^{02}+B^{02})\left\{(\omega^0-\widetilde{\omega})\sum_{t=1}^{n}t^2-\frac{1}{6}(\omega^0-\widetilde{\omega})^3\sum_{t=1}^{n}t^4\right\}}{(A^{02}+B^{02})+O_p\left(\frac{1}{n}\right)}$$

$$+\left[B^0\sum_{k=0}^{\infty}a(k)\sum_{t=1}^{n}e(t-k)t\cos(\omega^0 t)+A^0\sum_{k=0}^{\infty}a(k)\sum_{t=1}^{n}e(t-k)t\sin(\omega^0 t)\right]\times$$

$$\frac{6}{(A^{02}+B^{02})n^3+O_p\left(\frac{1}{n}\right)}+O_p(n^{-\frac{1}{2}-L_2\delta})+O_p(n^{-2})+O_p(n^{-1-4\delta})$$

$$= \omega^0 + (\omega^0-\widetilde{\omega})O_p(n^{-2\delta})$$

$$+\left[B^0\sum_{k=0}^{\infty}a(k)\sum_{t=1}^{n}e(t-k)t\cos(\omega^0 t)+A^0\sum_{k=0}^{\infty}a(k)\sum_{t=1}^{n}e(t-k)t\sin(\omega^0 t)\right]\times$$

$$\frac{6}{(A^{02}+B^{02})n^3+O_p\left(\frac{1}{n}\right)}+O_p(n^{-\frac{1}{2}-L_2\delta})+O_p(n^{-2})+O_p(n^{-1-4\delta}).$$

Finally, if $\delta \leq 1/6$, clearly $\widehat{\omega} - \omega^0 = O_p(n^{-1-3\delta})$, and if $\delta > 1/6$, then

$$n^{\frac{3}{2}}(\widehat{\omega}-\omega^0) \stackrel{d}{=} \frac{6n^{-\frac{3}{2}}}{(A^{02}+B^{02})}\left[B^0\sum_{k=0}^{\infty}a(k)\sum_{t=1}^{n}e(t-k)t\cos(\omega^0 t)\right.$$

$$\left.+A^0\sum_{k=0}^{\infty}a(k)\sum_{t=1}^{n}e(t-k)t\sin(\omega^0 t)\right]$$

$$= \frac{6n^{-\frac{3}{2}}}{(A^{02}+B^{02})}\left[B^0\sum_{t=1}^{n}X(t)t\cos(\omega^0 t)+A^0\sum_{t=1}^{n}X(t)t\sin(\omega^0 t)\right]$$

$$\left[\text{Using similar technique described in Appendix B.}\right]$$

$$\stackrel{d}{\longrightarrow} \mathcal{N}\left(0,\frac{6\sigma^2 c(\omega^0)}{A^{02}+B^{02}}\right).$$

That proves the theorem. ∎

Remark 4.3 In Eq. (4.27), write $(\widehat{A}, \widehat{B})^T = (A(\widehat{\omega}), B(\widehat{\omega}))^T$. Expanding $A(\widehat{\omega})$ around ω^0 by Taylor series,

$$A(\widehat{\omega}) - A(\omega^0) = (\widehat{\omega} - \omega^0)A'(\bar{\omega}) + o(n^2). \tag{4.35}$$

$A'(\bar{\omega})$ is the first-order derivative of $A(\omega)$ with respect to ω at $\bar{\omega}$; $\bar{\omega}$ is a point on the line joining $\widehat{\omega}$ and ω^0; $\widehat{\omega}$ can be either the LSE or the estimator obtained by modified Newton–Raphson method. Comparing the variances (asymptotic) of the two estimators of ω, note that the asymptotic variance of the corresponding estimator of A^0 is four times less than that of the LSE. The same is true for the estimator of B^0.

$$\text{Var}(A(\widehat{\omega}) - A(\omega^0)) \approx \text{Var}(\widehat{\omega} - \omega^0)[A'(\bar{\omega})]^2$$
$$\text{Var}(A(\widehat{\omega}) - A^0) = \frac{\text{Var}(\widehat{\omega}_{LSE})}{4}[A'(\bar{\omega})]^2 = \frac{\text{Var}(\widehat{A}_{LSE})}{4},$$

where $\widehat{\omega}_{LSE}$ and $\widehat{A}_{LSE}$ are the LSEs of ω^0 and A^0, respectively. Similarly, $\text{Var}(B(\widehat{\omega}) - B(\omega^0)) = \text{Var}(\widehat{B}_{LSE})/4$ and different notations involving B have similar meaning replacing A by B.

Remark 4.4 Theorem 4.5 suggests that an initial estimator of ω^0 having convergence rate equal to $O_p(n^{-1-\delta})$ is required to use it. But such an estimator is not easily available. Therefore, Kundu et al. [11] used an estimator having convergence rate $O_p(n^{-1})$ and using a fraction of the available data points to implement Theorem 4.5. The subset is chosen in such a way that the dependence structure of the error process is not destroyed. The argument maximum of $I(\omega)$, defined in (1.6), or $S(\omega)$, defined in (3.55), over Fourier frequencies, provides an estimator of the frequency with convergence rate $O_p(n^{-1})$.

4.6 Multiple Sinusoidal Model

The multicomponent frequency model, generally known as multiple sinusoidal model, has the following form in presence of p distinct frequency component:

$$y(t) = \sum_{k=1}^{p} \left[A_k^0 \cos(\omega_k^0 t) + B_k^0 \sin(\omega_k^0 t) \right] + X(t), \quad t = 1, \dots, n. \tag{4.36}$$

Here $\{y(1), \dots, y(n)\}$ is the observed data. For $k = 1, \dots, p$, A_k^0, and B_k^0 are amplitudes and ω_k^0 is the frequency. Similar to the single-component model (4.1), the additive error $\{X(t)\}$ is a sequence of random variables which satisfies different assumptions depending upon the problem at hand. In this chapter, p is assumed to be known. Problem of estimation of p is considered in Chap. 5.

Extensive work on multiple sinusoidal signal model has been done by several authors. Some references: Kundu and Mitra [13] studied the multiple sinusoidal model (4.36) and established the strong consistency and asymptotic normality of the LSEs under Assumption 3.1, that is, $\{X(t)\}$ is sequence of i.i.d. random variables with finite second moment. Kundu [10] proved the same results under Assumption 3.2. Irizarry [6] considered semi-parametric weighted LSEs of the parameters of model (4.36) and developed the asymptotic variance expression (discussed in the next section).

The LSEs of the unknown parameters are asymptotically normally distributed under Assumption 3.2 and are stated in the following theorem.

Theorem 4.6 *Under Assumption 3.2, as $n \to \infty$, $\{n^{\frac{1}{2}}(\widehat{A}_k - A_k^0), n^{\frac{1}{2}}(\widehat{B}_k - B_k^0), n^{\frac{3}{2}}(\widehat{\omega}_k - \omega_k^0)\}$ are jointly distributed as a 3-variate normal distribution with mean vector zero and dispersion matrix $2 \sigma^2 c(\omega_k^0) \Sigma_k^{-1}$, where $c(\omega_k^0)$ and Σ_k^{-1} are the same as $c(\omega^0)$ and Σ^{-1} with A^0, B^0, and ω^0, replaced by A_k^0, B_k^0, and ω_k^0, respectively. For $j \neq k$, $(n^{\frac{1}{2}}(\widehat{A}_j - A_j^0), n^{\frac{1}{2}}(\widehat{B}_j - B_j^0), n^{\frac{3}{2}}(\widehat{\omega}_j - \omega_j^0))$ and $(n^{\frac{1}{2}}(\widehat{A}_k - A_k^0), n^{\frac{1}{2}}(\widehat{B}_k - B_k^0), n^{\frac{3}{2}}(\widehat{\omega}_k - \omega_k^0))$ are asymptotically independently distributed. The quantities $c(\omega^0)$ and Σ^{-1} are defined in (4.10) and (4.11), respectively.*

4.7   Weighted Least Squares Estimators

The WLSEs are proposed by Irizarry [6] in order to analyze the harmonic components in musical sounds. The least squares results in case of multiple sinusoidal model are extended to the weighted least squares case for a class of weight functions. Under the assumption that the error process is stationary with certain other properties, the asymptotic results are developed. The WLSEs minimize the following criterion function

$$S(\boldsymbol{\omega}, \boldsymbol{\eta}) = \sum_{t=1}^{n} w\left(\frac{t}{n}\right) \left(y(t) - \sum_{k=1}^{p} \left\{A_k \cos(\omega_k t) + B_k \sin(\omega_k t)\right\}\right)^2, \quad (4.37)$$

with respect to $\boldsymbol{\omega} = (\omega_1, \ldots, \omega_p)$ and $\boldsymbol{\eta} = (A_1, \ldots A_p, B_1, \ldots, B_p)$. The error process $\{X(t)\}$ is stationary with autocovariance function $c_{xx}(h) = Cov(X(t), X(t + h))$ and spectral density function $f_x(\lambda)$ for $-\infty < \lambda < \infty$, satisfies Assumption 4.3 and the weight function $w(s)$ satisfies Assumption 4.4.

Assumption 4.3 The error process $\{X(t)\}$ is a strictly stationary real-valued random process with all moments exist, with zero mean and with $c_{xx...x}(h_1, \ldots, h_{L-1})$, the joint cumulant function of order L for $L = 2, 3, \ldots$. Also,

$$C_L = \sum_{h_1=-\infty}^{\infty} \cdots \sum_{h_{L-1}=-\infty}^{\infty} |c_{xx...x}(h_1, \ldots, h_{L-1})|$$

satisfy $\sum_k (C_k z^k)/k! < \infty$, for z in a neighborhood of 0.

Assumption 4.4 The weight function $w(s)$ is non-negative, bounded, of bounded variation, has support $[0, 1]$, and is such that $W_0 > 0$ and $W_1^2 - W_0 W_2 \neq 0$, where

$$W_n = \int_0^1 s^n w(s)ds. \tag{4.38}$$

Irizarry [6] uses the idea of Walker [23] discussed for the unweighted case by proving the following Lemma.

Lemma 4.5 *If $w(t)$ satisfies Assumption 4.4, then for $k = 0, 1, \ldots$*

$$\lim_{n\to\infty} \frac{1}{n^{k+1}} \sum_{t=1}^{n} w\left(\frac{t}{n}\right) t^k \exp(i\lambda t) = W_n, \qquad for \ \lambda = 0, 2\pi,$$

$$\sum_{t=1}^{n} w\left(\frac{t}{n}\right) t^k \exp(i\lambda t) = O(n^k), \quad for \ 0 < \lambda < 2\pi.$$

Similar to the unweighted case, that is the LSEs, the WLSEs of $(\boldsymbol{\omega}, \boldsymbol{\eta})$ is asymptotically equivalent to the weighted PEs which maximizes

$$I_{Wy}(\boldsymbol{\omega}) = \sum_{k=1}^{p} \left| \frac{1}{n} \sum_{t=1}^{n} w\left(\frac{t}{n}\right) y(t) \exp(it\omega_k) \right|^2$$

with respect to $\boldsymbol{\omega}$, and the estimators of $\boldsymbol{\eta}$ are given in Sect. 3.15. This is the sum of the periodogram functions of the tapered data $w\left(t/n\right) y(t)$. The following lemma is used to prove the consistency and asymptotic normality of the WLSEs.

Lemma 4.6 *Let the error process $\{X(t)\}$ satisfy Assumption 4.3 and the weight function $w(s)$ satisfy Assumption 4.4, then*

$$\lim_{n\to\infty} \sup_{0\leq\lambda\leq\pi} \left| \frac{1}{n^{k+1}} \sum_{t=n}^{\infty} w\left(\frac{t}{n}\right) t^k X(t) \exp(-it\lambda) \right| = 0, \quad in \ probability.$$

The consistency and asymptotic normality of the WLSEs are stated in Theorem 4.7 and Theorem 4.8, respectively.

Theorem 4.7 *Under Assumptions 4.3 and 4.4, for* $0 < \omega_k^0 < \pi$,

$$\widehat{A}_k \xrightarrow{p} A_k^0, \quad and \quad \widehat{B}_k \xrightarrow{p} B_k^0, \quad as \ n \to \infty$$

$$\lim_{n \to \infty} n|\widehat{\omega}_k - \omega_k^0| = 0, \quad for \ k = 1, \ldots, p.$$

Theorem 4.8 *Under the same assumptions as in Theorem 4.7,*

$$\left(n^{\frac{1}{2}}(\widehat{A}_k - A_k^0), n^{\frac{1}{2}}(\widehat{B}_k - B_k^0), n^{\frac{3}{2}}(\widehat{\omega}_k - \omega_k^0)\right) \xrightarrow{d} \mathcal{N}_3\left(\mathbf{0}, \frac{4\pi f_x(\omega_k^0)}{A_k^{0^2} + B_k^{0^2}}\mathbf{V_k}\right)$$

where

$$\mathbf{V_k} = \begin{bmatrix} c_1 A_k^{0^2} + c_2 B_k^{0^2} & -c_3 A_k^0 B_k^0 & -c_4 B_k^0 \\ -c_3 A_k^0 B_k^0 & c_2 A_k^{0^2} + c_1 B_k^{0^2} & c_4 A_k^0 \\ -c_4 B_k^0 & c_4 A_k^0 & c_0 \end{bmatrix}$$

with

$$c_0 = a_0 b_0$$
$$c_1 = U_0 W_0^{-1}$$
$$c_2 = a_0 b_1$$
$$c_3 = a_0 W_1 W_0^{-1}(W_o^2 W_1 U_2 - W_1^3 U_0 - 2W_0^2 + 2W_0 W_1 W_2 U_0)$$
$$c_4 = a_0(W_0 W_1 U_2 - W_1^2 U_1 - W_0 W_2 U_1 + W_1 W_2 U_0),$$

and

$$a_0 = (W_0 W_2 - W_1^2)^{-2}$$
$$b_i = W_i^2 U_2 + W_{i+1}(W_{i+1} U_0 - 2W_i U_1), \quad for \ i = 0, 1.$$

Here $W_i, i = 0, 1, 2$ *are defined in (4.38) and* $U_i, i = 0, 1, 2$ *are defined by*

$$U_i = \int_0^1 s^i w(s)^2 ds.$$

4.8 Conclusions

In this chapter, we mainly emphasize the asymptotic properties of LSEs of the unknown parameters of the sinusoidal signal model under different error assumptions. The theoretical properties of the super efficient estimator are discussed in with much detail because it has the same rate of convergence as the LSEs. At the same time, it has smaller asymptotic variance than the LSEs. Some of the other estimation procedures, presented in Chap. 3, have desirable theoretical properties. Bai et al. [1] proved the consistency of EVLP estimators whereas Kundu and Mitra [14] did it for NSD estimators. The proofs of convergence of the modified Prony algorithm and constrained MLE are found in Kundu [9] and Kundu and Kannan [12], respectively. Sequential estimators are strongly consistent and have the same limiting distribution as the LSEs, see Prasad, Kundu, and Mitra [17]. The frequency estimator obtained by using Quinn and Fernandes method is strongly consistent and is as efficient as the LSEs, Quinn and Fernandes [19]. Truong-Van [22] proved that the estimators of the frequencies obtained by amplitude harmonics method are strongly consistent, their bias converges to zero almost surely with rate $n^{-3/2}(\log n)^\delta$, $\delta > 1/2$, and they have the same asymptotic variances as Whittle estimators. Nandi and Kundu [16] proved that the algorithm presented in Sect. 3.16 is as efficient as the LSEs.

Appendix A

Proof of Lemma 4.1

In this proof, we denote $\widehat{\theta}$ by $\widehat{\theta}_n = (\widehat{A}_n, \widehat{B}_n, \widehat{\omega}_n)$ and $Q(\theta)$ by $Q_n(\theta)$ to emphasize that they depend on n. Suppose (4.3) is true and $\widehat{\theta}_n$ does not converge to θ^0 as $n \to \infty$, then there exists $c > 0$, $0 < M < \infty$, and a subsequence $\{n_k\}$ of $\{n\}$ such that $\widehat{\theta}_{n_k} \in S_{c,M}$ for all $k = 1, 2, \ldots$. Since $\widehat{\theta}_{n_k}$ is the LSE of θ^0 when $n = n_k$,

$$Q_{n_k}(\widehat{\theta}_{n_k}) \leq Q_{n_k}(\theta^0)$$
$$\Rightarrow \frac{1}{n_k}\left[Q_{n_k}(\widehat{\theta}_{n_k}) - Q_{n_k}(\theta^0)\right] \leq 0.$$

Therefore,

$$\underline{\lim}_{\widehat{\theta}_{n_k} \in S_{c,M}} \inf \frac{1}{n_k}[Q_{n_k}(\widehat{\theta}_{n_k}) - Q_{n_k}(\theta^0)] \leq 0,$$

which contradicts the inequality (4.3). Thus $\widehat{\theta}_n$ is a strongly consistent estimator of θ^0. ∎

Proof of Lemma 4.2 Under Assumption 4.1

We prove the result for $\cos(\omega t)$, the result for $\sin(\omega t)$ follows similarly. Let $Z(t) = X(t)I_{[|X(t)| \leq t^{\frac{1}{1+\delta}}]}$. Then

$$\sum_{t=1}^{\infty} P[Z(t) \neq X(t)] = \sum_{t=1}^{\infty} P[|X(t)| > t^{\frac{1}{1+\delta}}] = \sum_{t=1}^{\infty} \sum_{2^{t-1} \leq m < 2^t} P[|X(1)| > m^{\frac{1}{1+\delta}}]$$

$$\leq \sum_{t=1}^{\infty} 2^t P[2^{\frac{t-1}{1+\delta}} \leq |X(1)|] \leq \sum_{t=1}^{\infty} 2^t \sum_{k=t}^{\infty} P[2^{\frac{k-1}{1+\delta}} \leq |X(1)| < 2^{\frac{k}{1+\delta}}]$$

$$\leq \sum_{k=1}^{\infty} P[2^{\frac{k-1}{1+\delta}} \leq |X(1)| < 2^{\frac{k}{1+\delta}}] \sum_{t=1}^{k} 2^t$$

$$\leq C \sum_{k=1}^{\infty} 2^{k-1} P[2^{\frac{k-1}{1+\delta}} \leq |X(1)| < 2^{\frac{k}{1+\delta}}]$$

$$\leq C \sum_{k=1}^{\infty} E|X(1)|^{1+\delta} I_{[2^{\frac{k-1}{1+\delta}} \leq |X(1)| < 2^{\frac{k}{1+\delta}}]} \leq CE|X(1)|^{1+\delta} < \infty.$$

Therefore, $P[Z(t) \neq X(t) \text{ i.o.}] = 0$. Thus,

$$\sup_{0 \leq \omega \leq 2\pi} \frac{1}{n} \sum_{t=1}^{n} X(t) \cos(\omega t) \to 0 \quad \text{a.s} \Leftrightarrow \sup_{0 \leq \omega \leq 2\pi} \frac{1}{n} \sum_{t=1}^{n} Z(t) \cos(\omega t) \to 0 \quad \text{a.s.}$$

Let $U(t) = Z(t) - E(Z(t))$, then

$$\sup_{0 \leq \omega \leq 2\pi} \left| \frac{1}{n} \sum_{t=1}^{n} E(Z(t)) \cos(\omega t) \right| \leq \frac{1}{n} \sum_{t=1}^{n} |E(Z(t))| = \frac{1}{n} \sum_{t=1}^{n} \left| \int_{-t^{\frac{1}{1+\delta}}}^{t^{\frac{1}{1+\delta}}} x dF(x) \right| \to 0.$$

Thus, we only need to show that

$$\sup_{0 \leq \omega \leq 2\pi} \frac{1}{n} \sum_{t=1}^{n} U(t) \cos(\omega t) \to 0 \quad \text{a.s.}$$

For any fixed ω and $\varepsilon > 0$, let $0 \leq h \leq \frac{1}{2n^{\frac{1}{1+\delta}}}$, then we have

$$P\left\{ \left| \frac{1}{n} \sum_{t=1}^{n} U(t) \cos(\omega t) \right| \geq \varepsilon \right\} \leq 2e^{-hn\varepsilon} \prod_{t=1}^{n} E e^{hU(t) \cos(\omega t)} \leq 2e^{-hn\varepsilon} \prod_{t=1}^{n} \left(1 + 2Ch^{1+\delta} \right).$$

Since $|hU(t) \cos(\omega t)| \leq \frac{1}{2}$, $e^x \leq 1 + x + 2|x|^{1+\delta}$ for $|x| \leq \frac{1}{2}$ and $E|U(t)|^{1+\delta} < C$ for some $C > 0$. Clearly,

$$2e^{-hn\varepsilon} \prod_{t=1}^{n} \left(1 + 2Ch^{1+\delta} \right) \leq 2e^{-hn\varepsilon + 2nCh^{1+\delta}}.$$

Choose $h = 1/(2n^{-(1+\delta)})$, then for large n,

$$P\left\{\left|\frac{1}{n}\sum_{t=1}^{n}U(t)\cos(\omega t)\right|\geq\varepsilon\right\}\leq 2e^{-\frac{\varepsilon}{2}n^{\frac{\delta}{1+\delta}}+C}\leq Ce^{-\frac{\varepsilon}{2}n^{\frac{\delta}{1+\delta}}}.$$

Let $K=n^2$, choose $\omega_1,\ldots,\omega_K$, such that for each $\omega\in(0,2\pi)$, we have a ω_k satisfying $|\omega_k-\omega|\leq 2\pi/n^2$. Note that

$$\left|\frac{1}{n}\sum_{t=1}^{n}U(t)\left(\cos(\omega t)-\cos(\omega_k t)\right)\right|\leq C\frac{1}{n}\sum_{t=1}^{n}t^{\frac{1}{1+\delta}}.t.\left(\frac{2\pi}{n^2}\right)\leq C\pi n^{-\frac{\delta}{1+\delta}}\to 0.$$

Therefore for large n, we have

$$P\left\{\sup_{0\leq\omega\leq 2\pi}\left|\frac{1}{n}\sum_{t=1}^{n}U(t)\cos(\omega t)\right|\geq 2\varepsilon\right\}$$

$$\leq P\left\{\max_{k\leq n^2}\left|\frac{1}{n}\sum_{t=1}^{n}U(t)\cos(\omega t_k)\right|\geq\varepsilon\right\}\leq Cn^2 e^{-\frac{\varepsilon}{2}n^{\frac{\delta}{1+\delta}}}.$$

Since $\sum_{n=1}^{\infty}n^2 e^{-\frac{\varepsilon}{2}n^{\frac{\delta}{1+\delta}}}<\infty$, therefore,

$$\sup_{0\leq\omega\leq 2\pi}\left|\frac{1}{n}\sum_{t=1}^{n}U(t)\cos(\omega t)\right|\to 0\quad\text{a.s.}$$

by Borel Cantelli lemma.   ∎

Proof of Lemma 4.2 Under Assumption 3.2

Under Assumption 3.2, the error process $\{X(t)\}$ is a stationary linear process with absolutely summable coefficients. Observe that (Kundu [10])

$$\frac{1}{n}\sum_{t=1}^{n}X(t)\cos(\omega t)=\frac{1}{n}\sum_{t=1}^{n}\sum_{j=0}^{\infty}a(j)e(t-j)\cos(\omega t)$$

$$\frac{1}{n}\sum_{t=1}^{n}\sum_{j=0}^{\infty}a(j)e(t-t)\Big\{\cos((t-j)\omega)\cos(j\omega)-\sin((t-j)\omega)\sin(j\omega)\Big\}$$

$$=\frac{1}{n}\sum_{j=0}^{\infty}a(j)\cos(j\omega)\sum_{t=1}^{n}e(t-j)\cos((t-j)\omega)$$

$$-\frac{1}{n}\sum_{j=0}^{\infty}a(j)\sin(j\omega)\sum_{t=1}^{n}e(t-j)\sin((t-j)\omega). \tag{4.39}$$

Therefore,

$$\sup_{\omega} \frac{1}{n} \left| \sum_{t=1}^{n} X(t) \cos(\omega t) \right| \leq \sup_{\theta} \frac{1}{n} \left| \sum_{j=0}^{\infty} a(j) \cos(j\omega) \sum_{t=1}^{n} e(t-j) \cos((t-j)\omega) \right|$$

$$+ \sup_{\omega} \frac{1}{n} \left| \sum_{j=0}^{\infty} a(j) \sin(j\omega) \sum_{t=1}^{n} e(t-j) \sin((t-j)\omega) \right| \quad \text{a.s.}$$

$$\leq \frac{1}{n} \sum_{j=0}^{\infty} |a(j)| \sup_{\omega} \left| \sum_{t=1}^{n} e(t-j) \cos((t-j)\omega) \right|$$

$$+ \frac{1}{n} \sum_{j=0}^{\infty} |a(j)| \sup_{\omega} \left| \sum_{t=1}^{n} e(t-j) \sin((t-j)\omega) \right|.$$

Now taking expectation

$$E\left(\sup_{\omega} \frac{1}{n} \left| \sum_{t=1}^{n} X(t) \cos(\omega t) \right| \right)$$

$$+ \frac{1}{n} \sum_{j=0}^{\infty} |a(j)| E\left(\sup_{\omega} \left| \sum_{t=1}^{n} e(t-j) \sin((t-j)\omega) \right| \right)$$

$$\leq \frac{1}{n} \sum_{j=0}^{\infty} |a(j)| E\left(\sup_{\omega} \left| \sum_{t=1}^{n} e(t-j) \cos((t-j)\omega) \right| \right)$$

$$\leq \frac{1}{n} \sum_{j=0}^{\infty} |a(j)| \left\{ E \sup_{\theta} \left| \sum_{t=1}^{n} e(t-j) \cos((t-j)\omega) \right|^2 \right\}^{1/2}$$

$$+ \frac{1}{n} \sum_{j=0}^{\infty} |a(j)| \left\{ E \sup_{\theta} \left| \sum_{t=1}^{n} e(t-j) \sin((t-j)\omega) \right|^2 \right\}^{1/2}. \qquad (4.40)$$

The first term of the right hand side of (4.40)

$$\frac{1}{n} \sum_{j=0}^{\infty} |a(j)| \left\{ E \sup_{\theta} \left| \sum_{t=1}^{n} e(t-j) \cos((t-j)\omega) \right|^2 \right\}^{1/2}$$

$$\leq \frac{1}{n} \sum_{j=0}^{\infty} |a(j)| \left\{ n + \sum_{t=-(n-1)}^{n-1} E\left(\left| \sum_{m} e(m)e(m+t) \right| \right)^{1/2} \right\} \qquad (4.41)$$

where the sum $\sum_{t=-(n-1)}^{n-1}$ omits the term $t = 0$ and $\sum_m$ is over all such m such that $1 \le m + t \le n$, that is, $n - |t|$ terms (dependent on j). Similarly the second term of (4.40) can be bounded by the same. Since

$$E\left(\left|\sum_m e(m)e(m+t)\right|\right) \le E\left(\left|\sum_m e(m)e(m+t)\right|^2\right)^{1/2} = O(n^{1/2}), \qquad (4.42)$$

uniformly in j, the right-hand side of (4.41) is $O\left\{(n + n^{3/2})^{1/2}/n\right\} = O(n^{-1/4})$. Therefore, (4.40) is also $O(n^{-1/4})$. Let $M = n^3$, then $E\left(\sup_\omega \left|\sum_{t=1}^n X(t) \cos(\omega t)\right|/n\right) \le O(n^{-3/2})$. Therefore using Borel Cantelli Lemma, it follows that

$$\sup_\omega \frac{1}{n}\left|\sum_{t=1}^n X(t)\cos(\omega t)\right| \to 0, \quad \text{a.s.}$$

Now for J, $n^3 < J \le (n+1)^3$,

$$\sup_\omega \sup_{n^3 < J < (n+1)^3} \left|\frac{1}{n^3}\sum_{t=1}^{n^3} X(t)\cos(\omega t) - \frac{1}{J}\sum_{t=1}^J X(t)\cos(\omega t)\right|$$

$$= \sup_\omega \sup_{n^3 < J < (n+1)^3} \left|\frac{1}{n^3}\sum_{t=1}^{n^3} X(t)\cos(\omega t) - \frac{1}{n^3}\sum_{t=1}^J X(t)\cos(\omega t)\right.$$

$$\left. +\frac{1}{n^3}\sum_{t=1}^J X(t)\cos(\omega t) - \frac{1}{J}\sum_{t=1}^J X(t)\cos(\omega t)\right|$$

$$\le \frac{1}{n^3}\sum_{t=n^3+1}^{(n+1)^3} |X(t)| + \sum_{t=1}^{(n+1)^3} |X(t)|\left(\frac{1}{n^3} - \frac{1}{(n+1)^3}\right) \quad \text{a.s.} \qquad (4.43)$$

The mean squared error of the first term is of the order $O\left((1/n^6) \times ((n+1)^3 - n^3)^2\right) = O(n^{-2})$, and the mean squared error of the second term is of the order $O\left(n^6 \times \left(((n+1)^3 - n^3)/n^6\right)^2\right) = O(n^{-2})$. Therefore, both terms converge to zero almost surely. That proves the lemma. ∎

Proof of Lemma 4.4

Let $I'(\omega)$ and $I''(\omega)$ be the first and second derivatives of $I(\omega)$ with respect to ω. Expanding $I'(\widetilde{\omega})$ around ω^0 using Taylor series expansion:

$$I'(\widetilde{\omega}) - I'(\omega) = (\widetilde{\omega} - \omega^0)I''(\bar{\omega}), \qquad (4.44)$$

where $\bar{\omega}$ is a point on the line joining $\widetilde{\omega}$ and ω^0. Since $I'(\widetilde{\omega}) = 0$, (4.44) can be written as

$$n(\widetilde{\omega} - \omega^0) = \left[\frac{1}{n^2}I'(\omega^0)\right]\left[\frac{1}{n^3}I''(\bar{\omega})\right]^{-1}.$$

It can be shown that $\lim_{n\to\infty}\frac{1}{n^2}I'(\omega^0) = 0$ a.s. and since $I''(\omega)$ is a continuous function of ω and $\widetilde{\omega} \to \omega^0$ a.s.

$$\lim_{n\to\infty}\frac{1}{n^3}I''(\bar{\omega}) = \frac{1}{24}(A^{0^2} + B^{0^2}) \neq 0. \qquad (4.45)$$

Therefore, we have $n(\widetilde{\omega} - \omega^0) = 0$ a.s. ∎

Appendix B

We here calculate the variance covariance matrix of $Q'(\theta^0)\mathbf{D}$, present in (4.9). In this case, the error random variables $X(t)$ can be written as $\sum_{j=0}^{\infty} a(j)e(t - j)$. Note that

$$Q'(\theta^0)\mathbf{D} = \left(\frac{1}{n^{\frac{1}{2}}}\frac{\partial Q(\theta)}{\partial A}, \frac{1}{n^{\frac{1}{2}}}\frac{\partial Q(\theta)}{\partial B}, \frac{1}{n^{\frac{3}{2}}}\frac{\partial Q(\theta)}{\partial \omega}\right)\Bigg|_{\theta=\theta^0}$$

and $\lim_{n\to\infty} \text{Var}(Q'(\theta^0)\mathbf{D})) = \mathbf{\Sigma}$. In the following, we calculate Σ_{11} and Σ_{13}, where $\mathbf{\Sigma} = ((\Sigma_{ij}))$. Rest of the elements can be calculated similarly.

$$\Sigma_{11} = \lim_{n \to \infty} \text{Var}\left(\frac{1}{n^{\frac{1}{2}}}\frac{\partial Q(\boldsymbol{\theta})}{\partial A}\Big|_{\boldsymbol{\theta}=\boldsymbol{\theta}^0}\right) = \frac{1}{n}E\left[-2\sum_{t=1}^{n}X(t)\cos(\omega^0 t)\right]^2$$

$$= \lim_{n \to \infty}\frac{4}{n}E\left[\sum_{t=1}^{n}\sum_{j=0}^{\infty}a(j)e(t-j)\cos(\omega^0 t)\right]^3$$

$$= \lim_{n \to \infty}\frac{4}{n}E\left[\sum_{t=1}^{n}\sum_{j=0}^{\infty}a(j)e(t-j)\big(\cos(\omega^0(t-j))\cos(\omega^0 j)\right.$$

$$\left. - \sin(\omega^0(t-j))\sin(\omega^0 j)\big)\right]^2$$

$$= \lim_{n \to \infty}\frac{4}{n}E\left[\sum_{j=0}^{\infty}a(j)\cos(\omega^0 j)\sum_{t=1}^{n}e(t-j)\cos(\omega^0(t-j))\right.$$

$$\left. - \sum_{j=0}^{\infty}a(j)\sin(\omega^0 j)\sum_{t=1}^{n}e(t-j)\sin(\omega^0(t-j))\right]^2$$

$$= 4\sigma^2\left[\frac{1}{2}\left\{\sum_{j=0}^{\infty}a(j)\cos(\omega^0 j)\right\}^2 + \frac{1}{2}\left\{\sum_{j=0}^{\infty}a(j)\sin(\omega^0 j)\right\}^2\right]$$

$$= 2\sigma^2\left|\sum_{j=0}^{\infty}a(j)e^{-ij\omega^0}\right|^2 = 2c(\omega^0).$$

$$\Sigma_{13} = \lim_{n \to \infty}\text{Cov}\left(\frac{1}{n^{\frac{1}{2}}}\frac{\partial Q(\boldsymbol{\theta})}{\partial A}\Big|_{\boldsymbol{\theta}=\boldsymbol{\theta}^0}, \frac{1}{n^{\frac{3}{2}}}\frac{\partial Q(\boldsymbol{\theta})}{\partial \omega}\Big|_{\boldsymbol{\theta}=\boldsymbol{\theta}^0}\right)$$

$$= \lim_{n \to \infty}\frac{1}{n^2}E\left[-2\sum_{t=1}^{n}X(t)\cos(\omega^0 t)\right]\left[2\sum_{t=1}^{n}tX(t)\big(A^0\sin(\omega^0 t) - B^0\cos(\omega^0 t)\big)\right]$$

$$= \lim_{n \to \infty}-\frac{4}{n^2}E\left[\sum_{t=1}^{n}\sum_{j=0}^{\infty}a(j)e(t-j)\cos(\omega^0 t)\right]\times$$

$$\left[\sum_{t=1}^{n}t\sum_{j=0}^{\infty}a(j)e(t-j)\{A^0\sin(\omega^0 t) - B^0\cos(\omega^0 t)\}\right]$$

$$= \lim_{n \to \infty}-\frac{4}{n^2}E\left[\sum_{t=1}^{n}\sum_{j=0}^{\infty}a(j)e(t-j)\big(\cos(\omega^0(t-j))\cos(\omega^0 j)\right.$$

$$\left. - \sin(\omega^0(t-j))\sin(\omega^0 j)\big)\right]\times$$

$$\left[\sum_{t=1}^{n}\sum_{j=0}^{\infty}t\,a(j)e(t-j)\big(A^0\sin(\omega^0(t-j))\cos(\omega^0 j) + A^0\cos(\omega^0(t-j))\sin(\omega^0 j)\right.$$

$$\left. - B^0\cos(\omega^0(t-j))\cos(\omega^0 j) + B^0\sin(\omega^0(t-j))\sin(\omega^0 j)\big)\right]$$

$$
= \lim_{n \to \infty} -\frac{4}{n^2} E \left[\left(\sum_{j=0}^{\infty} a(j) \cos(\omega^0 j) \sum_{t=1}^{n} e(t-j) \cos(\omega^0 (t-j)) \right. \right.
$$

$$
\left. - \sum_{j=0}^{\infty} a(j) \sin(\omega^0 j) \sum_{t=1}^{n} e(t-j) \sin(\omega^0 (t-j)) \right)
$$

$$
\times \left(A^0 \sum_{j=0}^{\infty} a(j) \cos(\omega^0 j) \sum_{t=1}^{n} t \, e(t-j) \sin(\omega^0 (t-j)) \right.
$$

$$
+ A^0 \sum_{j=0}^{\infty} a(j) \sin(\omega^0 j) \sum_{t=1}^{n} t \, e(t-j) \cos(\omega^0 (t-j))
$$

$$
- B^0 \sum_{j=0}^{\infty} a(j) \cos(\omega^0 j) \sum_{t=1}^{n} t \, e(t-j) \cos(\omega^0 (t-j))
$$

$$
\left. \left. + B^0 \sum_{j=0}^{\infty} a(j) \sin(\omega^0 j) \sum_{t=1}^{n} t \, e(t-j) \sin(\omega^0 (t-j)) \right) \right]
$$

$$
= -4 \left[A^0 \frac{1}{4} \left\{ \sum_{j=0}^{\infty} a(j) \cos(\omega^0 j) \right\} \times \right.
$$

$$
\left\{ \sum_{j=0}^{\infty} a(j) \sin(\omega^0 j) \right\} \sigma^2 - B^0 \frac{1}{4} \left\{ \sum_{j=0}^{\infty} a(j) \cos(\omega^0 j) \right\}^2
$$

$$
\left. - A^0 \left\{ \sum_{j=0}^{\infty} a(j) \cos(\omega^0 j) \right\} \left\{ \sum_{j=0}^{\infty} a(j) \sin(\omega^0 j) \right\} - B^0 \frac{1}{4} \left\{ \sum_{j=0}^{\infty} a(j) \sin(\omega^0 j) \right\}^2 \right]
$$

$$
= \sigma^2 B^0 \left| \sum_{j=0}^{\infty} a(j) e^{ij\omega^0} \right|^2 = B^0 \sigma^2 c(\omega^0).
$$

∎

References

1. Bai, Z. D., Chen, X. R., Krishnaiah, P. R., & Zhao, L. C. (1987). Asymptotic properties of EVLP estimators for superimposed exponential signals in noise. Tech. Rep. 87-19, CMA, U. Pittsburgh.
2. Bai, Z. D., Rao, C. R., Chow, M., & Kundu, D. (2003). An efficient algorithm for estimating the parameters of superimposed exponential signals. *Journal of Statistical Planning and Inference*, *110*, 23–34.
3. Fuller, W. A. (1976). *Introduction to statistical time series*. New York: Wiley.
4. Hannan, E. J. (1971). Non-linear time series regression. *Journal of Applied Probability*, *8*, 767–780.
5. Hannan, E. J. (1973). The estimation of frequency. *Journal of Applied Probability*, *10*, 510–519.

6. Irizarry, R. A. (2002). Weighted estimation of harmonic components in a musical sound signal. *Journal of Time Series Analysis, 23*, 29–48.
7. Jennrich, R. I. (1969). Asymptotic properties of the non linear least squares estimators. *Annals of Mathematical Statistics, 40*, 633–643.
8. Kundu, D. (1993). Asymptotic theory of least squares estimators of a particular non-linear regression model. *Statistics & Probability Letters, 18*, 13–17.
9. Kundu, D. (1994). A modified Prony algorithm for sum of damped or undamped exponential signals. *Sankhya, 56*, 524–544.
10. Kundu, D. (1997). Estimating the number of sinusoids in additive white noise. *Signal Processing, 56*, 103–110.
11. Kundu, D., Bai, Z. D., Nandi, S., & Bai, L. (2011). Super efficient frequency estimation. *Journal of Statistical Planning and Inference, 141*(8), 2576–2588.
12. Kundu, D., & Kannan, N. (1994). On modified EVLP and ML methods for estimating superimposed exponential signals. *Signal Processing, 39*, 223–233.
13. Kundu, D., & Mitra, A. (1996). Asymptotic theory of least squares estimates of a non-linear time series regression model. *Communication in Statistics—Theory and Methods, 25*, 133–141.
14. Kundu, D., & Mitra, A. (1997). Consistent methods of estimating sinusoidal frequencies; a non iterative approach. *Journal of Statistical Computation and Simulation, 58*, 171–194.
15. Nandi, S., Iyer, S. K., & Kundu, D. (2002). Estimating the frequencies in presence of heavy tail errors. *Statistics & Probability Letters, 58*, 265–282.
16. Nandi, S., & Kundu, D. (2006). A fast and efficient algorithm for estimating the parameters of sum of sinusoidal model. *Sankhya, 68*, 283–306.
17. Prasad, A., Kundu, D., & Mitra, A. (2008). Sequential estimation of the sum of sinusoidal model parameters. *Journal of Statistical Planning and Inference, 138*, 1297–1313.
18. Quinn, B. G. (1994). Estimating frequency by interpolation using Fourier coefficients. *IEEE Transactions on Signal Processing, 42*, 1264–1268.
19. Quinn, B. G., & Fernandes, J. M. (1991). A fast efficient technique for the estimation of frequency. *Biometrika, 78*, 489–497.
20. Rice, J. A., & Rosenblatt, M. (1988). On frequency estimation. *Biometrika, 75*, 477–484.
21. Samorodnitsky, G., & Taqqu, M. S. (1994). *Stable non-Gaussian random processes; stochastic models with infinite variance*. New York: Chapman and Hall.
22. Truong-Van, B. (1990). A new approach to frequency analysis with amplified harmonics. *Journal of the Royal Statistical Society Series B, 52*, 203–221.
23. Walker, A. M. (1971). On the estimation of a harmonic component in a time series with stationary independent residuals. *Biometrika, 58*, 21–36.
24. Whittle, P. (1952). The simultaneous estimation of a time series Harmonic component and covariance structure. *Trabajos de estadástica, 3*, 43–57.
25. Wu, C. F. J. (1981). Asymptotic theory of non linear least squares estimation. *Annals of Statistics, 9*, 501–513.

Chapter 5
Estimating the Number of Components

5.1 Introduction

In the previous two chapters, we have discussed different estimation procedures of model (3.1) and the properties of these estimators. In all these developments, it has been assumed that the number of components "p" is known in advance. But in practice, estimation of p is also a very important problem. Although during the last 35–40 years an extensive amount of work has been done in estimating the frequencies of model (3.1), not that much of attention has been paid in estimating the number of components p.

The estimation of "p" can be considered as a model selection problem. Consider the class of models

$$\mathcal{M}_k = \left\{ \mu_k; \mu_k(t) = \sum_{j=1}^{k} A_j \left(\cos(\omega_j t) + B_j \sin(\omega_j t) \right) \right\}; \quad \text{for } k = 1, 2, \ldots.$$

$$(5.1)$$

Based on the data $\{y(t); t = 1, 2 \ldots, n\}$, estimating "$p$" is equivalent to finding $\widehat{p}$, so that $\mathcal{M}_{\widehat{p}}$ becomes the "best"-fitted model to the data. Therefore, any model selection method can be used in principle to choose p.

The most intuitive and natural estimator of p is the number of peaks of the periodogram function of the data as defined in (1.6). Consider the following examples.

Example 5.1 The data $\{y(t), t = 1, \ldots, n\}$ are obtained from model (3.1) with model parameters:

$$p = 2, \quad A_1 = A_2 = 1.0, \quad \omega_1 = 1.5, \quad \omega_2 = 2.0. \qquad (5.2)$$

The error random variables $X(1), \ldots, X(n)$ are i.i.d. normal random variables with mean 0 and variance 1. The periodogram function is plotted in Fig. 5.1, and it is immediate from the plot that the number of components is 2.

© Springer Nature Singapore Pte Ltd. 2020
S. Nandi and D. Kundu, *Statistical Signal Processing*,
https://doi.org/10.1007/978-981-15-6280-8_5

Fig. 5.1 The periodogram plot of the data obtained from model (5.2)

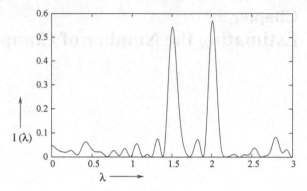

Fig. 5.2 The periodogram plot of the data obtained from model (5.3)

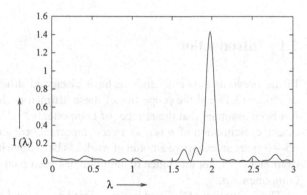

Example 5.2 The data $\{y(t), t = 1, \ldots, n\}$ are obtained from model (3.1) with model parameters:

$$p = 2, \quad A_1 = A_2 = 1.0, \quad \omega_1 = 1.95, \quad \omega_2 = 2.0. \tag{5.3}$$

The error random variables are the same as in Example 5.1. The periodogram is plotted in Fig. 5.2. It is not clear from Fig. 5.2 that $p = 2$.

Example 5.3 The data $\{y(t); t = 1, \ldots, n\}$ are obtained from model (3.1) with the same model parameters as in Example 5.1, but the errors are i.i.d. normal random variables with mean zero and variance 5. The periodogram function is presented in Fig. 5.3. It is not again clear from the periodogram plot that $p = 2$.

The above examples reveal that when frequencies are very close to each other or if the error variance is high, it may not be possible to detect the number of components from the periodogram plot of the data. Different methods have been proposed to detect the number of components of model (3.1). All the methods can be broadly classified into three different categories namely, (a) likelihood ratio approach, (b) cross-validation method, and (c) information theoretic criterion. In this chapter, we

Fig. 5.3 The periodogram plot of the data obtained from model (5.2) with error variance 5

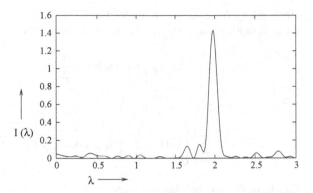

provide a brief review of different methods. Throughout this chapter without loss of generality, one can assume that

$$A_1^2 + B_1^2 > A_2^2 + B_2^2 > \cdots > A_p^2 + B_p^2.$$

5.2 Likelihood Ratio Approach

In estimating p of model (3.1), one of the natural procedures is to use a test of significance for each additional term as it is introduced in the model. Fisher [5] considered this as a simple testing of hypothesis problem. Such a test can be based on the well-known "maximum likelihood ratio", that is, the ratio of the maximized likelihood for k terms of model (3.1) to the maximized likelihood for $(k-1)$ terms of model (3.1). If this quantity is large, it provides evidence that the kth term is needed in the model.

The problem can be formulated as follows:

$$H_0 : p = p_0 \quad \text{against} \quad H_1 : p = p_1 \tag{5.4}$$

where $p_1 > p_0$. Based on the assumption that the error random variables follow i.i.d. normal distribution with mean 0 and variance σ^2, the maximized log-likelihood for fixed σ^2 can be written as

$$\text{constant} - \frac{n}{2}\ln\sigma^2 - \frac{1}{2\sigma^2}\sum_{t=1}^{n}\left[y(t) - \sum_{k=1}^{p_0}\{\widehat{A}_{k,p_0}\cos(\widehat{\omega}_{k,p_0}t) + \widehat{B}_{k,p_0}\sin(\widehat{\omega}_{k,p_0}t)\}\right]^2.$$

Here, $\widehat{A}_{k,p_0}, \widehat{B}_{k,p_0}, \widehat{\omega}_{k,p_0}$ are the MLEs of $A_{k,p_0}, B_{k,p_0}, \omega_{k,p_0}$, respectively, based on the assumption that $p = p_0$. The unconstrained maximized log-likelihood is then

$$l_{p_0} = \text{constant} - \frac{n}{2}\ln\widehat{\sigma}_{p_0}^2 - \frac{n}{2}, \tag{5.5}$$

where

$$\widehat{\sigma}_{p_0}^2 = \frac{1}{n}\sum_{t=1}^{n}\left[y(t) - \sum_{k=1}^{p_0}\left\{\widehat{A}_{k,p_0}\cos(\widehat{\omega}_{k,p_0}t) + \widehat{B}_{k,p_0}\sin(\widehat{\omega}_{k,p_0}t)\right\}\right]^2.$$

Similarly, $\widehat{\sigma}_{p_1}^2$ can be obtained when $p = p_1$.

Therefore, the likelihood ratio test takes the following form: rejects H_0 if L is large, where

$$L = \frac{\widehat{\sigma}_{p_0}^2}{\widehat{\sigma}_{p_1}^2} = \frac{\sum_{t=1}^{n}\left[y(t) - \sum_{k=1}^{p_0}\left\{\widehat{A}_{k,p_0}\cos(\widehat{\omega}_{k,p_0}t) + \widehat{B}_{k,p_0}\sin(\widehat{\omega}_{k,p_0}t)\right\}\right]^2}{\sum_{t=1}^{n}\left[y(t) - \sum_{k=1}^{p_1}\left\{\widehat{A}_{k,p_1}\cos(\widehat{\omega}_{k,p_1}t) + \widehat{B}_{k,p_1}\sin(\widehat{\omega}_{k,p_1}t)\right\}\right]^2}. \tag{5.6}$$

To find the critical point of the above test procedure, one needs to obtain the exact/asymptotic distribution of L under the null hypothesis. It seems that finding the exact/asymptotic distribution of L is a difficult problem.

Quinn [13] obtained the distribution of L as defined in (5.6) under the following assumptions: (a) errors are i.i.d. normal random variables, with mean 0 and variance σ^2, (b) frequencies are of the form $2\pi j/n$, where $1 \le j \le (n-1)/2$. If the frequencies are of the form (b), Quinn [13] showed that in this case L is of the form:

$$L = \frac{\sum_{t=1}^{n}y(t)^2 - J_{p_0}}{\sum_{t=1}^{n}y(t)^2 - J_{p_1}}, \tag{5.7}$$

where J_k is the sum of the k largest elements of $\{I(\omega_j); \omega_j = 2\pi j/n, 1 \le j \le (n-1)/2\}$, and $I(\omega)$ is the periodogram function of the data sequence $\{y(t); t = 1,\ldots,n\}$, as defined in (1.6). The likelihood ratio statistic L as defined in (5.7) can also be written as

$$L = \frac{\sum_{t=1}^{n}y(t)^2 - J_{p_0}}{\sum_{t=1}^{n}y(t)^2 - J_{p_1}} = \frac{1}{1 - G_{p_0,p_1}}, \tag{5.8}$$

where

$$G_{p_0,p_1} = \frac{J_{p_1} - J_{p_0}}{\sum_{t=1}^{n}y(t)^2 - J_{p_0}}. \tag{5.9}$$

When $p_0 = 0$ and $p_1 = 1$,

$$G_{0,1} = \frac{J_1}{\sum_{t=1}^{n}y(t)^2}, \tag{5.10}$$

and it is the well-known Fisher's g-statistic.

Finding the distribution of L is equivalent to finding the distribution of G_{p_0,p_1}. Quinn and Hannan [15] provided the approximate distribution of G_{p_0,p_1}, which is quite complicated in nature, and may not have much practical importance. The problem becomes more complicated when the frequencies are not in the form of (b) as defined above. Some attempts have been made to simplify the distribution of G_{p_0,p_1} by Quinn and Hannan [15] under the assumption of i.i.d. normal error. It is not further pursued here.

5.3 Cross-Validation Method

The cross-validation method is a model selection technique and it can be used in a fairly general setup. The basic assumption of the cross-validation technique is that there exists an M, such that $1 \leq k \leq M$ for the models defined in (5.1). The cross-validation method can be described as follows: For a given k, such that $1 \leq k \leq M$, remove jth observation from $\{y(1), \ldots, y(n)\}$, and estimate $y(j)$, say $\widehat{y}_k(j)$, based on the model assumption $\mathcal{M}_k$ and $\{y(1), \ldots, y(j-1), y(j+1), \ldots, y(n)\}$. Compute the cross-validation error for the kth model as

$$CV(k) = \sum_{t=1}^{n} (y(t) - \widehat{y}_k(t))^2; \quad \text{for} \quad k = 1, \ldots, M. \qquad (5.11)$$

Choose $\widehat{p}$ as an estimate of p, if

$$CV(\widehat{p}) < \{CV(1), \ldots, CV((\widehat{p}-1), CV((\widehat{p}+1), \ldots, CV(M)\}.$$

Cross-validation method has been used quite extensively in model selection as it is well-known that the small sample performance of cross-validation technique is very good, although it usually does not produce consistent estimator of the model order.

Rao [16] proposed to use the cross-validation technique to estimate the number of components in a sinusoidal model. The author did not provide any explicit method to compute $\widehat{y}_k(j)$ based on the observations $\{y(1), \ldots, y(j-1), y(j+1), \ldots, y(n)\}$, and most of the estimation methods are based on the fact that the data are equispaced.

Kundu and Kundu [11] first provided the modified EVLP method and some of its generalizations to estimate consistently the amplitudes and frequencies of a sinusoidal signal based on the assumption that the errors are i.i.d. random variables with mean zero and finite variance. The modified EVLP method has been further modified by Kundu and Mitra [12] by using both the forward and backward data, and it has been used quite effectively to estimate p of model (3.1) by using the cross-validation technique. Extensive simulation results suggest that the cross-validation technique works quite well for small sample sizes and for large error variances, although for large sample sizes the performance is not that satisfactory. For large samples, the cross-validation technique is computationally very demanding, hence it is not recommended.

In both the likelihood ratio approach and the cross-validation approach, it is important that the errors are i.i.d. random variables with mean zero and finite variance. It is not immediate how these methods can be modified for stationary errors.

5.4   Information Theoretic Criteria

Different information theoretic criteria such as AIC of Akaike [2, 3], BIC of Schwartz [19] or Rissanen [17], EDC of Bai, Krishnaiah, and Zhao [4] have been used quite successfully in different model selection problems. AIC, BIC, EDC, and their several modifications have been used to detect the number of components of model (3.1). All the information theoretic criteria are based on the following assumption that the maximum model order can be M, and they can be put in the general framework as follows: For the kth order model define

$$ITC(k) = f(\widehat{\sigma}_k^2) + N(k)\, c(n); \quad 1 \le k \le M. \tag{5.12}$$

Here, $\widehat{\sigma}_k^2$ is the estimated error variance and $N(k)$ denotes the number of parameters, both based on the assumption that the model order is k, $f(\cdot)$ is an increasing and $c(\cdot)$ is a monotone function. The quantity $N(k)c(n)$ is known as the penalty, and for fixed n, it increases with k. Depending on different information theoretic criteria, $f(\cdot)$ and $c(\cdot)$ change. Choose $\widehat{p}$ as an estimate of p, if

$$ITC(\widehat{p}) < \{ITC(1), \ldots, ITC((\widehat{p}-1), ITC((\widehat{p}+1), \ldots, ITC(M)\}.$$

The main focus of the different information theoretic criteria is to choose properly the functions $f(\cdot)$ and $c(\cdot)$.

As $\widehat{\sigma}_k^2$ is a decreasing function of k and $f(\cdot)$ is increasing, $f(\widehat{\sigma}_k^2)$ is a decreasing function of k. The function $N(k)\, c(n)$ acts as a penalty function, so that for small k, $f(\widehat{\sigma}_k^2)$ is the dominant part and $ITC(\cdot)$ behaves like a decreasing function. After a certain value of $k(\approx p)$, $N(k)\, c(n)$ becomes dominant and $ITC(\cdot)$ behaves like an increasing function.

5.4.1   Rao's Method

Rao [16] proposed different information theoretic criteria to detect the number of sinusoidal components based on the assumption that the errors are i.i.d. mean zero normal random variables. Based on the above error assumption, AIC takes the following form

$$AIC(k) = n \ln R_k + 2\,(3k+1), \tag{5.13}$$

where R_k denotes the minimum value of

$$\sum_{t=1}^{n} \left(y(t) - \sum_{j=1}^{k} \left(A_j \cos(\omega_j t) + B_j \sin(\omega_j t) \right) \right)^2 , \tag{5.14}$$

and the minimization of (5.14) is performed with respect to $A_1, \ldots, A_k, B_1, \ldots, B_k,$ $\omega_1, \ldots, \omega_k$. "$(3k + 1)$" denotes the number of parameters when the number of components is k.

Under the same assumption, BIC takes the form

$$BIC(k) = n \ln R_k + (3k + 1) \frac{1}{2} \ln n, \tag{5.15}$$

and EDC takes the form

$$EDC(k) = n \ln R_k + (3k + 1) \, c(n). \tag{5.16}$$

Here, $c(n)$ satisfies the following conditions:

$$\lim_{n \to \infty} \frac{c(n)}{n} = 0 \quad \text{and} \quad \lim_{n \to \infty} \frac{c(n)}{\ln \ln n} = \infty. \tag{5.17}$$

EDC is a very flexible criterion, and BIC is a special case of EDC. Several $c(n)$ satisfy (5.17). For example, $c(n) = n^a$, for $a < 1$ and $c(n) = (\ln \ln n)^b$, for $b > 1$ satisfy (5.17).

Although Rao [16] proposed to use information theoretic criteria to detect the number of components of model (3.1), he did not provide any practical implementation procedure, particularly the computation of R_k. He suggested to compute R_k by minimizing (5.14) with respect to the unknown parameters, which may not be very simple, as it has been observed in Chap. 3.

Kundu [8] suggested a practical implementation procedure of the method proposed by Rao [16], and performed extensive simulation studies to compare different methods for different models, different error variances, and for different choices of $c(n)$. It is further observed that AIC does not provide consistent estimate of the model order. Although for small sample sizes the performance of AIC is good, for large sample sizes it has a tendency to overestimate the model order. Among the different choices of $c(n)$ for EDC criterion, it is observed that BIC performs quite well.

5.4.2 Sakai's Method

Sakai [18] considered the problem of estimating p of model (3.1) under the assumptions that the errors are i.i.d. normal random variables with mean zero and the frequencies can be Fourier frequencies only. He has re-formulated the problem as follows. Consider the model

$$y(t) = \sum_{j=0}^{M} v_j (A_j \cos(\omega_j t) + B_j \sin(\omega_j t)) + X(t); \quad t = 1, \ldots, n, \qquad (5.18)$$

where $\omega_j = 2\pi j / n$, and $M = n/2$ or $M = (n - 1)/2$ depending on whether n is even or odd. The indicator function v_j is such that

$$v_j = \begin{cases} 1 \text{ if jth component is present} \\ \\ 0 \text{ if jth component is absent.} \end{cases}$$

Sakai [18] proposed the following information theoretic-like criterion as follows:

$$SIC(v_1, \ldots, v_M) = \ln \widehat{\sigma}^2 + \frac{2(\ln n + \gamma - \ln 2)}{n}(v_1 + \cdots + v_M). \qquad (5.19)$$

Here, $\gamma\, (\approx = 0.577)$ is the Euler's constant and

$$\widehat{\sigma}^2 = \frac{1}{n} \sum_{t=1}^{n} y(t)^2 - \frac{4}{n} \sum_{k=1}^{M} I(\omega_k) v_k,$$

where $I(\cdot)$ is the periodogram function of $\{y(t); t = 1, \ldots, n\}$. Now for all 2^M possible choices of $(v_1, \ldots, v_M)$, choose that combination for which $SIC(v_1, \ldots, v_M)$ is minimum.

Sakai [18] also suggested a Hopfield neural network for minimizing $SIC(v_1, \ldots, v_M)$, so that the problem can be solved without sorting operations. This model selection criterion is also applied to the order determination problem of an AR process.

5.4.3   Quinn's Method

Quinn [14] considered the same problem under the assumptions that the sequence of error random variable $\{X(t)\}$ is stationary and ergodic with mean zero and finite variance. It is further assumed that the frequencies are of the form of Fourier frequencies, $2\pi j / n$, for $1 \leq j \leq (n - 1)/2$. Under the above assumptions, Quinn [14] proposed an information theoretic-like criterion as follows. Let

$$QIC(k) = n \ln \widehat{\sigma}_k^2 + 2k\, c(n), \qquad (5.20)$$

where

$$\widehat{\sigma}_k^2 = \frac{1}{n} \left(\sum_{t=1}^{n} y(t)^2 - J_k \right),$$

and J_k is same as defined in (5.7). The penalty function $c(n)$ is such that it satisfies

$$\lim_{n \to \infty} \frac{c(n)}{n} = 0. \tag{5.21}$$

Then the number of sinusoids p is estimated as the smallest value of $k \geq 0$, for which $QIC(k) < QIC(k+1)$. Using the results of An, Chen, and Hannan [1], it has been shown that if $\widehat{p}$ is an estimate of p, then $\widehat{p}$ converges to p almost surely. Under the assumption that the sequence of error random variables $\{X(t)\}$ is stationary and ergodic, Quinn [14] proved that $\widehat{p} \to p$ almost surely if

$$\liminf_{n \to \infty} \frac{c(n)}{\log n} > 1.$$

Further, $\widehat{p}$ is greater than p infinitely often if

$$\limsup_{n \to \infty} \frac{c(n)}{\log n} < 1,$$

when $\{X(t)\}$ is an i.i.d. sequence with $E(|X(t)|^6) < \infty$ and

$$\sup_{|s| > s_0 > 0} |\psi(s)| = \beta(s_0) < 1,$$

ψ is the characteristic function of $\{X(t)\}$.

Quinn's method provides a strongly consistent estimate of the number of sinusoidal components, but it is not known how the method behaves for small sample sizes. Simulation experiments need to be done to verify the performance of this method.

5.4.4 Wang's Method

Wang [20] also considered this problem of estimating p under more general conditions than Rao [16] or Quinn [14]. Wang [20] assumed the same error assumptions as those of Quinn [14], but the frequencies need not be restricted to only Fourier frequencies. The method of Wang [20] is very similar to the method proposed by Rao [16], but the main difference is in the estimation procedure of the unknown parameters for the kth order model. The information criterion of Wang [20] can be described as follows. For the kth order model consider

$$WIC(k) = n \ln \widehat{\sigma}_k^2 + k \, c(n), \qquad (5.22)$$

where $\widehat{\sigma}_k^2$ is the estimated error variance and $c(n)$ satisfies the same condition as (5.21). Although Rao [16] did not provide any efficient estimation procedure of the unknown parameters for the kth order model, Wang [20] suggested to use the following estimation procedure of the unknown frequencies. Let $\Omega_1 = (-\pi, \pi]$, and $\widehat{\omega}_1$ is the argument maximizer of the periodogram function (1.6) over Ω_1. For $j > 1$, if Ω_{j-1} and $\widehat{\omega}_{j-1}$ are defined, then

$$\Omega_j = \Omega_{j-1} \backslash (\widehat{\omega}_{j-1} - u_n, \widehat{\omega}_{j-1} + u_n),$$

and $\widehat{\omega}_j$ is obtained as the argument maximizer of the periodogram function over Ω_j, where $u_n > 0$ and satisfies the conditions

$$\lim_{n \to \infty} u_n = 0, \quad \text{and} \quad \lim_{n \to \infty} (n \ln n)^{1/2} u_n = \infty.$$

Estimate p as the smallest value of $k \geq 0$, such that $WIC(k) < WIC(k+1)$. If

$$\liminf_{n \to \infty} \frac{c(n)}{\ln n} > C > 0,$$

then an estimator of p obtained by this method is a strongly consistent estimator of p, see Wang [20]. Although Wang's method is known to provide a consistent estimate of p, it is not known how this method performs for small sample sizes. Moreover, he did not mention about the practical implementation procedure of his method, mainly how to choose u_n or $c(n)$ for a specific case. It is very clear that the performance of this procedure heavily depends on them.

Kavalieries and Hannan [6] discussed some practical implementation procedure of Wang's method. They have suggested a slightly different estimation procedure of the unknown parameters than that of Wang [20]. It is not pursued further here, but interested readers may have a look at that paper.

5.4.5   Kundu's Method

Kundu [9] suggested the following simple estimation procedure of p. If M denotes the maximum possible model order, then for some fixed $L > 2M$, consider the data matrix $\mathbf{A}_L$:

$$\mathbf{A}_L = \begin{pmatrix} y(1) & \cdots & y(L) \\ \vdots & \ddots & \vdots \\ y(n-L+1) & \cdots & y(n) \end{pmatrix}. \tag{5.23}$$

Let $\widehat{\sigma}_1^2 > \cdots > \widehat{\sigma}_L^2$ be the L eigenvalues of the $L \times L$ matrix $\mathbf{R}_n = \mathbf{A}_L^T \mathbf{A}_L / n$. Consider

$$KIC(k) = \widehat{\sigma}_{2k+1}^2 + k\, c(n), \tag{5.24}$$

where $c(n) > 0$, satisfying the following two conditions:

$$\lim_{n\to\infty} c(n) = 0, \quad \text{and} \quad \lim_{n\to\infty} \frac{c(n)\sqrt{n}}{(\ln\ln n)^{1/2}} = \infty. \tag{5.25}$$

Now choose that value of k, as an estimate of p for which $KIC(k)$ is minimum.

Under the assumption of i.i.d. errors, the author proved the strong consistency of the above procedure. Moreover, the probability of wrong detection has also been obtained in terms of the linear combination of chi-square variables. Extensive simulations have been performed to check the effectiveness of the proposed method and to find proper $c(n)$. It is observed that $c(n) = 1/\sqrt{\ln n}$ works quite well, although no theoretical justification has been provided.

Kundu [10] in a subsequent paper discussed the choice of $c(n)$. He has considered a slightly different criterion than (5.24), and the new criterion takes the form

$$KIC(k) = \ln(\widehat{\sigma}_{2k+1}^2 + 1) + k\, c(n), \tag{5.26}$$

where $c(n)$ satisfies the same conditions as in (5.25).

5.4.5.1 Upper Bound of the Probability of Wrong Detection

In this section, we provide the upper bound of $P(\widehat{p} \neq p)$, the probability of wrong detection, where $\widehat{p}$ is obtained from the criterion (5.26). Now

$$
\begin{aligned}
P(\widehat{p} \neq p) &= P(\widehat{p} < p) + P(\widehat{p} > p) \\
&= \sum_{q=0}^{p-1} P(\widehat{p} = q) + \sum_{q=p+1}^{M} P(\widehat{p} = q) \\
&= \sum_{q=0}^{p-1} P(KIC(q) - KIC(p) < 0) + \sum_{q=p+1}^{M} P(KIC(q) - KIC(p) < 0).
\end{aligned}
$$

Let us consider two different cases:

Case I: $q < p$

$P(KIC(q) - KIC(p) < 0) =$

$P(\ln(\widehat{\sigma}^2_{2q+1} + 1) - \ln(\widehat{\sigma}^2_{2p+1} + 1) + (q - p)c(n) < 0) =$

$P(\ln(\sigma^2_{2q+1} + 1) - \ln(\sigma^2_{2p+1} + 1) + (q - p)c(n)$

$< \ln(\widehat{\sigma}^2_{2p+1} + 1) - \ln(\sigma^2_{2p+1} + 1) + \ln(\sigma^2_{2q+1} + 1) - \ln(\widehat{\sigma}^2_{2q+1} + 1)) <$

$P(\ln(\sigma^2_{2q+1} + 1) - \ln(\sigma^2_{2p+1} + 1) < (p - q)c(n) + |\ln(\widehat{\sigma}^2_{2p+1} + 1) - \ln(\sigma^2_{2p+1} + 1)| +$

$|\ln(\sigma^2_{2q+1} + 1) - \ln(\widehat{\sigma}^2_{2q+1} + 1)|).$

Note that there exists a $\delta > 0$, such that for large n,

$$\ln(\sigma^2_{2q+1} + 1) - \ln(\sigma^2_{2p+1} + 1) > (p - q)c(n) + \delta.$$

Therefore, for large n

$$P(KIC(q) - KIC(p) < 0) = 0,$$

as

$$\widehat{\sigma}^2_{2q+1} \xrightarrow{a.s.} \sigma^2_{2q+1} \quad \text{and} \quad \widehat{\sigma}^2_{2p+1} \xrightarrow{a.s.} \sigma^2_{2p+1}.$$

Case II: $q > p$

$$P(KIC(q) - KIC(p) < 0) =$$

$$P(\ln(\widehat{\sigma}^2_{2q+1} + 1) - \ln(\widehat{\sigma}^2_{2p+1} + 1) + (q - p)c(n) < 0) =$$

$$P(\ln(\widehat{\sigma}^2_{2p+1} + 1) - \ln(\widehat{\sigma}^2_{2q+1} + 1) > (q - p)c(n)). \qquad (5.27)$$

To compute (5.27), we need to know the joint distribution of $\widehat{\sigma}^2_k$, for $k = 2p + 1, \ldots, L$. Based on the perturbation theory of Wilkinson [21], the asymptotic joint distribution of the eigenvalues $\widehat{\sigma}^2_k$, for $k = 2p + 1, \ldots, L$ can be obtained. It has been shown that $\widehat{\sigma}^2_{2p+1}$ and $\widehat{\sigma}^2_{2q+1}$ for $q > p$ are jointly normal each with mean σ^2, the error variance, and it has a given variance covariance matrix. Based on this result, an upper bound of probability of wrong detection can be obtained. Extensive simulation results suggest that these theoretical bounds match very well with the simulated results.

5.4.5.2   Penalty Function Bank Approach

The natural question is how to choose a proper $c(n)$. It is observed, see Kundu [10], that the probability of over estimation and probability of under estimation depend on the asymptotic distribution of the eigenvalues of $\text{Vec}(\mathbf{R}_n)$. Here, $\text{Vec}(\cdot)$ of a $k \times k$

matrix is a $k^2 \times 1$ vector stacking the columns one below the other. The main idea to choose the proper penalty function from a class of penalty functions is to identify the penalty function which satisfies (5.25) and for which the theoretical bound of the probability of wrong selection is minimum. These theoretical bounds depend on the unknown parameters, and without knowing these parameters it is not possible to calculate these theoretical bounds. But even if the joint distribution of the eigenvalues are known, theoretically it is very difficult to estimate (5.27) from the given sample. Kundu [10] used bootstrap-like technique similarly as Kaveh, Wang, and Hung [7] to estimate (5.27) based on the observed sample, and then used it in choosing the proper penalty function from a class of penalty functions.

The second question naturally arises on how to choose the class of penalty functions. The suggestion is as follows. Take any particular class of reasonable size, maybe around ten or twelve, where all of them should satisfy (5.25) but they should converge to zero at varying rates (from very low to very high). Obtain the upper bound of the probability of wrong detection based on bootstrap-like technique as suggested above, for all these penalty functions, and compute the minimum of these values. If the minimum itself is high, it indicates that the class is not good, otherwise it is fine. Simulation results indicate that the method works very well for different sample sizes and for different error variances. The major drawback of this method is that it has been proposed when the errors are i.i.d. random variables, it is not immediate how the method can be modified for the correlated errors.

5.5 Conclusions

In this chapter, we have discussed different methods of estimating the number of components in a multiple sinusoidal model. This problem can be formulated as a model selection problem, hence any model selection procedure which is available in the literature can be used for this purpose. We have provided three different approaches; comparison of different methods is not available in the literature. We believe it is still an open problem, more work is needed along this direction.

References

1. An, H.-Z., Chen, Z.-G., & Hannan, E. J. (1983). The maximum of the periodogram. *Journal of Multivariate Analysis, 13*, 383–400.
2. Akaike, H. (1969). Fitting autoregressive models for prediction. *Annals of the Institute of Statistical Mathematics, 21*, 243–247.
3. Akaike, H. (1970). Statistical predictor identification. *Annals of the Institute of Statistical Mathematics, 22*, 203–217.
4. Bai, Z. D., Krishnaiah, P. R., & Zhao, L. C. (1986). On the detection of the number of signals in the presence of white noise. *Journal of Multivariate Analysis, 20*, 1–25.
5. Fisher, R. A. (1929). Tests of significance in harmonic analysis. *Proceedings of the Royal Society of London Series A, 125*, 54–59.

6. Kavalieris, L., & Hannan, E. J. (1994). Determining the number of terms in a trigonometric regression. *Journal of Time Series Analysis, 15*, 613–625.
7. Kaveh, M., Wang, H., & Hung, H. (1987). On the theoretical performance of a class of estimators of the number of narrow band sources. *IEEE Transactions on Acoustics, Speech, and Signal Processing, 35*, 1350–1352.
8. Kundu, D. (1992). Detecting the number of signals for undamped exponential models using information theoretic criteria. *Journal of Statistical Computation and Simulation, 44*, 117–131.
9. Kundu, D. (1997). Estimating the number of sinusoids in additive white noise. *Signal Processing, 56*, 103–110.
10. Kundu, D. (1998). Estimating the number of sinusoids and its performance analysis. *Journal of Statistical Computation and Simulation, 60*, 347–362.
11. Kundu, D., & Kundu, R. (1995). Consistent estimates of super imposed exponential signals when observations are missing. *Journal of Statistical Planning and Inference, 44*, 205–218.
12. Kundu, D., & Mitra, A. (1995). Consistent method of estimating the superimposed exponential signals. *Scandinavian Journal of Statistics, 22*, 73–82.
13. Quinn, B. G. (1986). Testing for the presence of sinusoidal components. *Journal of Applied Probability, Special Volume, 23 A*, 201–210.
14. Quinn, B. G. (1989). Estimating the number of terms in a sinusoidal regression. *Journal of Time Series Analysis, 10*, 71–75.
15. Quinn, B. G., & Hannan, E. J. (2001). *The estimation and tracking of frequency*. New York: Cambridge University Press.
16. Rao, C. R. (1988). Some results in signal detection. In S. S. Gupta & J. O. Berger (Eds.), *Decision theory and related topics, IV, 2* (pp. 319–332). New York: Springer.
17. Rissanen, J. (1978). Modeling by shortest data description. *Automatica, 14*, 465–471.
18. Sakai, H. (1990). An application of a BIC-type method to harmonic analysis and a new criterion for order determination of an error process. *IEEE Transactions of Acoustics, Speech and Signal Processing, 38*, 999–1004.
19. Schwartz, S. C. (1978). Estimating the dimension of a model. *Annals of Statistics, 6*, 461–464.
20. Wang, X. (1993). An AIC type estimator for the number of cosinusoids. *Journal of Time Series Analysis, 14*, 433–440.
21. Wilkinson, W. H. (1965). *The algebraic eigenvalue problem*. Oxford: Clarendon Press.

Chapter 6
Fundamental Frequency Model and Its Generalization

6.1 Introduction

The sinusoidal frequency model is a well-known model in different fields of science and technology and is a very useful model in explaining nearly periodic data. The fundamental frequency model (FFM) is practically the sinusoidal frequency model which exploits some special features present in the data. In case, the frequencies of the sinusoidal model appear at $\lambda, 2\lambda, \ldots, p\lambda$, the model that incorporates this extra information is the FFM corresponding to the fundamental frequency λ. The advantage of using this information in the model itself is that it reduces the total number of unknown parameters to be estimated to $2p + 1$ instead of $3p$. There is only one nonlinear parameter in FFM and the computational complexity reduces to a great extent. In FFM, the effective frequencies are at $\lambda, 2\lambda, \ldots, p\lambda$, it is said that the frequencies appear at the harmonics of the fundamental frequency λ and hence the name the FFM.

The presence of an exact periodicity may seem to be a convenient approximation, but many real-life phenomena can be described using FFM. This model has many applications in Signal Processing and Time Series literature. Bloomfield [3] considered the FFM when the fundamental frequency is a Fourier frequency. Baldwin and Thomson [2] used FFM to explain the visual observation of S. Carinae, a variable star in the Southern Hemisphere sky. Greenhouse, Kass, and Tsay [7] proposed higher order harmonics of one or more fundamental frequencies with stationary ARMA processes for the errors to study the biological rhythm data, illustrated by human core body temperature data. The harmonic regression plus correlated noise model has also been used in assessing static properties of the human circadian system, see Brown and Czeisler [5] and Brown and Liuthardt [6] and examples therein. Musical sound segments produced by certain musical instruments are mathematically explained using such models. Nandi and Kundu [15] and Kundu and Nandi [13] used FFM to analyze short duration speech data. These data sets were also analyzed using the multiple sinusoidal model.

© Springer Nature Singapore Pte Ltd. 2020
S. Nandi and D. Kundu, *Statistical Signal Processing*,
https://doi.org/10.1007/978-981-15-6280-8_6

Fig. 6.1 Plot of the mean
corrected "uuu" vowel sound

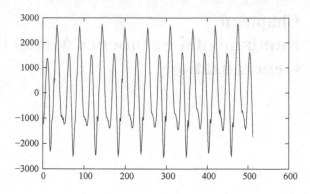

Fig. 6.2 Plot of the mean
corrected "ahh" sound

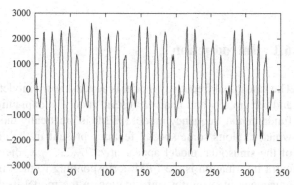

There are many signals like speech, where the data indicate the presence of harmonics with respect to a fundamental frequency. We provide the plot of two speech data sets. The Fig. 6.1 represents 'uuu' sound and Fig. 6.2 represents 'ahh' sound. The periodogram function of both the data sets are plotted in Figs. 6.3 and 6.4, respectively. From Figs. 6.3 and 6.4, it is clear that the harmonics of a particular frequency are present in both the cases. In these situations, it is better to use model (6.1) with one fundamental frequency than the multiple sinusoidal model (3.1) as model (6.1) has lesser number of nonlinear parameters than model (3.1) for fixed $p > 1$. Different short duration speech data and airline passenger data are analyzed in Chap. 7 as applications of models considered in this chapter.

In the same line as the FFM, a generalization has been proposed by Irizarry [10] when more than one fundamental frequency is present (See Sect. 6.3). The FFM is a special case of this generalized model with only one fundamental frequency. A further generalization is proposed by Nandi and Kundu [16], if the gap between two consecutive frequencies is approximately the same corresponding to one fundamental frequency. These models are referred to as generalized fundamental frequency models (GFFM). Motivation for FFM and GFFM came through some real data sets, two of which are already discussed above.

Fig. 6.3 Plot of the periodogram function of the "uuu" sound

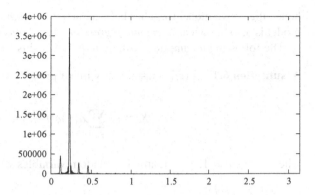

Fig. 6.4 Plot of the periodogram function of the "ahh" sound

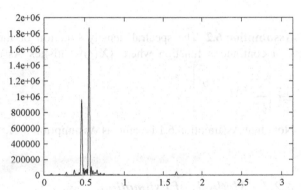

6.2 Fundamental Frequency Model

The FFM takes the following form;

$$y(t) = \sum_{j=1}^{p} \rho_j^0 \cos(tj\lambda^0 - \phi_j^0) + X(t); \tag{6.1}$$

$$= \sum_{j=1}^{p} \left[A_j^0 \cos(tj\lambda^0) + B_j^0 \sin(tj\lambda^0) \right] + X(t), \qquad t = 1, \ldots, n. \tag{6.2}$$

Here $\rho_j^0 > 0$, $j = 1, \ldots, p$ are unknown amplitudes; $\phi_j^0 \in (-\pi, \pi)$, $j = 1, \ldots, p$ are unknown phases; and $\lambda^0 \in (0, \pi/p)$ is the unknown frequency, the fundamental frequency; $\{X(t)\}$ is a sequence of error random variable; for $j = 1, \ldots, p$, $A_j^0 = \rho_j^0 \cos(\phi_j^0)$ and $B_j^0 = \rho_j^0 \sin(\phi_j^0)$. Models (6.1) and (6.2) are equivalent models with different parameterization. Most of the theoretical works involving fundamental frequency model, available in the literature, are derived under Assumption 3.2. The effective frequencies corresponding to the jth sinusoidal component present in $\{y(t)\}$ is $j\lambda^0$, which is of the form of harmonics of the fundamental frequency λ^0, hence the above model (6.1) has been named as *fundamental frequency model*. The model is also known as *harmonic regression signal plus noise model*.

The following assumptions will be required in this chapter.

Assumption 6.1 $\{X(t)\}$ is a stationary linear process with the following form

$$X(t) = \sum_{j=0}^{\infty} a(j)e(t - j), \tag{6.3}$$

where $\{e(t); t = 1, 2, \ldots\}$, are i.i.d. random variables with $E(e(t)) = 0$, $V(e(t)) = \sigma^2$, and $\sum_{j=0}^{\infty} |a(j)| < \infty$.

Assumption 6.2 The spectral density function or the spectrum $f(\lambda)$ of $\{X(t)\}$ is a continuous function where $\{X(t)\}$ satisfies Assumption 6.1. Then $f(\lambda) = \frac{\sigma^2}{2\pi} \left| \sum_{j=0}^{\infty} a(j)e^{-ij\lambda} \right|^2$.

Note that Assumption 6.1 is same as Assumption 3.2.

6.2.1 Methods of Estimation

6.2.1.1 Least Squares Estimator (LSE)

The LSEs of the unknown parameters of model (6.1) are obtained by minimizing the RSS;

$$Q(\theta) = \sum_{t=1}^{n} \left[y(t) - \sum_{j=1}^{p} \rho_j \cos(tj\lambda - \phi_j) \right]^2, \tag{6.4}$$

with respect to the parameter vector $\theta = (\rho_1, \ldots, \rho_p, \phi_1, \ldots, \phi_p, \lambda)^T$. Let $\widehat{\theta} = (\widehat{\rho}_1, \ldots, \widehat{\rho}_p, \widehat{\phi}_1, \ldots, \widehat{\phi}_p, \widehat{\lambda})^T$ be the LSE of $\theta^0 = (\rho_1^0, \ldots, \rho_p^0, \phi_1^0, \ldots, \phi_p^0, \lambda^0)^T$, that minimizes $Q(\theta)$ with respect to θ. We observe that λ is the only nonlinear parameter and $\rho_1, \ldots \rho_p$ and $\phi_1, \ldots, \phi_p$ are either linear parameters or can be expressed in terms of the linear parameters. Hence, using separable regression technique of Richards [19] (See Sect. 2.2 in Chap. 2), it is possible to explicitly write the LSEs of $\rho_1^0, \ldots \rho_p^0$ and $\phi_1^0, \ldots, \phi_p^0$ as functions of λ only. Therefore, it boils down to a 1-D minimization problem.

6.2.1.2   Approximate Least Squares Estimator (ALSE)

The ALSE of λ^0, say $\widetilde{\lambda}$, is obtained by maximizing $I_S(\lambda)$, the sum of the periodogram functions at $j\lambda$, $j = 1, \ldots, p$, defined as follows:

$$I_S(\lambda) = \frac{1}{n} \sum_{j=1}^{p} \left| \sum_{t=1}^{n} y(t) e^{itj\lambda} \right|^2 . \tag{6.5}$$

The ALSEs of the other parameters are estimated as;

$$\widetilde{\rho}_j = \frac{2}{n} \left| \sum_{t=1}^{n} y(t) e^{itj\widetilde{\lambda}} \right|, \qquad \widetilde{\phi}_j = arg \left\{ \frac{1}{n} \sum_{t=1}^{n} y(t) e^{itj\widetilde{\lambda}} \right\} \tag{6.6}$$

for $j = 1, \ldots, p$. Therefore, similar to LSE, estimating ALSE of λ^0 involves 1-D optimization and once $\widetilde{\lambda}$ is obtained, ALSEs of the other parameters are estimated using (6.6).

6.2.1.3   Weighted Least Squares Estimator (WLSE)

The weighted LSE was proposed by Hannan [9] for the alternative equivalent model (6.2). Write model (6.2) as $y(t) = \mu(t; \boldsymbol{\xi}^0) + X(t)$, where the parameter vector $\boldsymbol{\xi} = (A_1, \ldots, A_p, B_1, \ldots, B_p, \lambda)^T$ and $\boldsymbol{\xi}^0$ is the true value of $\boldsymbol{\xi}$. Define the following periodograms

$$I_y(\omega_j) = \frac{1}{2\pi n} \left| \sum_{t=1}^{n} y(t) \exp(it\omega_j) \right|^2 , \quad I_\mu(\omega_j, \boldsymbol{\xi}) = \frac{1}{2\pi n} \left| \sum_{t=1}^{n} \mu(t; \boldsymbol{\xi}) \exp(it\omega_j) \right|^2 ,$$

$$I_{y\mu}(\omega_j, \boldsymbol{\xi}) = \frac{1}{2\pi n} \left(\sum_{t=1}^{n} y(t) \exp(it\omega_j) \right) \overline{\left(\sum_{t=1}^{n} \mu(t; \boldsymbol{\xi}) \exp(it\omega_j) \right)} ,$$

where $\{\omega_j = \frac{2\pi j}{n}; j = 0, \ldots, n-1\}$ are Fourier frequencies. Then the WLSE of $\boldsymbol{\xi}^0$, say $\widehat{\boldsymbol{\xi}}$ minimizes the following objective function;

$$S_1(\boldsymbol{\xi}) = \frac{1}{n} \sum_{j=0}^{n-1} \left[\left\{ I_y(\omega_j) + I_\mu(\omega_j; \boldsymbol{\xi}) - 2Re(I_{y\mu}(\omega_j; \boldsymbol{\xi})) \right\} \psi(\omega_j) \right], \tag{6.7}$$

where $\psi(\omega)$ is a weight function and is a continuous even function of ω and it satisfies $\psi(\omega) \geq 0$ for $\omega \in [0, \pi]$.

6.2.1.4 Quinn and Thomson (QT) Estimator

The Quinn and Thomson estimator (QTE) of λ^0, proposed by Quinn and Thomson [18], say $\tilde{\tilde{\lambda}}$, maximizes

$$R(\lambda) = \frac{1}{n} \sum_{j=1}^{p} \frac{1}{f(j\lambda)} \left| \sum_{t=1}^{n} y(t) e^{itj\lambda} \right|^2, \tag{6.8}$$

where

$$f(\lambda) = \frac{1}{2\pi} \sum_{h=-\infty}^{\infty} e^{-ih\lambda} \gamma(h) \tag{6.9}$$

is the spectral density function or the spectrum of the error process $\{X(t)\}$ with auto-covariance function $\gamma(.)$. Under Assumption 6.1. the spectrum of $\{X(t)\}$ has the form $f(\lambda) = \frac{\sigma^2}{2\pi} \left| \sum_{j=0}^{\infty} a(j) e^{-ij\lambda} \right|^2$ and it is assumed that the spectrum of the error process is known and strictly positive on $[0, \pi]$. When the spectrum is unknown, $f(j\lambda)$ in (6.8) is replaced by its estimate. The QT estimators of ρ_j^0 and ϕ_j^0, $j = 1, \ldots, p$, have the same form as the ALSEs given in (6.6) with $\tilde{\lambda}$ replaced by $\tilde{\tilde{\lambda}}$. Denote them as $\tilde{\tilde{\rho}}_j$ and $\tilde{\tilde{\phi}}_j$, $j = 1, \ldots, p$ and the vector of QTE as $\tilde{\tilde{\theta}} = (\tilde{\tilde{\rho}}_1, \ldots, \tilde{\tilde{\rho}}_p, \tilde{\tilde{\phi}}_1, \ldots, \tilde{\tilde{\phi}}_p, \tilde{\tilde{\lambda}})^T$.

In case of QTE, $R(\lambda)$ is a weighted sum of the squared amplitude estimators of model (6.1) at the harmonics $j\lambda$, $j = 1, \ldots, p$ and the weights are inversely proportional to the spectral density of the error random variables at these frequencies. Hence, $R(\lambda)$ coincides with $I_S(\lambda)$ when $\{X(t)\}$ is a sequence of uncorrelated random variables as in this case $f(\lambda)$ is a constant function. Similar to WLSE, the QT estimator is nothing but a Weighted ALSE.

Note: In case of WLSE, Hannan [8] showed that the optimal frequency domain weight function $\psi(\omega) = f^{-1}(\omega)$, where $f(\omega)$ is the spectrum of $\{X(t)\}$. This choice of weight function is asymptotically equivalent to using the inverse of the covariance matrix of $\{X(t)\}$ as in generalized least squares objective function.

The QT estimator of λ is equivalent to generalized least squares estimator for large n, Quinn and Thomson [18]. The QT method incorporates the variances of the error process $\{X(t)\}$ through the spectrum $f(\omega)$. It is expected that the QT method and the weighted least squares method provide estimators with lower asymptotic variances than the LSEs, at least theoretically.

6.2.2 Bayes Estimates

In order to discuss Bayes estimates, write $X(t) = e(t)$, $t = 1, \ldots, n$ in model (6.2), then in matrix notation it can be written as

$$\mathbf{Y} = \mathbf{Z}_p \boldsymbol{\alpha}_p + \mathbf{e},$$

where $\mathbf{Y} = (y(1), \ldots, y(n))^T, \mathbf{e} = (e(1), \ldots, e(n))^T, \boldsymbol{\alpha}_p = (A_1, B_1, \ldots, A_p, B_p)^T$
and $\mathbf{Z}_p = (\mathbf{X}_1, \ldots, \mathbf{X}_p)$ with

$$\mathbf{X}_k = \begin{bmatrix} \cos(k\lambda) & \sin(k\lambda) \\ \cos(2k\lambda) & \sin(2k\lambda) \\ \vdots & \vdots \\ \cos(nk\lambda) & \sin(nk\lambda) \end{bmatrix}.$$

In most of the literature on fundamental frequency model, the observation model considered is that $f(\mathbf{y}|\boldsymbol{\alpha}_p, \sigma^2, \lambda, \mathscr{I}_p)$ is Gaussian with mean $\mathbf{Z}_p \boldsymbol{\alpha}_p$ and covariance matrix $\sigma^2 \mathbf{I}_n$; $\mathscr{I}_p$ denotes the prior information available for a p-component model. The parameters are $\boldsymbol{\alpha}_p$, the amplitude vector of order $2p$; λ, the fundamental frequency; σ^2, the noise variance; p, the model order. The fundamental frequency is clearly connected with p as $\lambda < \pi/p$ and from the prior information, the signal is bandlimited to $[\lambda_a, \lambda_b], \lambda_a > 0$. Therefore, λ must lie on the set $\Omega_p = [\lambda_a, \lambda_b/p]$. Assume that the prior probability density function (PDF) is factored as

$$f(\boldsymbol{\alpha}_p, \sigma^2, \lambda, p|\mathscr{I}_p) = f(\boldsymbol{\alpha}_p|\mathscr{I}_p)f(\sigma^2|\mathscr{I}_p)f(\lambda|\mathscr{I}_p)f(p|\mathscr{I}_p). \tag{6.10}$$

Usually, prior distribution on the noise variance is taken as the improper prior PDF $f(\sigma^2|\mathscr{I}_p) \propto \sigma^{-1}$ and is known as Jeffreys' prior. This prior PDF is improper as it does not integrate to one. In practice, the noise variance cannot be zero and is always upper bounded. So a normalized prior PDF on σ^2 is (Nielsen, Christensen, and Jensen [17])

$$f(\sigma^2|\mathscr{I}_p) = \begin{cases} \left[\ln(w/v)\sigma^2\right]^{-1} & v < \sigma^2 < w \\ 0 & \text{otherwise.} \end{cases}$$

The bounds have negligible influence on the inference, so often they are selected as $v \to 0$ and $w \to \infty$ (see Bretthorst [4]).

The prior PDF given the model order p on the fundamental frequency λ is

$$f(\lambda|p, \mathscr{I}_p) = \begin{cases} (F_p\lambda)^{-1} & \lambda \in \Omega_p \\ 0 & \text{otherwise,} \end{cases}$$

where $F_p = \ln(\lambda_b) - \ln(p\lambda_a)$. The ith amplitude pair (A_i, B_i) is assumed to have Gaussian prior $\mathscr{N}_2 \left(\mathbf{0}, \frac{\sigma_\alpha^2}{2}\mathbf{I}_2\right)$. Therefore, the joint prior PDF of $\boldsymbol{\alpha}_p$ is

$$f(\boldsymbol{\alpha}_p|\sigma_\alpha^2, \mathscr{I}_p) = \mathscr{N}_{2p} \left(\mathbf{0}, \frac{\sigma_\alpha^2}{2}\mathbf{I}_{2p}\right).$$

The prior variance of a single amplitude is $\frac{\sigma_\alpha^2}{2}$ and this is unknown. The hyper parameters are treated as random variables and like the original noise variance, the following hyper prior on σ_α^2 is considered.

$$f(\sigma_\alpha^2|\mathscr{I}_p) = \begin{cases} \left[\ln(w/v)\sigma_\alpha^2\right]^{-1} & v < \sigma_\alpha^2 < w \\ 0 & \text{otherwise.} \end{cases}$$

The model order p is a discrete parameter. Under the constraint that the prior PMF of the model order must sum to one, hence the discrete uniform PMF on $\{1, 2, \ldots, L\}$ is the prior for p. The model order cannot be larger than $\left[\frac{\lambda_b}{\lambda_a}\right]$, hence, L should not be chosen larger than this value.

The prior information $\mathscr{I}_p$ is a default probability model and the inference becomes analytically intractable. One can use Markov Chain Monte Carlo (MCMC) method to solve this problem numerically. However, Nielsen, Christensen, and Jensen [17] proposed a reparameterization with a few minor approximations and obtained a prior of the same form as the Zellner's g-prior (Zellner [20]) which has some tractable analytical properties. Interested readers are referred to Nielsen, Christensen, and Jensen [17] for details.

6.2.3 Theoretical Results

All the methods, discussed in Sects. 6.2.1.1–6.2.1.4, are nonlinear in nature. Therefore, the theoretical results concerning these estimators are asymptotic. All the four estimators are consistent and asymptotically normally distributed. The LSE and ALSE are asymptotically equivalent under Assumption 6.1 whereas under Assumption 6.2, WLSE and QTE have the same asymptotic distribution when the weight function $\psi(\omega) = f^{-1}(\omega)$ in case of WLSE.

Define a diagonal matrix of order $2p + 1$ as follows:

$$\mathbf{D}(n) = \begin{pmatrix} n^{\frac{1}{2}}\mathbf{I}_p & \mathbf{0} & \mathbf{0} \\ \mathbf{0} & n^{\frac{1}{2}}\mathbf{I}_p & \mathbf{0} \\ \mathbf{0} & \mathbf{0} & n^{\frac{3}{2}} \end{pmatrix},$$

where $\mathbf{I}_p$ is the identity matrix of order p. The diagonal entries of $\mathbf{D}(n)$ correspond to the rates of convergence of the LSEs of the parameters. The following theorem states the asymptotic distribution of the LSE and ALSE of $\boldsymbol{\theta}^0$.

Theorem 6.1 *Under Assumption 6.1,*

$$\mathbf{D}(n)(\widehat{\boldsymbol{\theta}} - \boldsymbol{\theta}^0) \xrightarrow{d} \mathscr{N}_{2p+1}(\mathbf{0}, 2\sigma^2\mathbf{V}),$$

as n tends to infinity. The dispersion matrix $\mathbf{V}$ *is as follows:*

$$\mathbf{V} = \begin{bmatrix} \mathbf{C} & \mathbf{0} & \mathbf{0} \\ \mathbf{0} & \mathbf{CD}_{\rho^0}^{-1} + \frac{3\delta_G \mathbf{LL}^T}{(\sum_{k=1}^p k^2 \rho_k^{0^2})^2} & \frac{6\delta_G \mathbf{L}}{(\sum_{k=1}^p k^2 \rho_k^{0^2})^2} \\ \mathbf{0} & \frac{6\delta_G \mathbf{L}^T}{(\sum_{k=1}^p k^2 \rho_k^{0^2})^2} & \frac{12\delta_G}{(\sum_{k=1}^p k^2 \rho_k^{0^2})^2} \end{bmatrix},$$

where

$$\mathbf{D}_{\rho^0} = diag\{\rho_1^{0^2}, \ldots, \rho_p^{0^2}\}, \quad \mathbf{L} = (1, 2, \ldots, p)^T,$$

$$\delta_G = \mathbf{L}^T \mathbf{D}_{\rho^0} \mathbf{CL} = \sum_{k=1}^p k^2 \rho_k^{0^2} c(k), \quad \mathbf{C} = diag\{c(1), \ldots, c(p)\},$$

$$c(k) = \left\{\sum_{i=0}^\infty a(i) \cos(ki\lambda^0)\right\}^2 + \left\{\sum_{i=0}^\infty a(i) \sin(ki\lambda^0)\right\}^2.$$

The asymptotic distribution of $\widetilde{\boldsymbol{\theta}}$*, the ALSE of* $\boldsymbol{\theta}^0$ *is same as* $\widehat{\boldsymbol{\theta}}$*, the LSE of* $\boldsymbol{\theta}^0$*.* ∎

The following theorem states the asymptotic distribution of the WLSE of $\boldsymbol{\theta}$.

Theorem 6.2 *Let* $\boldsymbol{\xi}$ *and* $\boldsymbol{\xi}^0$ *be same as defined in Sect. 6.2.1.3. Then,*

$$\mathbf{D}(n)(\widehat{\boldsymbol{\xi}} - \boldsymbol{\xi}^0) \xrightarrow{d} \mathcal{N}_{2p+1}(\mathbf{0}, \mathbf{W}_2^{-1}\mathbf{W}_1\mathbf{W}_2^{-1}),$$

where

$$\mathbf{W}_1 = \begin{pmatrix} \frac{1}{4\pi}\mathbf{U} & \mathbf{0} & \frac{1}{8\pi}\mathbf{u}_1 \\ \mathbf{0} & \frac{1}{4\pi}\mathbf{U} & \frac{1}{8\pi}\mathbf{u}_2 \\ \frac{1}{8\pi}\mathbf{u}_1' & \frac{1}{8\pi}\mathbf{u}_2' & \frac{1}{12\pi}\sum_{j=1}^p j^2(A_j^{0^2} + B_j^{0^2})\psi(j\lambda^0)^2 f(j\lambda^0) \end{pmatrix}$$

$\mathbf{U} = diag\{\psi(\lambda^0)^2 f(\lambda^0), \psi(2\lambda^0)^2 f(2\lambda^0), \ldots, \psi(p\lambda^0)^2 f(p\lambda^0)\}$,
$\mathbf{u}_1 = (B_1^0 \psi(\lambda^0)^2 f(\lambda^0), 2B_2^0 \psi(2\lambda^0)^2 f(2\lambda^0), \ldots, pB_p^0 \psi(p\lambda^0)^2 f(p\lambda^0))^T$,
$\mathbf{u}_2 = (-A_1^0 \psi(\lambda^0)^2 f(\lambda^0), -2A_2^0 \psi(2\lambda^0)^2 f(2\lambda^0), \ldots, -pA_p^0 \psi(p\lambda^0)^2 f(p\lambda^0))^T$, *and*

$$\mathbf{W}_2 = \begin{pmatrix} \frac{1}{4\pi}\mathbf{V} & \mathbf{0} & \frac{1}{8\pi}\mathbf{v}_1 \\ \mathbf{0} & \frac{1}{4\pi} & \frac{1}{8\pi}\mathbf{v}_2 \\ \frac{1}{8\pi}\mathbf{v}_1' & \frac{1}{8\pi}\mathbf{v}_2' & \frac{1}{12\pi}\sum_{j=1}^p j^2(A_j^{0^2} + B_j^{0^2})\psi(j\lambda^0)^2 f(j\lambda^0) \end{pmatrix},$$

$$\mathbf{V} = diag\{\psi(\lambda^0), \psi(2\lambda^0), \dots, \psi(p\lambda^0)\}$$
$$\mathbf{v}_1 = (B_1^0\psi(\lambda^0), 2B_2\psi(2\lambda^0), \dots, pp B_p^0\psi(p\lambda^0))^T,$$
$$\mathbf{v}_2 = (-A_1^0\psi(\lambda^0), -2A_2^0\psi(2\lambda^0), \dots, -A_p^0\psi(p\lambda^0))^T.$$ ∎

The asymptotic distribution of the QTE was explicitly given by Quinn and Thomson [18]. We have already discussed that QTE is a special case of WLSE for large n. When the weight function $\psi(\cdot)$ in WLSE is the inverse of the spectral density $f(\cdot)$, the asymptotic distribution of QTE and WLSE are the same.

Quinn and Thomson [18] assumed that the error process $\{X(t)\}$ is an ergodic and zero mean strictly stationary process. It is also required that the spectrum $f(\omega)$ is positive on $[0, \pi]$ and $\rho_j^0 > 0$.

Theorem 6.3 *Under the above assumptions, if $\{X(t)\}$ is a weakly mixing process with twice differentiable spectral density function, then*

$$\mathbf{D}(n)(\widetilde{\widetilde{\boldsymbol{\theta}}} - \boldsymbol{\theta}^0) \xrightarrow{d} \mathcal{N}_{2p+1}(\mathbf{0}, \boldsymbol{\Gamma}),$$

$$\boldsymbol{\Gamma} = \begin{bmatrix} \mathbf{C} & \mathbf{0} & \mathbf{0} \\ \mathbf{0} & \mathbf{C}\mathbf{D}_{\rho^0}^{-1} + 3\beta^*\mathbf{L}\mathbf{L}^T & 6\beta^*\mathbf{L} \\ \mathbf{0} & 6\beta^*\mathbf{L}^T & 12\beta^* \end{bmatrix},$$

where

$$\beta^* = \left(\sum_{k=1}^{p} \frac{k^2\rho_k^{0^2}}{c(k)}\right)^{-1},$$

and the matrices $\mathbf{C}$, $\mathbf{D}_{\rho^0}$, the vector $\mathbf{L}$ and $c(k)$ are same as defined in Theorem 6.1. ∎

Comparing Theorems 6.1 and 6.3, it is obvious that the asymptotic variances of $\widehat{\rho}_j$ and $\widetilde{\widetilde{\rho}}_j$, the LSE and the QTE of ρ_j, respectively, are same whereas the asymptotic variances of $\widehat{\phi}_j$ and $\widetilde{\widetilde{\phi}}_j$ are different; the same is true for the LSE and QTE of λ^0. Asymptotic covariances of estimators of ρ_j^0 and (ϕ_j^0, λ^0) corresponding to both the methods are same. A further comparison reveals that if $\dfrac{\delta_G}{(\sum_{k=1}^{p} k^2\rho_k^{0^2})^2} - \beta^* > 0$, then asymptotic variances of LSEs of ϕ_j^0 as well as λ^0 are bigger than those of the QTEs. If $f(\omega) > 0$ and known, theoretically QTEs have lower asymptotic variances for the fundamental frequency and the phases.

6.2.4   A Test for Harmonics

Fundamental frequency model (6.1) is a special case of the sinusoidal frequency model discussed in previous chapters. If frequencies are harmonics of a fundamental frequency then model (6.1) is a more suitable model instead of the multiple sinusoidal model. Therefore, it is required to test for presence of harmonics of a fundamental frequency. Consider the following general model

$$y(t) = \sum_{j=1}^{p} \rho_j^0 \cos(t\lambda_j^0 - \phi_j^0) + X(t); \qquad (6.11)$$

where $\lambda_1^0 < \lambda_2^0 < \cdots < \lambda_p^0$ are unknown frequencies, not necessarily harmonics of a fundamental frequency and ρ_j^0, ϕ_j^0, $j = 1, \ldots, p$ are same as before. The error sequence $\{X(t)\}$ satisfies Assumption 6.2. The estimation procedure for this model has been thoroughly discussed in Chap. 3. The test procedure, we describe in the following, uses the QTE. Similar tests involving LSE and ALSE are also available, see Nandi and Kundu [15].

Consider testing $H_0 : \lambda_j^0 = j\lambda^0$ against H_A : not H_0, where λ^0 is known. Quinn and Thomson [18] proposed the following test statistic for testing H_0 against H_A;

$$\chi^2 = \frac{n}{\pi} \left\{ \sum_{j=1}^{p} \frac{J(\widehat{\lambda}_j)}{f(\widehat{\lambda}_j)} - \sum_{j=1}^{p} \frac{J(j\widehat{\lambda})}{f(j\widehat{\lambda})} \right\}, \quad J(\lambda) = \left| \frac{1}{n} \sum_{t=1}^{n} y(t) e^{it\lambda} \right|^2, \qquad (6.12)$$

where $\widehat{\lambda}$ is the QTE of λ^0 under H_0 and $\widehat{\lambda}_j$ is the QTE of λ_j^0, $j = 1, \ldots, p$ under H_A, respectively. Under the assumption of normality on the error sequence $\{X(t)\}$, the likelihood ratio test statistic is asymptotically equivalent to χ^2. It has been assumed that $f(\omega)$ is known, but asymptotically nothing is lost if a consistent estimator is used. The asymptotic distribution of χ^2 under H_0 is known and stated in the following theorem. These results hold irrespective of the data being normal or not.

Theorem 6.4 *Under assumptions of Theorem 6.3, χ^2 is distributed as χ_{p-1}^2 random variable for large n and it is asymptotically equivalent to*

$$\chi_*^2 = \frac{n^3}{48\pi} \sum_{j=1}^{p} \widehat{\rho}_j^2 (\widehat{\lambda}_j - j\widehat{\lambda})^2 / f(j\widehat{\lambda}),$$

when H_0 is true. Here $\widehat{\rho}_j$ is the QTE of ρ_j^0 under H_0. ∎

Write χ_{LSE}^2 in (6.12) when QT estimators of λ^0 and λ_j^0 are replaced by LSE or ALSE of λ^0 and λ_j^0. Then $\chi_{LSE}^2 = \chi^2 + o(1)$, see Nandi and Kundu [15].

6.2.5　Estimation of Number of Harmonics

In this chapter, so far, we have considered the problem of estimation of λ_0, ρ_j^0, ϕ_j^0, $j = 1, \ldots, p$ assuming p, the number of harmonics to be known. In practice, the number of harmonics is unknown and has to be estimated. In this section, we discuss the problem of estimation of p under Assumption 6.1. Kundu and Nandi [13] addressed this problem using penalty function approach in the line of information theoretic criteria. Instead of using a fixed penalty function, Kundu and Nandi [13], proposed using a class of penalty functions satisfying some conditions. In this section, we denote the true number of harmonics as p^0. The proposed method provides a strongly consistent estimator of p^0. In the following, we first describe the estimation procedure and then how to implement it in practice.

6.2.5.1　Estimation Procedure: Number of Harmonics

Assume that the number of harmonics can be at most K, a fixed number. Let L denote the possible range of p^0, that is, $L \in \{0, 1, \ldots, K\}$ and M_L denotes the fundamental frequency model (6.1) of order L, $L = 0, 1, \ldots, K$. The problem is a model selection problem from a class of models $\{M_0, M_1, \ldots, M_K\}$. Define

$$R(L) = \min_{\substack{\lambda, \rho_j, \phi_j \\ j=1,\ldots,L}} \frac{1}{n} \sum_{t=1}^{n} \left\{ y(t) - \sum_{j=1}^{L} \rho_j \cos(tj\lambda - \phi_j) \right\}^2. \qquad (6.13)$$

Denote, $\widehat{\lambda}_L$, $\widehat{\rho}_{jL}$ and $\widehat{\phi}_{jL}$ as the LSEs of λ, ρ_j and ϕ_j, respectively, if the model order is L. As these estimators depend on L, we make it explicit. Consider

$$IC(L) = n \log R(L) + 2LC_n, \qquad (6.14)$$

here C_n is a function of n and following the terminology of AIC and BIC, it is termed as penalty function. It satisfies the following conditions;

$$(1) \quad \lim_{n \to \infty} \frac{C_n}{n} = 0 \quad \text{and} \quad (2) \quad \lim_{n \to \infty} \frac{C_n}{\log n} > 1. \qquad (6.15)$$

The number of harmonics p^0 is estimated by the smallest value $\widehat{p}$ such that;

$$IC(\widehat{p} + 1) > IC(\widehat{p}). \qquad (6.16)$$

Similarly as in Chap. 5, it is like other information theoretic criteria, but unlike AIC or BIC, it does not have a fixed penalty function. It can be anything provided it satisfies (6.15). The first term of $IC(L)$ is a decreasing function of L, whereas the second term is an increasing function of L. As model order L increases, $n \log R(L)$

decreases and the penalty function increases and discourages to add more and more terms in the model. Kundu and Nandi [13] proved the strong consistency of $\widehat{p}$.

Theorem 6.5 *Let C_n be any function of n satisfying (6.15) and $\widehat{p}$ is the smallest value satisfying (6.16). If $\{X(t)\}$ satisfies Assumption 6.1, then $\widehat{p}$ is a strongly consistent estimator of p^0.* ∎

See the appendix of this chapter for the proof of this theorem.

6.2.5.2 How to Obtain $R(L)$?

Write

$$\mu_t^L = \sum_{j=1}^{L} \rho_j \cos(j\lambda t - \phi_j) = \sum_{j=1}^{L} \left[\rho_j \cos(\phi_j) \cos(j\lambda t) + \rho_j \sin(\phi_j) \sin(j\lambda t) \right].$$

Then, in matrix notation

$$
\begin{bmatrix} \mu_1^L \\ \vdots \\ \mu_n^L \end{bmatrix} = \begin{bmatrix} \cos(\lambda) & \sin(\lambda) & \cdots & \cos(L\lambda) & \sin(L\lambda) \\ \vdots & \vdots & \vdots & \vdots & \vdots \\ \cos(n\lambda) & \sin(n\lambda) & \cdots & \cos(nL\lambda) & \sin(nL\lambda) \end{bmatrix} \begin{bmatrix} \rho_1 \cos(\phi_1) \\ \rho_1 \sin(\phi_1) \\ \vdots \\ \rho_L \cos(\phi_L) \\ \rho_L \sin(\phi_L) \end{bmatrix}
$$

$$= \mathbf{A}_L(\lambda)\mathbf{b}_L \quad \text{(say)}.$$

Now,

$$\sum_{t=1}^{n} \left(y(t) - \sum_{j=1}^{L} \rho_j \cos(j\lambda t - \phi_j) \right)^2 = (\mathbf{Y} - \mathbf{A}_L(\lambda)\mathbf{b}_L)^T (\mathbf{Y} - \mathbf{A}_L(\lambda)\mathbf{b}_L),$$

$$(6.17)$$

where $\mathbf{Y} = (y(1), \ldots, y(n))^T$. We can use separable regression technique of Richards [19] in this case, see Sect. 2.2. For a given λ, (6.17) is minimized when $\widehat{\mathbf{b}}_L = \left[\mathbf{A}_L(\lambda)^T \mathbf{A}_L(\lambda) \right]^{-1} \mathbf{A}_L(\lambda)^T \mathbf{Y}$. Replacing $\mathbf{b}_L$ by $\widehat{\mathbf{b}}_L$ in (6.17), we have

$$\left(\mathbf{Y} - \mathbf{A}_L(\lambda)\widehat{\mathbf{b}}_L \right)^T \left(\mathbf{Y} - \mathbf{A}_L(\lambda)\widehat{\mathbf{b}}_L \right) = \mathbf{Y}^T \left(\mathbf{I} - \mathbf{P}_{\mathbf{A}_L(\lambda)} \right) \mathbf{Y} = Q(\lambda, L) \quad \text{(say)},$$

$$(6.18)$$

with $\mathbf{P}_{\mathbf{A}_L(\lambda)} = \mathbf{A}_L(\lambda) \left(\mathbf{A}_L(\lambda)^T \mathbf{A}_L(\lambda) \right)^{-1} \mathbf{A}_L(\lambda)^T$ is the projection matrix on the column space of $\mathbf{A}_L(\lambda)$. Therefore, λ can be estimated by minimizing $Q(\lambda, L)$ with respect to λ when the number of harmonics is L. We denote it $\widehat{\lambda}$, then $R(L) = Q(\widehat{\lambda}, L)$. Additionally, note that for large n

$$\frac{1}{n}\mathbf{Y}^T\left(\mathbf{P}_{\mathbf{A}_L(\lambda)}\right)\mathbf{Y} = \left(\frac{1}{n}\mathbf{Y}^T\mathbf{A}_L(\lambda)\right)\left(\frac{1}{n}\mathbf{A}_L(\lambda)^T\mathbf{A}_L(\lambda)\right)^{-1}\left(\frac{1}{n}\mathbf{A}_L(\lambda)^T\mathbf{Y}\right)$$

$$\approx 2\sum_{j=1}^{L}\left[\left(\frac{1}{n}\sum_{t=1}^{n}y(t)\cos(j\lambda t)\right)^2 + \left(\frac{1}{n}\sum_{t=1}^{n}y(t)\sin(j\lambda t)\right)^2\right]$$

$$= 2\sum_{j=1}^{L}\left|\frac{1}{n}\sum_{t=1}^{n}y(t)e^{ijt\lambda}\right|^2 = 2I_F(\lambda, L), \quad (say)$$

because

$$\lim_{n\to\infty}\frac{1}{n}\mathbf{A}_L(\lambda)^T\mathbf{A}_L(\lambda) = \frac{1}{2}\mathbf{I}_{2L}, \tag{6.19}$$

where $\mathbf{I}_{2L}$ is the identity matrix of order $2L$. Thus, for large n the estimator of λ obtained by minimizing $Q(\lambda, L)$, can also be obtained by maximizing $I_F(\lambda, L)$. For a given L, $Q(\lambda, L)$ is a function of λ only and thus minimization of $Q(\lambda, L)$ is an 1-D optimization problem and can be solved easily.

6.2.5.3 Probability of Wrong Estimates—Bootstrap

Kundu and Nandi [13] used bootstrap to obtain the probability of wrong estimates. Consider a dataset $\{y(1), \dots, y(n)\}$ from model (6.1). Here p^0 is unknown, but it is assumed that $p^0 \le K$ for some known fixed integer. Compute $\widehat{\lambda}$ by maximizing $I_F(\lambda, L)$, or minimizing $Q(\lambda, L)$ with respect to λ, for a given L. Using $\widehat{\lambda}$, obtain an estimate of σ^2 as

$$\widehat{\sigma}^2 = \frac{1}{n}\mathbf{Y}^T\left(\mathbf{I} - \mathbf{P}_{\mathbf{A}_K(\widehat{\lambda})}\right)\mathbf{Y} = \frac{1}{n}Q(\widehat{\lambda}, L) = R(L).$$

For a given choice of the penalty function C_n, compute $IC(L)$, for different values of $L = 1, \dots K$ and choose $\widehat{p}$ an estimate of p as suggested in Sect. 6.2.5.1. In practical implementation we use a collection of different penalty functions, called penalty function bank. We would like to choose that C_n for which $P[\widehat{p} \ne p^0]$ is minimum. Write

$$P[\widehat{p} \ne p^0] = P[\widehat{p} < p^0] + P[\widehat{p} > p^0] = \sum_{q=0}^{p^0-1}P[\widehat{p} = q] + \sum_{q=p^0+1}^{K}P[\widehat{p} = q].$$

Case I: $q < p^0$

$$P(\widehat{p} = q) = P[IC(0) > IC(1) > \cdots > IC(q) < IC(q+1)] \le P\left[\log\frac{R(q)}{R(q+1)} < 2\frac{C_n}{n}\right].$$

Note that there exists a $\delta > 0$ such that for large n,

$$\log \frac{R(q)}{R(q+1)} > \delta \quad \text{a.s.}$$

This implies that for large n,

$$\log \frac{R(q)}{R(q+1)} > 2\frac{C_n}{n} \quad \text{a.s.}$$

Therefore, when n is large, for $q < p^0$,

$$P[\widehat{p} = q] = 0. \tag{6.20}$$

So, for large n, the probability of underestimation is zero.

Case II: $q > p^0$

$$P(\widehat{p} = q) = P[IC(0) > IC(1) > \cdots > IC(q) < IC(q+1)].$$

Note that for large n,

$$P(\widehat{p} = q) = P[IC(p^0) > \cdots > IC(q) < IC(q+1)],$$

as for large n

$$P[IC(0) > \cdots > IC(p^0)] = 1.$$

Therefore,

$$P[\widehat{p} = q] = P\left[\log \frac{R(q+1)}{R(q)} + 2\frac{C_n}{n} > 0, \log \frac{R(j)}{R(j-1)} + 2\frac{C_n}{n} < 0, j = p^0 + 1, \ldots, q\right]. \tag{6.21}$$

To compute (6.21) we need to find the joint distribution of $R(p^0), \ldots, R(K)$, which is not straight forward to obtain and it depends on the unknown parameters. One can use resampling or bootstrap technique to estimate the probability of wrong detection as follows:

Given a realization from model (6.1), estimate the order of the model as $M(C_n)$, using the penalty function C_n. Then, estimate $\widehat{\sigma}^2$ using $L = M(C_n)$ and normalize the data so that the error variance is $\frac{1}{2}$. Generate n i.i.d. Gaussian random variables with mean zero and variance $\frac{1}{2}$, say $\epsilon(1), \ldots, \epsilon(n)$. Obtain the bootstrap sample as

$$y(t)^B = y(t) + \epsilon(t); \quad \text{for} \quad t = 1, \ldots, n.$$

Assuming $M(C_n)$ is the correct order model, check for $q < M(C_n)$, whether

$$\log \frac{R(q)}{R(q+1)} < 2\frac{C_n}{n}$$

and for $q > M(C_n)$, check whether

$$\log \frac{R(q+1)}{R(q)} + 2\frac{C_n}{n} > 0, \quad \log \frac{R(j)}{R(j-1)} + 2\frac{C_n}{n} < 0, \quad j = p^0 + 1, \ldots q.$$

Repeating the process, say B times, estimate $P(\widehat{p} \neq p^0)$. Finally, choose that C_n for which the estimated $P(\widehat{p} \neq p^0)$ is minimum.
Note: The bootstrap sample $\{y(1)^B, \ldots, y(n)^B\}$ can be thought of coming from a model similar to (6.1) with error variance equal to one.

6.2.5.4 Discussion and Illustration

Consider data generated using the following model and parameter values

$$y(t) = \sum_{j=1}^{4} \rho_j^0 \cos(tj\lambda^0 - \phi_j^0) + X(t); \quad t = 1, \ldots, 50. \tag{6.22}$$

Here $\rho_1^0 = 2.5, \rho_2^0 = 2.0, \rho_3^0 = 3.5, \rho_4^0 = 1.0; \phi_1^0 = 0.5, \phi_2^0 = 0.9, \phi_3^0 = 0.75, \phi_4^0 = 0.5; \lambda^0 = 0.75398; X(t) = 0.5e(t-1) + e(t)$.

Here $\{e(t)\}$ is a sequence of i.i.d. Gaussian random variables with mean zero and variance τ^2. The generated data with $\tau^2 = 1.0$ are plotted in Fig. 6.5. The corresponding periodogram function is plotted in Fig. 6.6. It is well-known that the number of peaks in the periodogram function plot roughly gives an estimate of the number of components p^0. But it depends on the magnitude of the amplitude associated with each effective frequency and the error variance. If a particular amplitude is relatively small as compared to others, that component may not be significantly visible in the periodogram plot. Figure 6.6 exhibits three peaks properly and it is not clear that the data are generated using FFM with four harmonics.

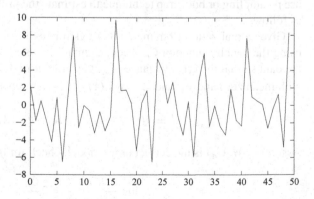

Fig. 6.5 Plot of the data generated by model (6.22) with $n = 50$ and error variance $= 1.0$

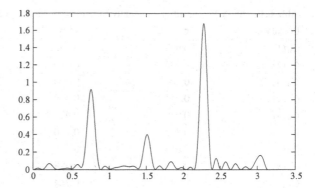

Fig. 6.6 Plot of the periodogram function of the data plotted in Fig. 6.5

6.2.5.5 Penalty Function Bank Approach

The probability of wrong estimate, discussed in Sect. 6.2.5.3, can be used to choose the best possible penalty function from a particular class of penalty functions. Instead of using a fixed penalty function, a class of penalty functions, satisfying some special properties can be used. Any penalty function from that particular class will provide consistent estimates of the unknown parameters. But, it is observed as expected that a particular penalty function may not work well for all possible error variances or for all possible parameter values. Therefore, it is suggested to compute the probability of correct estimates (PCE's) of a particular penalty function. Based on the resampling technique discussed in Sect. 6.2.5.3, (see also Kundu and Mitra [12]), compute an estimate of the PCE for each penalty function. Once an estimate of PCE is obtained, one can use that penalty function for which the estimated PCE is maximum.

In order to understand the procedure, take a penalty function bank which contains the following twelve different penalty functions satisfying (6.15). $C_n(1) = n^{.2} \log \log n$, $C_n(2) = n^{.3}$, $C_n(3) = \frac{(\log n)^3}{\log \log n}$, $C_n(4) = n^{.4}$, $C_n(5) = (\log n)^{1.1}$, $C_n(6) = (\log n)^{1.2}$, $C_n(7) = (\log n)^{1.3}$, $C_n(8) = \frac{n^{.5}}{(\log n)^{.9}}$, $C_n(9) = (\log n)^{1.4}$, $C_n(10) = \frac{n^{.4}}{\log n}$, $C_n(11) = \frac{n^{.5}}{\log n}$ and $C_n(12) = \frac{n^{.6}}{\log n}$. The main idea is to choose a wide variety of C_n's. Our analysis in this section is based on model (6.22). It is assumed that the number of components can be at most 6, i.e. $K = 6$. For each simulated data vector from model (6.22), p was estimated for all C_n's. Then this procedure is replicated 1000 times to obtain the PCE's. The results are reported in Table 6.1 for each penalty function in the bank and for different error variances. Points to be noted from Table 6.1:

1. The performances of all $C_n(j)$'s improve as the error variance decreases, which is expected.
2. The performances of the different $C_n(j)$'s vary from one extreme to the other. Some of the $C_n(j)$'s can detect the correct order model all the times considered here, whereas some of the $C_n(j)$'s cannot detect the correct order model at all, although all of them satisfy (6.15).

Table 6.1 The PCE's for different penalty functions for model (6.22)

Penalty	$\tau^2 = 0.5$	$\tau^2 = 0.75$	$\tau^2 = 1.0$
$C_n(1)$	0.998	0.998	0.996
$C_n(2)$	0.999	0.999	0.994
$C_n(3)$	0.000	0.000	0.000
$C_n(4)$	0.941	0.869	0.799
$C_n(5)$	0.978	0.923	0.865
$C_n(6)$	0.873	0.780	0.702
$C_n(7)$	0.614	0.525	0.459
$C_n(8)$	0.988	0.988	0.988
$C_n(9)$	0.286	0.264	0.235
$C_n(10)$	0.928	0.926	0.926
$C_n(11)$	0.979	0.979	0.979
$C_n(12)$	0.995	0.995	0.995

3. Typically, a particular $C_n(j)$ may not work for a particular model but may work for some others.

On the other hand when we use above procedure, the PCE becomes 1.000, 1.000, 0.998 for $\tau^2 = 0.5, 0.75, 1.0$, respectively.

6.3 Generalized Fundamental Frequency Model

The fundamental frequency model has been generalized by the authors (i) Irizarry [10] and (ii) Nandi and Kundu [16] and call these models as generalized fundamental frequency model (GFFM). Such models can be used in case there is more than one fundamental frequency.

Irizarry [10] proposed the signal plus noise model with M periodic components; for $j = 1, \ldots, M$, $s_j(t; \boldsymbol{\xi}_j^0)$ is the contribution of the jth fundamental frequency, and is a sum of K_j sinusoidal components of the following form:

$$y(t) = \sum_{j=1}^{M} s_j(t; \boldsymbol{\xi}_j^0) + X(t), \quad t = 1, \ldots, n. \tag{6.23}$$

$$s_j(t; \boldsymbol{\xi}_j^0) = \sum_{k=1}^{K_j} \left\{ A_{j,k}^0 \cos(k\theta_j^0 t) + B_{j,k}^0 \sin(k\theta_j^0 t) \right\}, \tag{6.24}$$

$$\boldsymbol{\xi}_j^0 = (A_{j,1}^0, B_{j,1}^0, \ldots, A_{j,K_j}^0, B_{j,K_j}^0, \theta_j^0).$$

This model is a harmonic model with multiple fundamental frequencies $\theta_1^0, \ldots, \theta_M^0$ and coincide with FFM (6.2) when $M = 1$. The amplitudes $A_{j,k}^0$ and $B_{j,k}^0$ are corresponding to the kth component of the jth fundamental frequency.

Nandi and Kundu [16] considered a further generalization with M fundamental frequencies of the form;

$$s_j(t; \eta_j^0) = \sum_{k=1}^{q_j} \rho_{jk}^0 \cos\left\{[\lambda_j^0 + (k-1)\omega_j^0]t - \phi_{jk}^0\right\}, \qquad (6.25)$$

$$\eta_j^0 = (\rho_{j_1}^0, \ldots, \rho_{j_{q_k}}^0, \phi_{j_1}^0, \ldots, \phi_{j_{q_k}}^0, \lambda_j^0, \omega_j^0), \qquad (6.26)$$

where λ_j^0, $j = 1, \ldots, M$ are fundamental frequencies and the other frequencies associated with λ_j^0 are occurring at λ_j^0, $\lambda_j^0 + \omega_j^0$, $\ldots$, $\lambda_j^0 + (q_j - 1)\omega_j^0$. Corresponding to the jth fundamental frequency, there are q_j bunch of effective frequencies. If $\lambda_j^0 = \omega_j^0$, then frequencies effectively appear at the harmonics of λ_j^0. Corresponding to the frequency $\lambda_j^0 + (k-1)\omega_j^0$, ρ_{jk}^0 and ϕ_{jk}^0, $k = 1, \ldots, q_j$, $j = 1, \ldots, M$ represent the amplitude and phase components, respectively and they are also unknown. Observe that when $\lambda_j^0 = \omega_j^0$, model (6.25) is nothing but model (6.23) where frequencies appear at the harmonics of λ_j^0 corresponding to the jth fundamental frequency. The number of fundamental frequencies is M whereas K_j and q_j refer to number of harmonics associated with the jth fundamental frequency in case of models (6.23) and (6.25), respectively.

The presence of this kind of periodicity is a convenient approximation, but many real-life phenomena can be described quite effectively, using models (6.23) and (6.25).

Irizarry [10] assumed that the noise sequence $\{X(t)\}$ is a strictly stationary real-valued random process with autocovariance function $c_{xx}(u) = \text{Cov}(X(t+u), X(t))$ and power spectrum

$$f_{xx}(\lambda) = \frac{1}{2\pi} \sum_u c_{xx}(u) \exp(-i\lambda u), \quad -\infty < \lambda < \infty$$

and satisfies the following assumption;

Assumption 6.3 The sequence $\{X(t)\}$ is a zero mean process such that all of its moments exist along with the joint cumulant function of order L, $c_{x\ldots x}(u_1, \ldots, u_{L-1})$, $L = 2, 3, \ldots$ Furthermore,

$$C_L = \sum_{u_1=-\infty}^{\infty} \cdots \sum_{u_{L-1}=-\infty}^{\infty} |c_{x\ldots x}(u_1, \ldots, u_{L-1})|$$

satisfy $\sum_k C_k z^k / k < \infty$ for z in a neighborhood of 0.

6.3.1 Weighted Least Squares Estimator Proposed by Irizarry

Irizarry [10] proposed a window based weighted least squares method and developed asymptotic variance expression of the proposed estimators. Kundu and Nandi [16] studied the theoretical properties of the LSEs of the unknown parameters. Given n observations, the weighted least squares method proposed by Irizarry [10] to estimate $\xi_1^0, \ldots, \xi_M^0$, minimizes the following criterion function

$$s_n(\xi_1, \ldots, \xi_M) = \sum_{t=1}^{n} w\left(\frac{t}{n}\right)\left\{y(t) - \sum_{j=1}^{M} s_j(t; \xi_j)\right\}^2, \tag{6.27}$$

where $w(\cdot)$ is a weight function and satisfies Assumption 6.4.

Assumption 6.4 The weight function $w(s)$ is non-negative, bounded, of bounded variation, has support $[0, 1]$ and it is such that $W_0 > 0$ and $W_1^2 - W_0 W_2 \neq 0$, where

$$W_n = \int_0^1 s^n w(s) ds. \tag{6.28}$$

Note that Assumption 6.4 is same as Assumption 4.4 in Chap. 4. Let $\widehat{\xi}_j^I$ be the WLSE of ξ_j, $j = 1, \ldots, M$ for the model given in (6.23)–(6.24). For each $j = 1, \ldots, M$, let $\mathbf{D}_j^I(n)$ be a $(2K_j + 1) \times (2K_j + 1)$ diagonal matrix whose first $2K_j$ diagonal entries are $n^{\frac{1}{2}}$ and $(2K_j + 1)$-st diagonal entry is $n^{\frac{3}{2}}$. Then we have the following theorem (Irizarry [10]).

Theorem 6.6 *Under Assumption 6.3 on the error process $\{X(t)\}$ and Assumption 6.4 on the weight function $w(\cdot)$, for each $j = 1, \ldots, M$, $\widehat{\xi}_j^I$ is a consistent estimator of ξ_j^0 and $\mathbf{D}_j^I(n)(\widehat{\xi}_j^I - \xi_j^0) \xrightarrow{d} \mathcal{N}_{2K_j+1}\left(\mathbf{0}, \Sigma_{jI}\right)$, where*

$$\Sigma_{jI} = 4\pi \left\{\sum_{k=1}^{K_j} k^2 \left(A_{j,k}^{0^2} + B_{j,k}^{0^2}\right) / f_{xx}(k\theta_j^0)\right\}^{-1} \begin{pmatrix} \mathbf{D}_j + c_0^{-1} \mathbf{E}_j \mathbf{E}_j' & \mathbf{E}_j \\ \mathbf{E}_j' & c_0 \end{pmatrix}$$

$$\mathbf{D}_j = \left\{\sum_{k=1}^{K_j} k^2 \left(A_{j,k}^{0^2} + B_{j,k}^{0^2}\right) / f_{xx}(k\theta_j^0)\right\} \begin{pmatrix} \mathbf{D}_{j,1} & \mathbf{0} & \cdots & \mathbf{0} \\ \mathbf{0} & \mathbf{D}_{j,2} & \cdots & \mathbf{0} \\ \vdots & \vdots & \vdots & \vdots \\ \mathbf{0} & \mathbf{0} & \cdot & \mathbf{D}_{j,K_j} \end{pmatrix}$$

$$\mathbf{D}_{j,k} = \frac{f_{xx}(k\theta_j^0)}{b_0 \left(A_{j,k}^{0^2} + B_{j,k}^{0^2}\right)} \begin{pmatrix} c_1 b_0 A_{j,k}^{0^2} + a_1 B_{j,k}^{0^2} & a_2 A_{j,k}^0 B_{j,k}^0 \\ a_2 A_{j,k}^0 B_{j,k}^0 & a_1 A_{j,k}^{0^2} + c_1 b_0 B_{j,k}^{0^2} \end{pmatrix}$$

$$\mathbf{E}_j = c_4\left(-B_{j,1}^0, A_{j,1}^0, -2B_{j,2}^0, 2A_{j,2}^0, \ldots, -K_j B_{j,K_j}^0, K_j A_{j,K_j}^0\right)$$

$$c_0 = a_0 b_0, \quad c_1 = U_0 W_0^{-1}$$
$$c_4 = a_0 (W_0 W_1 U_2 - W_1^2 U_1 - W_0 W_2 U_1 + W_1 W_2 U_0),$$

and

$$a_0 = (W_0 W_2 - W_1^2)^{-2}$$
$$b_i = W_i^2 U_2 + W_{i+1}(W_{i+1} U_0 - 2 W_i U_1), \quad for \ i = 0, 1.$$

Here W_i, $i = 0, 1, 2$ are defined in (6.28) and U_i, $i = 0, 1, 2$ are defined by

$$U_i = \int_0^1 s^i w(s)^2 ds.$$

Also, $\mathbf{D}_j^I(n)(\widehat{\boldsymbol{\xi}}_j^I - \boldsymbol{\xi}_j^0)$ and $\mathbf{D}_k^I(n)(\widehat{\boldsymbol{\xi}}_k^I - \boldsymbol{\xi}_k^0)$ are independently distributed. ∎

The quantities a_0, b_i, c_i, W_i and U_i are same as defined in Theorem 4.8 in Chap. 4.

The asymptotic variances of the estimates of the fundamental frequencies obtained in Theorem 6.6 provides an useful approximation of the variance of the estimate of the jth fundamental frequency

$$\text{Var}\left(\widehat{\theta}_j^I\right) \approx 4\pi c_0 \left\{ \sum_{k=1}^{K_j} k^2 \left(A_{j,k}^{0^2} + B_{j,k}^{0^2} \right) / f_{xx}(k\theta_j^0) \right\}^{-1}, \quad j = 1, \ldots, M.$$

The denominator is the sum of weighted squares of amplitudes corresponding to the harmonics. Therefore, the precision of the estimate increases (that is, variance decreases) with the total magnitude of the respective harmonic components.

Nandi and Kundu [16] studied the usual least squares method to estimate the unknown parameter of the model given in (6.23) and (6.25). Apart from Assumption 6.1, the following assumptions on the true values of the parameters are also required.

Assumption 6.5

$$\rho_{j_k}^0 > 0, \quad \phi_{j_k}^0 \in (-\pi, \pi), \quad \lambda_j^0, \omega_j^0 \in (0, \pi), \quad k = 1, \ldots, q_j, \quad j = 1, \ldots, M.$$
$$(6.29)$$

Assumption 6.6 λ_j^0 and ω_j^0, $j = 1, \ldots, M$ are such that

$$\lambda_j^0 + (i_1 - 1)\omega_j^0 \neq \lambda_l^0 + (i_2 - 1)\omega_l^0$$

for $i_1 = 1, \ldots, q_j$; $i_2 = 1, \ldots, q_l$ and $j \neq l = 1, \ldots, M$.

The least squares method consists of choosing $\widehat{\boldsymbol{\eta}}_1^L, \ldots, \widehat{\boldsymbol{\eta}}_M^L$ by minimizing the criterion function

$$Q_n(\boldsymbol{\eta}_1, \ldots, \boldsymbol{\eta}_M) = \sum_{t=1}^n \left(y(t) - \sum_{j=1}^M \sum_{k=1}^{q_j} \rho_{jk} \cos\left\{ (\lambda_j + (k-1)\omega_j)t - \phi_{jk} \right\} \right)^2 .$$

(6.30)

with respect to $\boldsymbol{\eta}_1, \ldots, \boldsymbol{\eta}_M$ when M and $q_1, \ldots, q_M$ are known in advance. Note that obtaining the LSEs involves a $2\sum_{j=1}^M q_j + 2M$ dimensional minimization search. When M and q_j, $j = 1, \ldots, M$ are large the LSEs may be very expensive. But ρ_{jk}s and ϕ_{jk}s can be expressed as functions of the frequencies λ_js and ω_js, so using the separable regression technique of Richards [19], it boils down to a $2M$ dimensional search.

In order to find the theoretical properties of $\widehat{\boldsymbol{\eta}}_1^L, \ldots, \widehat{\boldsymbol{\eta}}_M^L$, define a diagonal matrix $\mathbf{D}_j^L(n)$ of order $2q_j + 2$ such that first $2q_j$ diagonal entries are $n^{\frac{1}{2}}$ and last two diagonal entries are $n^{\frac{3}{2}}$. Then, the LSEs $\widehat{\boldsymbol{\eta}}_1^L, \ldots, \widehat{\boldsymbol{\eta}}_M^L$ enjoy the following result.

Theorem 6.7 *Under Assumptions 6.1, 6.5 and 6.6, when M and q_j, $j = 1, \ldots, M$ are known, $\widehat{\boldsymbol{\eta}}_j^L$ is strongly consistent for $\boldsymbol{\eta}_j^0$. For each $j = 1, \ldots, M$, $\mathbf{D}_j^L(n)(\widehat{\boldsymbol{\eta}}_j^L - \boldsymbol{\eta}_j^0) \xrightarrow{d} \mathcal{N}_{2q_j+2}\left(\mathbf{0}, 2\sigma^2 \boldsymbol{\Sigma}_{jL}^{-1} \mathbf{G}_{jL} \boldsymbol{\Sigma}_{jL}^{-1} \right)$ where for $j = 1, \ldots, M$*

$$\boldsymbol{\Sigma}_{jL} = \begin{pmatrix} \mathbf{I}_{q_j} & \mathbf{0} & \mathbf{0} & \mathbf{0} \\ \mathbf{0} & \mathbf{P}_j & -\frac{1}{2}\mathbf{P}_j\mathbf{J}_j & -\frac{1}{2}\mathbf{P}_j\mathbf{L}_j \\ \mathbf{0} & -\frac{1}{2}\mathbf{J}_j^T\mathbf{P}_k & \frac{1}{3}\mathbf{J}_j^T\mathbf{P}_j\mathbf{J}_j & \frac{1}{3}\mathbf{J}_j^T\mathbf{P}_j\mathbf{L}_j \\ \mathbf{0} & -\frac{1}{2}\mathbf{L}_j^T\mathbf{P}_j & \frac{1}{3}\mathbf{L}_j^T\mathbf{P}_j\mathbf{J}_j & \frac{1}{3}\mathbf{L}_j^T\mathbf{P}_j\mathbf{L}_j \end{pmatrix},$$

$$\mathbf{G}_{jL} = \begin{pmatrix} \mathbf{C}_j & \mathbf{0} & \mathbf{0} & \mathbf{0} \\ \mathbf{0} & \mathbf{P}_j\mathbf{C}_j & -\frac{1}{2}\mathbf{P}_j\mathbf{C}_j\mathbf{J}_j & -\frac{1}{2}\mathbf{P}_j\mathbf{C}_j\mathbf{L}_j \\ \mathbf{0} & -\frac{1}{2}\mathbf{J}_j^T\mathbf{P}_j\mathbf{C}_j & \frac{1}{3}\mathbf{J}_j^T\mathbf{P}_j\mathbf{C}_j\mathbf{J}_j & \frac{1}{3}\mathbf{J}_k^T\mathbf{P}_j\mathbf{C}_j\mathbf{L}_j \\ \mathbf{0} & -\frac{1}{2}\mathbf{L}_j^T\mathbf{P}_j\mathbf{C}_j & \frac{1}{3}\mathbf{L}_j^T\mathbf{P}_j\mathbf{C}_j\mathbf{J}_j & \frac{1}{3}\mathbf{L}_j^T\mathbf{P}_j\mathbf{C}_j\mathbf{L}_j \end{pmatrix}$$

Here

$$\mathbf{P}_j = diag\{\rho_{j_1}^{0}{}^2, \ldots, \rho_{j_{q_j}}^{0}{}^2\}, \quad \mathbf{J}_j = (1, 1, \ldots, 1)_{q_j \times 1}^T,$$
$$\mathbf{L}_j = (0, 1, \ldots, q_j - 1)_{q_j \times 1}^T, \quad \mathbf{C}_j = diag\{c_j(1), \ldots, c_j(q_j)\}, \quad j = 1, \ldots, M.$$

Additionally, for large n, $\mathbf{D}_j^L(n)(\widehat{\eta}_j^L - \eta_j^0)$ and $\mathbf{D}_k^L(n)(\widehat{\eta}_k^L - \eta_k^0)$, $j \neq k = 1, \ldots, M$ are asymptotically independently distributed. ∎

Remark 6.1 The form of $\mathbf{D}_k^L(n)$ indicates that for a given sample size n, the frequencies can be estimated more accurately than the other parameters and the rate of convergence is much higher in case of estimators of λ_j^0 and ω_j^0.

Remark 6.2 The matrices $\mathbf{\Sigma}_{jL}$ and $\mathbf{G}_{jL}$ are of the form $\mathbf{\Sigma}_{jl} = \begin{pmatrix} \mathbf{I}_{q_j} & 0 \\ 0 & \mathbf{F_j} \end{pmatrix}$ and $\mathbf{G}_{jL} = \begin{pmatrix} \mathbf{C_j} & 0 \\ 0 & \mathbf{H_j} \end{pmatrix}$. So $\mathbf{\Sigma}_{jL}^{-1}\mathbf{G}_{jL}\mathbf{\Sigma}_{jL}^{-1} = \begin{pmatrix} \mathbf{C_j} & 0 \\ 0 & \mathbf{F_j^{-1}H_jF_j^{-1}} \end{pmatrix}$ which implies that, the amplitude estimators are asymptotically independent of the corresponding phase and frequency parameter estimators.

Remark 6.3 It can be seen from Theorem 6.7 that the asymptotic distribution of the LSEs of the unknown parameters is independent of the true values of the phases.

6.3.2 Estimating Number of Fundamental Frequencies and Number of Harmonics

When one uses model (6.23), the number of fundamental frequencies and the number of harmonics corresponding to each fundamental frequencies are usually unknown. In practice, it is required to estimate them to analyze a data set using GFFM. First consider the following sum of sinusoidal model with p^0 sinusoidal components.

$$y(t) = \sum_{k=1}^{p^0}\left[A_k \cos(\omega_k t) + B_k \sin(\omega_k t)\right] + e(t), t = 1, \ldots, n.$$

Let $\widehat{p}_n$ be a strongly consistent estimate of the true number of sinusoids p^0 which is obtained using any of the order estimation method discussed in Chap. 5, say Wang's method. So, for large n, $\widehat{p}_n = p^0$ with probability one.

Given a strongly consistent estimate $\widehat{p}_n$ of the total number of sinusoidal components $p^0 = \sum_{j=1}^{M} K_j$ in model (6.23), the number of fundamentals, M and respective number of harmonics $K_1, \ldots, K_M$ can be estimated in the following way.

Take $J_0 = \left\{\widehat{\omega}_{(k)}, k = 1, \ldots, \widehat{p}_n\right\}$ as the set of ordered maximum periodogram frequencies. Consider $\widehat{\omega}_{1,1} = \widehat{\omega}_{(1)}$ to be the estimate of the first fundamental frequency. The set $J_1 = \left\{\omega \in J_0 : |\omega - k\widehat{\omega}_{1,1}| \leq n^{\frac{1}{2}} \text{ tor some } k = 1, 2, \ldots\right\}$ is considered to be the set containing the harmonics related to $\widehat{\omega}_{1,1}$. Suppose that the fundamental frequencies for components $1, \ldots, j-1$ and their respective harmonics contained in the sets $J_1, \ldots, J_{j-1}$ are identified. Then, define the jth fundamental frequency $\widehat{\omega}_{j,1}$ as the smallest frequency in $J_0 \setminus \cup_{l=1}^{j-1} J_l$ and

$$J_j = \left\{ \omega \in J_0 \setminus \cup_{l=1}^{j-1} J_l : |\widehat{\omega}_{j,1} - k\widehat{\omega}_{j,1}| \leq n^{\frac{1}{2}}, \text{ for some } k = 1, 2, \ldots \right\}.$$

This process is continued until all $\widehat{p}_n$ frequencies in J_0 are exhausted. The number of fundamental frequencies found in the process gives an estimate of M, say $\widehat{M}$ and the number of elements in J_j provides an estimate of the number of harmonics corresponding to the jth fundamental frequency K_j, say $\widehat{K}_j$, $j = 1, \ldots, M$. It can be shown that for large n, $\widehat{M}$ and $\widehat{K}_j$, $j = 1, \ldots, M$ are consistent estimators of M and K_j, $j = 1, \ldots, M$, respectively using the fact that $\widehat{p}_n$ is a strongly consistent estimator of p^0.

6.4 Conclusions

The FFM and different forms of the GFFM are special cases of the multiple sinusoidal model. But many real-life phenomena can be analyzed using such special models. In estimating unknown parameters of multiple sinusoidal model, there are several algorithms available, but the computational loads of these algorithms are usually quite high. Therefore, the FFM and the GFFM are very convenient approximations where inherent frequencies are harmonics of a fundamental frequency. Human circadian system originates data which can be analyzed using FFM. Nielsen, Christensen, and Jensen [17] analyzed a human voice signal which was originated from a female voice uttering "*Why were you away a year, Roy?*" uniformly sampled at 8 kHz frequency. Therefore, it is quite understandable that the models discussed in this chapter, have an ample number of applications in data from different spheres of nature.

Appendix A

In order to prove Theorem 6.5, we need the following lemmas.

Lemma 6.1 (An, Chen, and Hannan [1]) *Define,*

$$I_X(\lambda) = \left| \frac{1}{n} \sum_{t=1}^{n} X(t) e^{it\lambda} \right|^2.$$

If $\{X(t)\}$ *satisfies Assumption 6.1, then*

$$\limsup_{n \to \infty} \max_{\lambda} \frac{n I_X(\lambda)}{\sigma^2 \log n} \leq 1 \quad a.s. \tag{6.31}$$

Lemma 6.2 (Kundu [11]) *If $\{X(t)\}$ satisfies Assumption 6.1, then*

$$\lim_{n\to\infty} \sup_{\lambda} \frac{1}{n} \left| \sum_{t=1}^{n} X(t)e^{i\lambda t} \right| = 0 \quad a.s.$$

Proof of Theorem 6.5

Observe that, we need to show

$$IC(0) > IC(1) > \cdots > IC(p^0 - 1) > IC(p^0) < IC(p^0 + 1).$$

Consider two cases separately.
Case I: $L < p^0$

$$\lim_{n\to\infty} R(L) = \lim_{n\to\infty} \frac{1}{n} \sum_{t=1}^{n} \left(y(t) - \sum_{j=1}^{L} \widehat{\rho}_j \cos(j\widehat{\lambda}t - \widehat{\phi}_j) \right)^2$$

$$= \lim_{n\to\infty} \left[\frac{1}{n} \mathbf{Y}^T \mathbf{Y} - 2 \sum_{j=1}^{L} \left| \frac{1}{n} \sum_{t=1}^{n} y(t)e^{ijt\widehat{\lambda}} \right|^2 \right] = \sigma^2 + \sum_{j=L+1}^{p^0} \rho_j^{0^2} \quad a.s.$$

Therefore, for $0 \le L < p^0 - 1$,

$$\lim_{n\to\infty} \frac{1}{n}[IC(L) - IC(L+1)] =$$

$$\lim_{n\to\infty} \left[\log\left(\sigma^2 + \sum_{j=L+1}^{p^0} \rho_j^{0^2} \right) - \log\left(\sigma^2 + \sum_{j=L+2}^{p^0} \rho_j^{0^2} \right) - \frac{2C_n}{n} \right] \quad a.s.$$

(6.32)

and for $L = p^0 - 1$,

$$\lim_{n\to\infty} \frac{1}{n}\left[IC(p^0 - 1) - IC(p^0) \right] = \lim_{n\to\infty} \left[\log\left(\sigma^2 + \rho_{p^0}^{0^2} \right) - \log\sigma^2 - \frac{2C_n}{n} \right] \quad a.s.$$

(6.33)

Since $\frac{C_n}{n} \to 0$, therefore as $n \to \infty$ for $0 \le L \le p^0 - 1$,

$$\lim_{n\to\infty} \frac{1}{n}[IC(L) - IC(L+1)] > 0.$$

It implies that for large n, $IC(L) > IC(L+1)$, when $0 \le L \le p^0 - 1$.
Case II: $L = p^0 + 1$.
Now consider

$$R(p^0 + 1) = \frac{1}{n}\mathbf{Y}^T\mathbf{Y} - 2\sum_{j=1}^{p^0}\left|\frac{1}{n}\sum_{t=1}^{n}y(t)e^{i\widehat{\lambda}_jt}\right|^2 - 2\left|\frac{1}{n}\sum_{t=1}^{n}y(t)e^{i\widehat{\lambda}(p^0+1)t}\right|^2.$$

(6.34)

Note that $\widehat{\lambda} \to \lambda^0$ $a.s.$ as $n \to \infty$ (Nandi [14]). Therefore, for large n

$$IC(p^0 + 1) - IC(p^0)$$

$$= n\left(\log R(p^0 + 1) - \log R(p^0)\right) + 2C_n = n\left[\log \frac{R(p^0 + 1)}{R(p^0)}\right] + 2C_n$$

$$\approx n\left[\log\left(1 - \frac{2\left|\frac{1}{n}\sum_{t=1}^{n}y(t)e^{i\lambda^0(p^0+1)t}\right|^2}{\sigma^2}\right)\right] + 2C_n \text{ (using lemma 6.2)}$$

$$\approx 2\log n\left[\frac{C_n}{\log n} - \frac{n\left|\frac{1}{n}\sum_{t=1}^{n}X(t)e^{i\lambda^0(p^0+1)t}\right|^2}{\sigma^2\log n}\right]$$

$$= 2\log n\left[\frac{C_n}{\log n} - \frac{nI_X(\lambda^0(p^0+1))}{\sigma^2\log n}\right] > 0 \quad a.s.$$

Note that the last inequality follows because of the properties of C_n and due to Lemma 6.1.   ∎

References

1. An, H.-Z., Chen, Z.-G., & Hannan, E. J. (1983). The maximum of the periodogram. *Journal of Multi-Criteria Decision Analysis, 13*, 383–400.
2. Baldwin, A. J., & Thomson, P. J. (1978). Periodogram analysis of S. Carinae. *Royal Astronomical Society of New Zealand (Variable Star Section), 6*, 31–38.
3. Bloomfiled, P. (1976). *Fourier analysis of time series. An introduction*. New York: Wiley.
4. Bretthorst, G. L. (1988). *Bayesian spectrum analysis and parameter estimation*. Berlin: Springer.
5. Brown, E. N., & Czeisler, C. A. (1992). The statistical analysis of circadian phase and amplitude in constant-routine core-temperature data. *Journal of Biological Rhythms, 7*, 177–202.
6. Brown, E. N., & Liuthardt, H. (1999). Statistical model building and model criticism for human circadian data. *Journal of Biological Rhythms, 14*, 609–616.
7. Greenhouse, J. B., Kass, R. E., & Tsay, R. S. (1987). Fitting nonlinear models with ARMA errors to biological rhythm data. *Statistics in Medicine, 6*, 167–183.
8. Hannan, E. J. (1971). Non-linear time series regression. *Journal of Applied Probability, 8*, 767–780.
9. Hannan, E. J. (1973). The estimation of frequency. *Journal of Applied Probability, 10*, 510–519.
10. Irizarry, R. A. (2000). Asymptotic distribution of estimates for a time-varying parameter in a harmonic model with multiple fundamentals. *Statistica Sinica, 10*, 1041–1067.
11. Kundu, D. (1997). Asymptotic properties of the least squares estimators of sinusoidal signals. *Statistics, 30*, 221–238.
12. Kundu, D., & Mitra, A. (2001). Estimating the number of signals of the damped exponential models. *Computational Statistics & Data Analysis, 36*, 245–256.

13. Kundu, D., & Nandi, S. (2005). Estimating the number of components of the fundamental frequency model. *Journal of the Japan Statistical Society, 35*(1), 41–59.

14. Nandi, S. (2002). Analyzing some non-stationary signal processing models. Ph.D. Thesis, Indian Institute of Technology Kanpur.

15. Nandi, S., & Kundu, D. (2003). Estimating the fundamental frequency of a periodic function. *Statistical Methods and Applications, 12*, 341–360.

16. Nandi, S., & Kundu, D. (2006). Analyzing non-stationary signals using a cluster type model. *Journal of Statistical Planning and Inference, 136*, 3871–3903.

17. Nielsen, J. K., Christensen, M. G. and Jensen, S. H. (2013). Default Bayesian estimation of the fundamental frequency. *IEEE Transactions Audio, Speech and Language Processing 21*(3), 598–610.

18. Quinn, B. G., & Thomson, P. J. (1991). Estimating the frequency of a periodic function. *Biometrika, 78*, 65–74.

19. Richards, F. S. G. (1961). A method of maximum likelihood estimation. *Journal of the Royal Statistical Society, B*, 469–475.

20. Zellner, A. (1986). On assessing prior distributions and Bayesian regression analysis with g-prior distributions. In P. K. Goel & A. Zellner (Eds.) *Bayesian inference and decision techniques*. The Netherlands: Elsevier.

Chapter 7
Real Data Example Using Sinusoidal-Like Models

7.1 Introduction

In this chapter, we analyze some real data sets using the models we have already discussed, namely, the multiple sinusoidal model, the fundamental frequency model (FFM), and the generalized fundamental frequency model (GFFM). It is observed that the FFM has only one nonlinear parameter, namely the fundamental frequency. On the other hand, the GFFM is a generalization of the FFM where more than one fundamental frequencies are present.

In this chapter, we discuss the analyses of the following data sets

(a) a segment of Electrocardiogram (ECG) signal of a healthy human being,
(b) the well-known "variable star" data,
(c) three short duration voiced speech signals: "uuu", "aaa" and "eee", and
(d) airline passenger data.

Periodogram function $I(\omega)$, in (1.6), as well as $R_1(\omega)$ defined in (3.39), can be used for initial identification of the number of components. The simple (raw) periodogram function, not the smoothed periodogram, which is commonly known as a spectrogram in time series literature, is used to find the initial estimates. The periodogram or $R_1(\omega)$ is calculated at each grid point of a fine enough grid of $(0, \pi)$. These are not calculated only at the so-called Fourier frequencies $\left\{ \frac{2\pi j}{n}, j = 0, 1, \ldots, \left[\frac{n}{2} \right] \right\}$. So the number of peaks in the plot of the periodogram function gives an estimate of the number of effective frequencies to be considered in the underlying model to analyze a particular data set. It may be quite subjective sometimes, depending on the error variance and magnitude of the amplitudes. The periodogram may show only the more dominant frequencies, dominant in terms of larger amplitudes. In such cases, when the effects of these frequencies are removed from the observed series and the periodogram function of the residual series is plotted, then it may show some peaks corresponding to other frequencies. If the error variance is too high, periodogram plot may not exhibit a significant distinct peak at ω^*, even if this ω^* has a significant contribution to the data. Also, in case, two frequencies are "close enough" then periodogram may

© Springer Nature Singapore Pte Ltd. 2020
S. Nandi and D. Kundu, *Statistical Signal Processing*,
https://doi.org/10.1007/978-981-15-6280-8_7

show only one peak. In such cases, when two frequencies are closer than $O(\frac{1}{n})$, it is recommended to use of larger sample size, if it is possible and use of a finer grid may provide some more information about the presence of another frequency. In some cases, information theoretic criterion discussed in Chap. 5, have also been used. Once the initial estimates are found, the unknown parameters are estimated using either LSE, ALSE or sequential estimation procedure.

In most of the data sets considered in this chapter, the additive error random variables satisfy Assumption 3.2, the assumption of a stationary linear process. Under this assumption, in order to use the asymptotic distribution, it is required to estimate $\sigma^2 c(\omega_k^0)$, where

$$c(\omega_k^0) = \left| \sum_{j=0}^{\infty} a(j)e^{-ij\omega_k^0} \right|^2, \quad k = 1, \ldots, p,$$

when either LSE, ALSE or sequential procedure are used to estimate the unknown parameters in the sinusoidal frequency model. It is important to note that we cannot estimate σ^2 and $c(\omega_k^0)$ separately. But, it is possible to estimate $\sigma^2 c(\omega_k^0)$ which is needed to use the asymptotic distribution in case of finite samples. It can be shown that

$$\sigma^2 c(\omega_k^0) = E\left(\frac{1}{n} \left| \sum_{t=1}^{n} X(t)e^{-i\omega_k^0 t} \right|^2 \right).$$

Since $\sigma^2 c(\omega_k^0)$ is the expected value of the periodogram function of the error random variable $X(t)$ at ω_k^0, estimate $\sigma^2 c(\omega_k^0)$ by local averaging of the periodogram function of the estimated error over a window $(-L, L)$, for a suitable choice of L, across the point estimate of ω_k^0, see, for example, Quinn and Thomson [12].

7.2 ECG Data

The data set represents a segment of the ECG signal of a healthy human being and is plotted in Fig. 7.1. The data set contains 512 data points. The data are first mean corrected and scaled by the square root of the estimated variance of $\{y(t)\}$. The $S(\omega)$ function in the interval $(0, \pi)$ is plotted in Fig. 7.2. This $S(\omega)$ is defined in (3.55) in Chap. 3. This gives an idea about the number of frequencies present in the data. We observe that the total number of frequencies is quite large. It is not easy to obtain an estimate of p from this figure. Apart from that, all the frequencies may not be visible, depending on the magnitudes of some of the dominant frequencies and error variance. In fact, $\widehat{p}$ is much larger than what Fig. 7.2 reveals. The BIC has been used to estimate p. We estimate the unknown parameters of the multiple sinusoidal model sequentially for $k = 1, \ldots, 100$. For each k, the residual series is approximated by an AR process and the corresponding parameters are estimated. Here k represents the number of sinusoidal components. Let ar_k be the number of AR parameters in the AR model fitted to the residuals when k sinusoidal components are estimated and

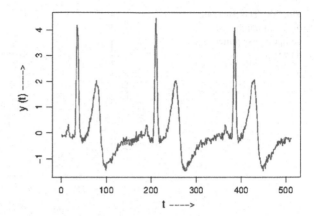

Fig. 7.1 The plot of the observed ECG signal

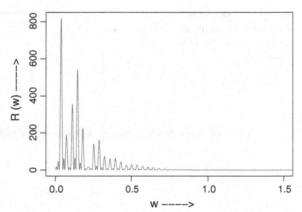

Fig. 7.2 The plot of the $S(\omega)$ function of the ECG signal data

$\widehat{\sigma}_k^2$ be the estimated error variance. Then minimize $\mathrm{BIC}(k)$ for this class of models for estimating p which takes the following form in this case;

$$BIC(k) = n \log \widehat{\sigma}_k^2 + \frac{1}{2}(3k + ar_k + 1) \log n.$$

The $\mathrm{BIC}(k)$ values are plotted in Fig. 7.3 for $k = 75, \ldots, 85$ and at $k = 78$, the $\mathrm{BIC}(k)$ takes its minimum value, therefore, we estimate p as $\widehat{p} = 78$. Then, the other unknown parameters are estimated using sequential method described in Sect. 3.12 as $\widehat{p}$ is quite large and simultaneous estimation might be a problem. With the estimated $\widehat{p}$, we plug-in the other estimates of the linear parameters and frequencies and obtain the fitted values $\{\widehat{y}(t); t = 1, \ldots, n\}$. They are plotted in Fig. 7.4 along with their observed values. The fitted values match reasonably well with the observed one. The residual sequence satisfies the assumption of stationarity, see also Prasad, Kundu, and Mitra [11] and Kundu et al. [3] for more details.

Fig. 7.3 The BIC values for different number of components applied to ECG data

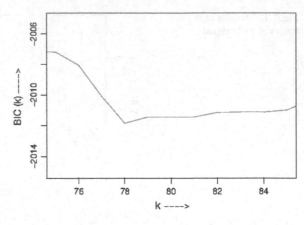

Fig. 7.4 The plot of the observed (red) and fitted values (green) of the ECG signal

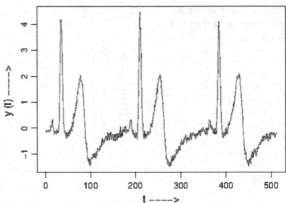

7.3 Variable Star Data

The variable star data is an astronomical data and widely used in time series literature. This data set represents the daily brightness of a variable star on 600 successive midnights. The data is collected from Time Series Library of StatLib (http://www.stat. cmu.edu; Source: Rob J. Hyndman). The observed data is displayed in Fig. 7.5 and its periodogram function in Fig. 7.6. Initial inspection of the periodogram function gives an idea of the presence of two frequency components in the data, resulting in two sharp separate peaks in the periodogram plot. With $\widehat{p} = 2$, once we estimate the frequencies and the amplitudes and obtain the residual series, the periodogram plot of the resulting residual series gives evidence of the presence of another significant frequency. The not so dominating third component is not visible in the periodogram plot of the original series. This is due to the fact that the first two components are dominant in terms of the large absolute values of their associated amplitudes. In addition, the first one is very close to the third one as compared to the available data points to distinguish them. Therefore, we take $\widehat{p} = 3$ and estimate the unknown

Fig. 7.5 The plot of the brightness of a variable star

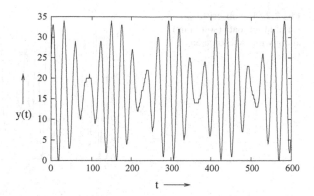

Fig. 7.6 The plot of the periodogram function of the variable star data

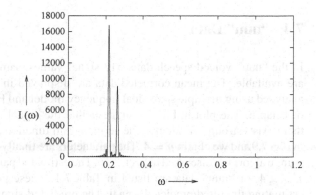

Table 7.1 The point estimates of the unknown parameters for variable star data and "uuu" data

Data set: "Variable Star"					
A_1	7.482622	B_1	7.462884	ω_1	0.216232
A_2	−1.851167	B_2	6.750621	ω_2	0.261818
A_3	−0.807286	B_3	0.068806	ω_2	0.213608
Data set: "uuu"					
A_1	8.927288	B_1	1698.65247	ω_1	0.228173
A_2	−584.679871	B_2	−263.790344	ω_2	0.112698
A_3	−341.408905	B_3	−282.075409	ω_3	0.343263
A_4	−193.936096	B_4	−300.509613	ω_4	0.457702

parameters. The estimated point estimates are listed in Table 7.1. The observed (solid line) and the estimated values (dotted line) are plotted in Fig. 7.7, and it is not possible to distinguish them. So the performance of the multiple sinusoidal model is quite good in analyzing variable star data. See also Nandi and Kundu [6–8] in this respect.

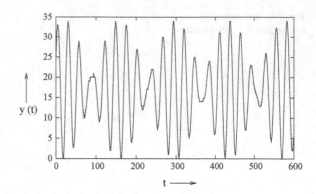

Fig. 7.7 The observed variable star data (solid line) along with the estimated values

7.4 "uuu" Data

In the "uuu" voiced speech data, 512 signal values, sampled at 10 kHz frequency are available. The mean corrected data are displayed in Fig. 7.8. This data can be analyzed using multiple sinusoidal frequency model and FFM, discussed in Sect. 6.2 of Chap. 6. The plot in Fig. 7.8 suggests that the signal is mean nonstationary and there exists strong periodicity. The periodogram function of the "uuu" data is plotted in Fig. 7.9 and we obtain $\widehat{p} = 4$. The parameters are finally estimated using Nandi and Kundu algorithm discussed in Sect. 3.16. The estimated parameters $(\widehat{A}_k, \widetilde{B}_k, \widehat{\omega}_k)$, $k = 1, \ldots, 4$ for "uuu" data are listed in Table 7.1. These point estimates are used in estimating the fitted/predicted signal. The predicted signal (solid line) of the mean corrected data along with the mean corrected observed "uuu" data (dotted line) are presented in Fig. 7.10. The fitted values match quite well with the mean corrected observed data.

We observe that the parameter estimates of ω_1, ω_3, and ω_4 are approximately integer multiples of ω_2. Therefore, "uuu" data can also be analyzed using FFM. The number of components p in case of the sinusoidal model was estimated from

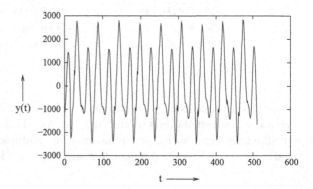

Fig. 7.8 The plot of the mean corrected "uuu" vowel sound data

Fig. 7.9 The plot of the
periodogram function of
"uuu" vowel data

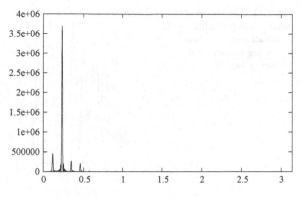

Fig. 7.10 The plot of the
fitted values (solid line) and
the mean corrected "uuu"
data

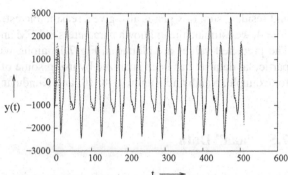

the periodogram function plot given in Fig. 7.9. But when we are using FFM, then
the method discussed in Sect. 6.2.5 of Chap. 6 can be used to estimate the number
of harmonics, that is, the number of components. We take the maximum possible
harmonics as $K = 9$ and the class of penalty functions is the same as mentioned
in Sect. 6.2.5.5. Then, p is estimated for each penalty function using the method
discussed in Sect. 6.2.5.1. The penalty functions $C_n(4)$, $C_n(5)$, $C_n(6)$, $C_n(7)$, $C_n(9)$
and $C_n(12)$ estimate the number of harmonics to be 4, the penalty function $C_n(1)$,
$C_n(2)$, $C_n(8)$, $C_n(10)$ and $C_n(11)$ estimate it to be 5, whereas $C_n(3)$ estimates it
to be 2 only. Therefore, it is not clear which one is to be taken. Now using the
second step based on maximum PCE obtained using re-sampling technique discussed
in Sect. 6.2.5.3, we obtain the estimate of the number of harmonics to be 4. The
residual sums of squares for different orders are as follows: 0.454586, 0.081761,
0.061992, 0.051479, 0.050173, 0.050033, 0.049870, 0.049866, and 0.049606 for
the model orders 1, 2, ... and 9, respectively. The residual sum of squares is a
decreasing function of p and it almost stabilizes at $\widehat{p}$. So it also indicates that the
number of harmonics should be 4. So for 'uuu' data set the periodogram function

Fig. 7.11 The plot of the
fitted values (solid line) and
the mean corrected "uuu"
data using fundamental
frequency model

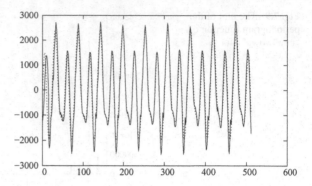

and residual sum of squares also give a reasonable estimate of the order. Now using $\widehat{p} = 4$, we estimate the unknown parameters in FFM and obtain the predicted values. The predicted values are plotted in Fig. 7.11 along with the observed values. This particular data set has been analyzed by using some of the related models also, see, for example, Nandi and Kundu [6–8, 10] and Kundu and Nandi [4, 5] in this respect.

7.5 "aaa" Data

The "aaa" data set, similar to "uuu" data, contains 512 signal values sampled at 10 kHz frequency. This data set can be analyzed using the sinusoidal frequency model and the GFFM. The data are first scaled by 1000 and are plotted in Fig. 7.12 and the periodogram function of the scaled data is plotted in Fig. 7.13. When the sinusoidal frequency model is used, the number of components p is estimated from

Fig. 7.12 The plot of the
"aaa" data

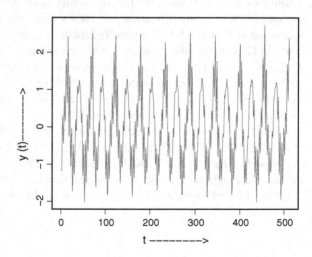

Fig. 7.13 The plot of the
periodogram function of
"aaa" data

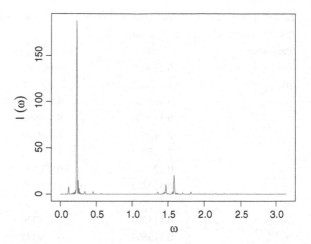

the periodogram plot. In case the GFFM is used, then M and q_j, $j = 1, \ldots, M$ are
also estimated from the periodogram. Here M is the total number of fundamental
frequencies and corresponding to the jth fundamental frequency, q_j frequencies are
associated. Preliminary analysis of the periodogram function indicates that for "aaa"
data set $\widehat{p} = 9$ for sinusoidal model and for GFFM, $\widehat{M} = 2$ and $\widehat{q}_1 = 4$ and $\widehat{q}_2 = 5$.

The initial estimates of the frequencies are obtained from the periodogram. Using
these initial estimates as starting values, the LSEs of the unknown parameters are
obtained using the sinusoidal frequency model as well as the GFFM. Theorem 4.6 can
be used to find confidence intervals of the LSEs in case of sinusoidal frequency model
and Theorem 6.7 in case of GFFM. The estimated error has been used to estimate
the term of the form $\sigma^2 c(\widehat{\omega})$ where $\widehat{\omega}$ is the estimated effective frequency. Then
$100(1 - \alpha)\%$ confidence interval can be obtained for both the models. The point
estimates, and 95% confidence bounds for each parameter estimates are given in
Table 7.2 in case of GFFM, and in Table 7.3 for multiple sinusoidal model. The error
has been modeled as a stationary AR(3) process for GFFM, whereas for the sinusoidal
model it is a stationary AR(1) process. The predicted values using the GFFM along
with the observed data are plotted in Fig. 7.14 and in Fig. 7.15, the predicted values
using the multiple sinusoidal frequency model along with the observed data are
plotted. Both the models provide reasonable fits in this case. It should be mentioned
that Grover, Kundu, and Mitra [2], analyzed this data set using the chirp model, and
the fit is quite satisfactory.

Table 7.2 LSEs and 95% confidence bounds for parameters of **'aaa'** data set when GFFM is used
Data set: 'aaa'. $M = 2$, $q_1 = 4$, $q_2 = 5$

Parameter	Estimate	Lower bound	Upper bound
ρ_{1_1}	0.220683	0.168701	0.272666
ρ_{1_2}	1.211622	1.160225	1.263019
ρ_{1_3}	0.160045	0.109573	0.210516
ρ_{1_4}	0.138261	0.088990	0.187531
ϕ_{1_1}	2.112660	1.358938	2.866383
ϕ_{1_2}	2.748044	2.530074	2.966015
ϕ_{1_3}	−0.129586	−0.896532	0.637359
ϕ_{1_4}	0.377879	−1.025237	1.780995
λ_1	0.113897	0.111100	0.116694
ω_1	0.113964	0.111330	0.116599
ρ_{2_1}	0.146805	0.109372	0.184237
ρ_{2_2}	0.245368	0.209165	0.281571
ρ_{2_3}	0.377789	0.342714	0.412865
ρ_{2_4}	0.096959	0.062910	0.131008
ρ_{2_5}	0.133374	0.100253	0.166495
ϕ_{2_1}	2.949186	2.424410	3.473963
ϕ_{2_2}	0.051864	−0.357301	0.461030
ϕ_{2_3}	1.364091	0.981884	1.746297
ϕ_{2_4}	2.106739	1.550737	2.662742
ϕ_{2_5}	−0.706613	−1.300238	−0.112988
λ_2	1.360523	1.358731	1.362314
ω_2	0.113994	0.113308	0.114621

$\widehat{X}(t) = 0.696458\,\widehat{X}(t-1) - 0.701408\,\widehat{X}(t-2) + 0.618664\,\widehat{X}(t-3) + e(t)$
Run Test: z for $\widehat{X}(t) = -5.873135$,
        z for $\widehat{e}(t) = 0.628699$
Residual sum of squares: 0.101041

7.6 "eee" Data

The speech data "eee" contains 512 signal values sampled at 10 kHz frequency, similar to "uuu" and "aaa" data. This data set can be analyzed using the multiple sinusoidal frequency model and also using the GFFM, see, for example, Nandi and Kundu [7, 8]. The data are first scaled by 1000 and are plotted in Fig. 7.16 and the periodogram function of the scaled data is plotted in Fig. 7.17. The number of components, p, is estimated from the periodogram function is 9 and for the GFFM there are two clusters with 3 and 6 frequencies. This implies that $\widehat{M} = 2$, $\widehat{q_1} = 3$ and $\widehat{q_2} = 6$. LSEs are obtained using the initial estimates of the frequencies from the periodogram function in case of both the models. The point estimates and 95% confidence bounds are presented in Tables 7.4 and 7.5 for the GFFM and the

Table 7.3 LSE and confidence intervals for 'aaa' data with sinusoidal frequency model

Parameters	LSE	Lower limit	Upper limit
a_1	0.226719	0.190962	0.262477
a_2	1.213009	1.177251	1.248766
a_3	0.163903	0.128146	0.199660
a_4	0.137947	0.102190	0.173705
a_5	0.143389	0.107632	0.179147
a_6	0.266157	0.230400	0.301915
a_7	0.389306	0.353548	0.425063
a_8	0.101906	0.066149	0.137663
a_9	0.134323	0.098566	0.170080
ϕ_1	2.161910	1.846478	2.477343
ϕ_2	2.548266	2.489310	2.607223
ϕ_3	−0.069081	−0.505405	0.367242
ϕ_4	0.371035	−0.147385	0.889456
ϕ_5	3.045470	2.546726	3.544215
ϕ_6	−0.602790	−0.871483	−0.334097
ϕ_7	0.905654	0.721956	1.089352
ϕ_8	1.586090	0.884319	2.287860
ϕ_9	−1.121016	−1.653425	−0.588607
β_1	0.114094	0.113027	0.115161
β_2	0.227092	0.226893	0.227292
β_3	0.342003	0.340527	0.343479
β_4	0.455799	0.454046	0.457553
β_5	1.360839	1.359152	1.362526
β_6	1.472003	1.471094	1.472912
β_7	1.586924	1.586302	1.587545
β_8	1.700715	1.698341	1.703089
β_9	1.815016	1.813215	1.816817

$\widehat{X}(t) = 0.315771\widehat{X}(t-1) + e(t)$
Run Test: z for $\widehat{X}(t) = -3.130015$,
z for $\widehat{e}(t) = 0.262751$
Residual sum of squares: 0.087489

multiple sinusoidal frequency model, respectively. The estimated errors are modeled as i.i.d. random variables in case of both the models. The predicted values using the GFFM along with the observed data are plotted in Fig. 7.18 and in Fig. 7.19 using the multiple sinusoidal frequency model. The predicted values match quite well in both

Fig. 7.14 The plot of the predicted values along with the observed 'aaa' data when GFFM is used

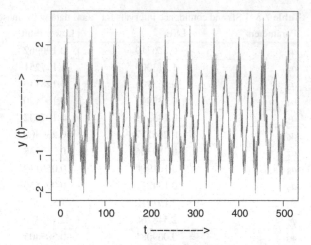

Fig. 7.15 The plot of the predicted values along with the observed 'aaa' data when multiple sinusoidal frequency model is used

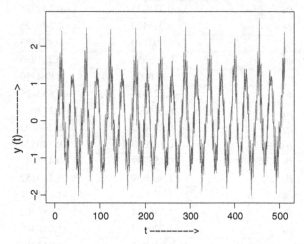

the cases. But in GFFM, the number of nonlinear parameters is less than the number of nonlinear parameters in case of sinusoidal model, therefore, it is preferable to use GFFM in case of "eee" data set.

7.7   Airline Passenger Data

This is a classical data set in time series analysis. The data represent the number of monthly international airline passenger during January 1953 to December 1960 and are collected from the Time Series Data Library of Hyndman (n.d.). The raw data are plotted in Fig. 7.20. It is clear from the plot that the variability increases with time, so it cannot be considered as a constant variance case. The log transform

Fig. 7.16 The plot of the
"eee" data

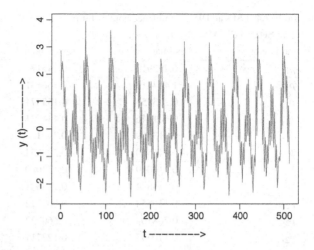

Fig. 7.17 The plot of the
periodogram function of
"eee" data

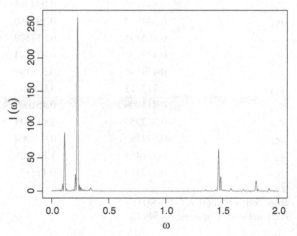

of the data is plotted in Fig. 7.21 to stabilize the variance. The variance seems to
be approximately constant now, but at the same time there is a significant linear
trend component present along with multiple periodic components. Therefore, a
transformation of the form $a + bt$ or application of the difference operator to the log
transform data is required. We choose to use the difference operator and finally we
have $y(t) = \log x(t + 1) - \log x(t)$ which can be analyzed using multiple sinusoidal
model. The transformed data are plotted in Fig. 7.22. It now appears that there is no
trend component with approximately constant variance. Here $\{x(t)\}$ represents the
observed data. Now to estimate the frequency components and to get an idea about
the number of frequency components present in $\{y(t)\}$, we plot the periodogram
function of $\{y(t)\}$ in the interval $(0, \pi)$ in Fig. 7.23.

There are six peaks corresponding to dominating frequency components in the
periodogram plot. The initial estimates of $\omega_1, \ldots, \omega_6$ are obtained one by one using

Table 7.4 LSEs and 95% confidence bounds for parameters of **"eee"** data set when GFFM is used

Data Set: "eee". $M = 2$, $q_1 = 3$, $q_2 = 6$

Parameter	Estimate	Lower bound	Upper bound
ρ_{1_1}	0.792913	0.745131	0.840695
ρ_{1_2}	1.414413	1.366631	1.462195
ρ_{1_3}	0.203153	0.155371	0.250935
ϕ_{1_1}	0.681846	0.434021	0.929672
ϕ_{1_2}	0.913759	0.747117	1.080401
ϕ_{1_3}	-0.041550	-0.298854	0.215753
λ_1	0.114019	0.113080	0.114958
ω_1	0.113874	0.113531	0.114217
ρ_{2_1}	0.110554	0.062772	0.158336
ρ_{2_2}	0.708104	0.660322	0.755886
ρ_{2_3}	0.170493	0.122711	0.218275
ρ_{2_4}	0.141166	0.093384	0.188948
ρ_{2_5}	0.344862	0.297080	0.392644
ρ_{2_6}	0.165291	0.117509	0.213073
ϕ_{2_1}	0.427151	-0.034474	0.888776
ϕ_{2_2}	0.479346	0.348149	0.610542
ϕ_{2_3}	-2.712411	-3.010046	-2.414776
ϕ_{2_4}	0.815808	0.451097	1.180520
ϕ_{2_5}	-0.773955	-1.012844	-0.535067
ϕ_{2_6}	-0.577183	-0.966904	-0.187463
λ_2	1.351453	1.350820	1.352087
ω_2	0.113463	0.113173	0.113753

$\widehat{X}(t) = e(t)$

Run Test: z for $\widehat{e}(t) = -1.109031$

Residual sum of squares: 0.154122

the sequential procedures. After taking out the effects of these six frequency components, we again study the periodogram function of the residual plot, similarly as in case of the variable star data. We observe that there is an additional significant frequency component present. Hence, we estimate p as seven and accordingly estimate the other parameters. Finally, plugging in the estimated parameters we have the fitted series which is plotted in Fig. 7.24 along with the log difference data. They match quite well. The sum of squares of the residuals of this sinusoidal fit is 5.54×10^{-4}. As mentioned earlier, the monthly international airline passenger data is a well studied data set in time series literature. Usually a seasonal ARIMA (multiplicative) model is used to analyze it. A reasonable fit using this class of models is a seasonal ARIMA of order $(0, 1, 1) \times (0, 1, 1)_{12}$ to the log of the data which is same as the model $(0, 0, 1) \times (0, 1, 1)_{12}$ applied to the difference of the log data (discussed by Box,

Table 7.5 LSE and confidence intervals for **"eee"** data with multiple frequency model

Parameters	LSE	Lower limit	Upper limit
a_1	0.792754	0.745823	0.839686
a_2	1.413811	1.366879	1.460743
a_3	0.203779	0.156848	0.250711
a_4	0.111959	0.065028	0.158891
a_5	0.703978	0.657047	0.750910
a_6	0.172563	0.125632	0.219495
a_7	0.145034	0.098103	0.191966
a_8	0.355013	0.308082	0.401945
a_9	0.176292	0.129360	0.223223
ϕ_1	0.687658	0.569257	0.806059
ϕ_2	0.912467	0.846077	0.978857
ϕ_3	0.031799	−0.428813	0.492411
ϕ_4	0.628177	−0.210193	1.466547
ϕ_5	0.495832	0.362499	0.629164
ϕ_6	2.828465	2.284531	3.372400
ϕ_7	0.847214	0.200035	1.494393
ϕ_8	−0.360236	−0.624629	−0.095842
ϕ_9	0.002140	−0.530290	0.534571
β_1	0.114038	0.113638	0.114439
β_2	0.227886	0.227661	0.228110
β_3	0.342098	0.340540	0.343656
β_4	1.352172	1.349336	1.355008
β_5	1.464954	1.464503	1.465405
β_6	1.575381	1.573541	1.577222
β_7	1.691908	1.689718	1.694097
β_8	1.807006	1.806111	1.807900
β_9	1.920891	1.919090	1.922692

$\widehat{X}(t) = e(t)$
Run Test: z for $\widehat{e}(t) = -1.032388$
Residual sum of squares: 0.148745

Jenkins, and Reinsel [1] in details). The ARIMA model to the difference of the log data $\{y(t)\}$

$$y(t) = y(t - 12) + z(t) + 0.3577z(t - 1) + .4419z(t - 12) + 0.1581z(t - 13),$$

where $\{z(t)\}$ is a white noise sequence with mean zero and estimated variance 0.001052. In this case, we observe that the residual sum of squares is 9.19×10^{-4}, which is greater than the residual sum of squares when multiple sinusoidal frequency model is used. See Nandi and Kundu [9] for more details.

Fig. 7.18 The plot of the predicted values along with the observed "eee" data when generalized fundamental frequency model is used

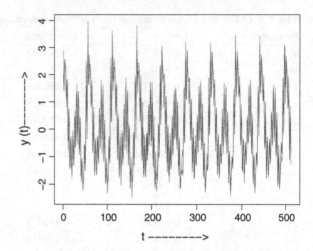

Fig. 7.19 The plot of the predicted values along with the observed "eee" data when multiple sinusoidal frequency model is used

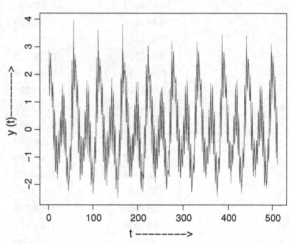

Fig. 7.20 Monthly international airline passengers from January 1953 to December 1960

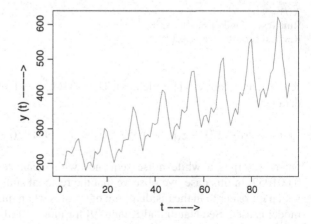

Fig. 7.21 The logarithm of the airline passenger data

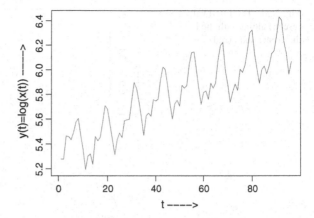

Fig. 7.22 The first difference values of the logarithmic data

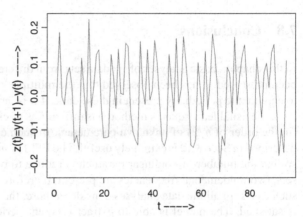

Fig. 7.23 The periodogram function of the log difference data

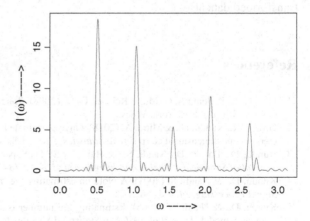

Fig. 7.24 The fitted values
(green) along with the log
difference data (red)

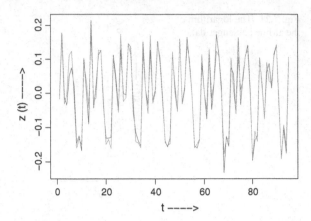

7.8   Conclusions

In this chapter, basic analyses of six data sets from different fields of application have
been presented. It has been observed that the multiple sinusoidal model is effective
to capture the periodicity present in these data. If p is large, as in case of ECG signal
data, the usual least squares method has difficulties to estimate such a large number
(in the order of $n/2$) of unknown parameters. In such cases, the sequential method
described in Sect. 3.12 is extremely useful. The FFM is suitable to analyze "uuu" data
so that the number of nonlinear parameters reduce to one. In "aaa" and "eee" data
sets, two fundamental frequencies are present, therefore, GFFM is a useful model in
such cases. In all the data analyses considered here, the stationary error assumption
is satisfied. The model is able to extract inherent periodicity if present, from the
transformed data also.

References

1. Box, G. E. P., Jenkins, G. M., & Reinsel, G. C. (2008). *Time series analysis, forecasting and
   control* (4th ed.). New York: Wiley.
2. Grover, R., Kundu, D., & Mitra, A. (2018). On approximate least squares estimators of parameters on one dimensional chirp signal. *Statistics*, *52*(5), 1060–1085.
3. Kundu, D., Bai, Z. D., Nandi, S., & Bai, L. (2011). Super efficient frequency Estimation.
   Journal of Statistical Planning and Inference, *141*(8), 2576–2588.
4. Kundu, D., & Nandi, S. (2004). A Note on estimating the frequency of a periodic function.
   Signal Processing, *84*, 653–661.
5. Kundu, D., & Nandi, S. (2005). Estimating the number of components of the fundamental
   frequency model. *Journal of the Japan Statistical Society*, *35*(1), 41–59.
6. Nandi, S., & Kundu, D. (2003). Estimating the fundamental frequency of a periodic function.
   Statistical Methods and Applications, *12*, 341–360.
7. Nandi, S., & Kundu, D. (2006). A fast and efficient algorithm for estimating the parameters of
   sum of sinusoidal model. *Sankhya*, *68*, 283–306.

8. Nandi, S., & Kundu, D. (2006). Analyzing non-stationary signals using a cluster type model. *Journal of Statistical Planning and Inference, 136*, 3871–3903.
9. Nandi, S., & Kundu, D. (2013). Estimation of parameters of partially sinusoidal frequency model. *Statistics, 47*, 45–60.
10. Nandi, S., & Kundu, D. (2019). Estimating the fundamental frequency using modified Newton-Raphson algorithm. *Statistics, 53*, 440–458.
11. Prasad, A., Kundu, D., & Mitra, A. (2008). Sequential estimation of the sum of sinusoidal model parameters. *Journal of Statistical Planning and Inference, 138*, 1297–1313.
12. Quinn, B. G., & Thomson, P. J. (1991). Estimating the frequency of a periodic function. *Biometrika, 78*, 65–74.

Chapter 8
Multidimensional Models

8.1 Introduction

In the last few chapters, we have discussed different aspects of the one-dimensional (1-D) sinusoidal frequency model. In this chapter, our aim is to introduce two dimensional (2-D) and three dimensional (3-D) frequency models, and discuss several issues related to them. The 2-D and 3-D sinusoidal frequency models are natural generalizations of the 1-D sinusoidal frequency model (3.1), and these models have various applications.

The 2-D sinusoidal frequency model has the following form:

$$y(m, n) = \sum_{k=1}^{p} [A_k \cos(m\lambda_k + n\mu_k) + B_k \sin(m\lambda_k + n\mu_k)] + X(m, n);$$

$$\text{for} \quad m = 1, \ldots, M, \ n = 1, \ldots, N. \tag{8.1}$$

Here for $k = 1, \ldots, p$, A_k, B_k are unknown real numbers, λ_k, μ_k are unknown frequencies, $\{X(m, n)\}$ is a 2-D sequence of error random variables with mean zero and finite variance. Several correlation structures have been assumed in the literature, and they are explicitly mentioned later. Two problems are of major interest associated to model (8.1). One is the estimation of A_k, B_k, λ_k, μ_k for $k = 1, \ldots, p$, and the other is the estimation of the number of components namely p. The first problem has received considerable amount of attention in the statistical signal processing literature.

In the particular case when $\{X(m, n); m = 1, \ldots, M, n = 1, \ldots, N\}$ are i.i.d. random variables, this problem can be interpreted as a "signal detection" problem, and it has different applications in "Multidimensional Signal Processing". This is a basic model in many fields like antenna array processing, geophysical perception, biomedical spectral analysis, etc., see for example, the works of Barbieri and Barone [1], Cabrera and Bose [2], Chun and Bose [3], Hua [5], and the references cited

© Springer Nature Singapore Pte Ltd. 2020
S. Nandi and D. Kundu, *Statistical Signal Processing*,
https://doi.org/10.1007/978-981-15-6280-8_8

therein. This problem has a special interest in spectrography, and it has been studied using the group theoretic method by Malliavan [16, 17].

Zhang and Mandrekar [28] used the 2-D sinusoidal frequency model for analyzing symmetric grayscale textures. Any grayscale picture, stored in digital format in a computer, is composed of tiny dots or pixels. In digital representation of grayscale or black and white pictures, the gray shade at each pixel is determined by the grayscale intensity at that pixel. For instance, if 0 represents black and 1 represents white, then any real number $\in [0, 1]$ corresponds to a particular intensity of gray. If a picture is composed of 2-D array of M pixels arranged in rows and N pixels arranged in columns, the size of the picture is $M \times N$ pixels. The grayscale intensities corresponding to the gray shades of various pixels can be stored in an $M \times N$ matrix. This transformation from a picture to a matrix and again back to a picture can easily be performed by image processing tools of any standard mathematical software, say R or MATLAB.

Consider the following synthesized grayscale texture data for $m = 1, \ldots, 40$ and $n = 1, \ldots, 40$.

$$
\begin{aligned}
y(m, n) = {} & 4.0 \cos(1.8m + 1.1n) + 4.0 \sin(1.8m + 1.1n) + \\
& 1.0 \cos(1.7m + 1.0n) + 1.0 \sin(1.7m + 1.0n) + X(m, n), \quad (8.2)
\end{aligned}
$$

where $X(m, n) = e(m, n) + 0.25e(m - 1, n) + 0.25e(m, n - 1)$, and $e(m, n)$ for $m = 1, \ldots, 40$ and $n = 1, \ldots, 40$ are i.i.d. normal random variables with mean 0 and variance 2.0. This texture is displayed in Fig. 8.1. It is clear from Fig. 8.1 that model (8.1) can be used quite effectively to generate symmetric textures.

In this chapter, we briefly discuss different estimators of the unknown parameters of the 2-D model given in (8.1) and their properties. It can be seen that model (8.1) is a nonlinear regression model, and therefore, the LSEs seem to be the most reasonable estimators. It is observed that the 1-D periodogram method can be extended to the 2-D model, but it also has similar problems as the 1-D periodogram method. The

Fig. 8.1 The image plot of a simulated data from model (8.2)

Fig. 8.2 The image plot of a simulated data from model (8.3)

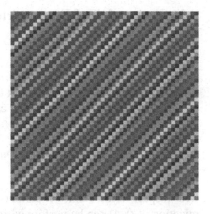

LSEs are the most intuitive natural estimators in this case, but it is well-known that finding the LSEs is a computationally challenging problem, particularly when p is large. It is observed that the sequential method, proposed by Prasad, Kundu, and Mitra [23], can be used quite effectively in producing estimators which are equivalent to the LSEs. For $p = 1$, an efficient algorithm has been proposed by Nandi, Prasad, and Kundu [21] which produces estimators which are equivalent to the LSEs in three steps starting from the PEs.

Similarly, the 3-D sinusoidal frequency model takes the following form:

$$y(m, n, s) = \sum_{k=1}^{p} \left[A_k \cos(m\lambda_k + n\mu_k + s\nu_k) + B_k \sin(m\lambda_k + n\mu_k + s\nu_k) \right] + X(m, n, s);$$

$$\text{for } m = 1, \ldots, M, \; n = 1, \ldots, N, \; s = 1, \ldots, S. \tag{8.3}$$

Here for $k = 1, \ldots, p$, A_k, B_k are unknown real numbers, λ_k, μ_k, ν_k are unknown frequencies, $\{X(m, n, s)\}$ is a 3-D sequence of error random variables with mean zero, finite variance and "p" denotes the number of 3-D sinusoidal components. Model (8.3) has been used for describing color textures by Prasad and Kundu [22], see Fig. 8.2. The third dimension represents different color schemes. In digital representation, any color picture is stored digitally in RGB format. Almost any color can be represented by a unique combination of red, green, and blue color intensities. In RGB format, $S = 3$. A color picture can be stored digitally in an $M \times N \times S$ array. Similar to a black and white picture, any image processing tool of a mathematical software can be used to convert a color picture to a 3-D array and vice versa. For detailed description on how a color picture is stored digitally, the readers are referred to Prasad and Kundu [22].

8.2 Estimation of Frequencies for 2-D Sinusoidal Model

8.2.1 LSEs

It has been mentioned earlier that in the presence of i.i.d. errors, the LSEs seem to be the most natural choice, and they can be obtained by minimizing

$$\sum_{m=1}^{M}\sum_{n=1}^{N}\left(y(m,n)-\sum_{k=1}^{p}[A_k\cos(m\lambda_k+n\mu_k)+B_k\sin(m\lambda_k+n\mu_k)]\right)^2, \quad (8.4)$$

with respect to the unknown parameters, A_k, B_k λ_k, and μ_k, $k=1,\ldots,p$. Minimization of (8.4) can be performed in two steps by using the separable regression technique of Richards [25]. For $k=1,\ldots,p$, when λ_k and μ_k are fixed, the LSEs of A_k and B_k can be obtained as

$$\left[\widehat{A}_1(\lambda,\mu),\widehat{B}_1(\lambda,\mu):\cdots:\widehat{A}_p(\lambda,\mu),\widehat{B}_p(\lambda,\mu)\right]^T=\left(\mathbf{U}^T\mathbf{U}\right)^{-1}\mathbf{U}^T\mathbf{Y}. \quad (8.5)$$

Here, $\lambda=(\lambda_1,\ldots,\lambda_p)^T$, $\mu=(\mu_1,\ldots,\mu_p)^T$, $\mathbf{U}$ is an $MN\times 2p$ matrix, and $\mathbf{Y}$ is an $MN\times 1$ vector as follows:

$$\mathbf{U}=[\mathbf{U}_1:\cdots:\mathbf{U}_p]$$

$$\mathbf{U}_k=\begin{bmatrix}\cos(\lambda_k+\mu_k)\cdots\cos(\lambda_k+N\mu_k)\cos(2\lambda_k+\mu_k)\cdots\cos(M\lambda_k+N\mu_k)\\\sin(\lambda_k+\mu_k)\cdots\sin(\lambda_k+N\mu_k)\sin(2\lambda_k+\mu_k)\cdots\sin(M\lambda_k+N\mu_k)\end{bmatrix}^T,$$
$$(8.6)$$

for $k=1,\ldots,p$, and

$$\mathbf{Y}=\left(y(1,1)\cdots y(1,N)\,y(2,1)\cdots y(2,N)\cdots y(M,1)\cdots y(M,N)\right)^T. \quad (8.7)$$

Once $\widehat{A}_k(\lambda,\mu)$ and $\widehat{B}_k(\lambda,\mu)$ are obtained for $k=1,\ldots,p$, the LSEs of λ_k and μ_k are obtained by minimizing

$$\sum_{m=1}^{M}\sum_{n=1}^{N}\left(y(m,n)-\sum_{k=1}^{p}[\widehat{A}_k(\lambda,\mu)\cos(m\lambda_k+n\mu_k)+\widehat{B}_k(\lambda,\mu)\sin(m\lambda_k+n\mu_k)]\right)^2,$$
$$(8.8)$$

with respect to $\lambda_1,\ldots,\lambda_p$ and $\mu_1,\ldots,\mu_p$. Denote $\widehat{\lambda}_k$ and $\widehat{\mu}_k$ as the LSEs of λ_k and μ_k, respectively. Once $\widehat{\lambda}_k$ and $\widehat{\mu}_k$ are obtained, the LSEs of A_k and B_k can be obtained as $\widehat{A}_k(\widehat{\lambda},\widehat{\mu})$ and $\widehat{B}_k(\widehat{\lambda},\widehat{\mu})$, respectively, where $\widehat{\lambda}=(\widehat{\lambda}_1,\ldots,\widehat{\lambda}_p)^T$ and $\widehat{\mu}=(\widehat{\mu}_1,\ldots,\widehat{\mu}_p)^T$. The minimization of (8.8) can be obtained by solving a $2p$ dimensional optimization problem, which can be computationally quite challenging if p is large.

Rao, Zhao, and Zhou [24] obtained the theoretical properties of the LSEs of the parameters of a similar model namely 2-D superimposed complex exponential model, that is,

$$y(m, n) = \sum_{k=1}^{p} C_k e^{i(m\lambda_k + n\mu_k)} + Z(m, n); \qquad (8.9)$$

here for $k = 1, \ldots, p$, C_k are unknown complex-valued amplitudes, λ_k and μ_k are same as defined before, $\{Z(m, n)\}$ is a 2-D sequence of complex-valued random variables. Rao, Zhao, and Zhou [24] obtained the consistency and asymptotic normality properties of the LSEs under the assumptions that $\{Z(m, n); m = 1, \ldots, M, n = 1, \ldots, N\}$ is a 2-D sequence i.i.d. complex normal random variables. Mitra and Stoica [19] provided the asymptotic Cramer–Rao bound of the MLEs of the frequencies. In case of complex models, LSEs are obtained by minimizing the sum of least norm squares. Kundu and Gupta [10] proved the consistency and asymptotic normality properties of the LSEs of model (8.1) under the assumption that the error random variables $\{X(m, n); m = 1, \ldots, M, n = 1, \ldots, N\}$ are i.i.d. with mean zero and finite variance, see also Kundu and Mitra [12] in this respect. Later, Kundu, and Nandi [13] provided the consistency and asymptotic normality properties of the LSEs under the following stationary assumptions on $\{X(m, n)\}$.

Assumption 8.1 The 2-D sequence of random variables $\{X(m, n)\}$ can be represented as follows:

$$X(m, n) = \sum_{j=-\infty}^{\infty} \sum_{k=-\infty}^{\infty} a(j, k)e(m - j, n - k),$$

where $a(j, k)$ are real constants such that

$$\sum_{j=-\infty}^{\infty} \sum_{k=-\infty}^{\infty} |a(j, k)| < \infty,$$

and $\{e(m, n)\}$ is a 2-D sequence of i.i.d. random variables with mean zero and variance σ^2.

We use the following notation to provide the asymptotic distribution of the LSEs of the parameters of model (8.1) obtained by Kundu and Nandi [13].

$$\boldsymbol{\theta}_1 = (A_1, B_1, \lambda_1, \mu_1)^T, \ldots, \boldsymbol{\theta}_p = (A_p, B_p, \lambda_p, \mu_p)^T$$

$$\mathbf{D} = \text{diag}\{M^{1/2}N^{1/2}, M^{1/2}N^{1/2}, M^{3/2}N^{1/2}, M^{1/2}N^{3/2}\},$$

and for $k = 1, \ldots, p$,

$$\Sigma_k = \begin{bmatrix} 1 & 0 & \frac{1}{2}B_k & \frac{1}{2}B_k \\ 0 & 1 & -\frac{1}{2}A_k & -\frac{1}{2}A_k \\ \frac{1}{2}B_k & -\frac{1}{2}A_k & \frac{1}{3}(A_k^2+B_k^2) & \frac{1}{4}(A_k^2+B_k^2) \\ \frac{1}{2}B_k & -\frac{1}{2}A_k & \frac{1}{4}(A_k^2+B_k^2) & \frac{1}{3}(A_k^2+B_k^2) \end{bmatrix}.$$

Theorem 8.1 *Under Assumption 8.1, $\widehat{\theta}_1, \ldots, \widehat{\theta}_p$, the LSEs of $\theta_1, \ldots, \theta_p$, respectively, are consistent estimators, and as $\min\{M, N\} \to \infty$*

$$\left(\mathbf{D}(\widehat{\theta}_1 - \theta_1), \ldots, \mathbf{D}(\widehat{\theta}_p - \theta_p)\right) \xrightarrow{d} \mathcal{N}_{4p}\left(\mathbf{0}, 2\sigma^2 \mathbf{\Delta}^{-1}\right);$$

here

$$\mathbf{\Delta}^{-1} = \begin{bmatrix} c_1 \mathbf{\Sigma}_1^{-1} & 0 & \cdots & 0 \\ 0 & c_2 \mathbf{\Sigma}_2^{-1} & \cdots & 0 \\ \vdots & & \ddots & \vdots \\ 0 & 0 & \cdots & c_p \mathbf{\Sigma}_p^{-1} \end{bmatrix}$$

and for $k = 1, \ldots, p$,

$$c_k = \left| \sum_{u=-\infty}^{\infty} \sum_{v=-\infty}^{\infty} a(u, v) e^{-i(u\lambda_k + v\mu_k)} \right|^2.$$

Theorem 8.1 indicates that even for the 2-D model, the LSEs of the frequencies have much faster convergence rates than that of the linear parameters.

8.2.2 Sequential Method

It has been observed that the LSEs are the most efficient estimators, but computing the LSEs is a difficult problem. It involves solving a $2p$ dimensional optimization problem as it has been mentioned in the previous section. It might be quite difficult particularly if p is large. To avoid this problem, Prasad, Kundu, and Mitra [23] proposed a sequential estimation procedure of the unknown parameters which have the same rate of convergence as the LSEs. Moreover, the sequential estimators can be obtained by solving p separate 2-D optimization problems sequentially. Therefore, even if p is large, sequential estimators can be obtained quite easily compared to the LSEs. It can be described as follows. At the first step, minimize the quantity

$$Q_1(A, B, \lambda, \mu) = \sum_{m=1}^{M} \sum_{n=1}^{N} (y(m, n) - A\cos(m\lambda + n\mu) - B\sin(m\lambda + n\mu))^2,$$

$$(8.10)$$

with respect to A, B, λ, μ. Here, λ and μ are scaler parameters. It can be seen using separable regression technique of Richards [25] that for fixed λ and μ, $\widetilde{A}(\lambda, \mu)$ and $\widetilde{B}(\lambda, \mu)$ minimize (8.10), where

$$\left[\widetilde{A}(\lambda, \mu) \ \widetilde{B}(\lambda, \mu) \right]^T = \left(\mathbf{U}_1^T \mathbf{U}_1 \right)^{-1} \mathbf{U}_1^T \mathbf{Y}, \tag{8.11}$$

$\mathbf{U}_1$ is an $MN \times 2$ and $\mathbf{Y}$ is an $MN \times 1$ data vector as defined in (8.6) and (8.7), respectively. Replacing $\widetilde{A}(\lambda, \mu)$ and $\widetilde{B}(\lambda, \mu)$ in (8.10), we obtain

$$R_1(\lambda, \mu) = Q_1(\widetilde{A}(\lambda, \mu), \widetilde{B}(\lambda, \mu), \lambda, \mu). \tag{8.12}$$

If $\widetilde{\lambda}$ and $\widetilde{\mu}$ minimize $R_1(\lambda, \mu)$, $\left(\widetilde{A}(\widetilde{\lambda}, \widetilde{\mu}), \widetilde{B}(\widetilde{\lambda}, \widetilde{\mu}), \widetilde{\lambda}, \widetilde{\mu} \right)$ minimizes (8.10). Denote these estimators as $\widetilde{\boldsymbol{\theta}}_1 = \left(\widetilde{A}_1, \widetilde{B}_1, \widetilde{\lambda}_1, \widetilde{\mu}_1 \right)^T$.

Consider $\{y_1(m, n); \ m = 1, \ldots, M, n = 1, \ldots, N\}$ where

$$y_1(m, n) = y(m, n) - \widetilde{A}_1 \cos(m\widetilde{\lambda}_1 + n\widetilde{\mu}_1) - \widetilde{B}_1 \sin(m\widetilde{\lambda}_1 + n\widetilde{\mu}_1). \tag{8.13}$$

Repeating the whole procedure as described above by replacing $y(m, n)$ with $y_1(m, n)$, obtain $\widetilde{\boldsymbol{\theta}}_2 = \left(\widetilde{A}_2, \widetilde{B}_2, \widetilde{\lambda}_2, \widetilde{\mu}_2 \right)^T$. Along the same line, one can obtain $\widetilde{\boldsymbol{\theta}}_1, \ldots, \widetilde{\boldsymbol{\theta}}_p$. Further, $\widetilde{\boldsymbol{\theta}}_j$ and $\widetilde{\boldsymbol{\theta}}_k$ for $j \neq k$ are independently distributed. Prasad, Kundu, and Mitra [23] proved that Theorem 8.1 also holds for $\widetilde{\boldsymbol{\theta}}_1, \ldots, \widetilde{\boldsymbol{\theta}}_p$. It implies that the LSEs and the estimators obtained by using the sequential method are asymptotically equivalent estimators.

8.2.3   Periodogram Estimators

The 2-D periodogram function of any 2-D sequence of observations $\{y(m, n); \ m = 1, \ldots, M, n = 1, \ldots, N\}$ is defined as follows:

$$I(\lambda, \mu) = \frac{1}{MN} \left| \sum_{m=1}^{M} \sum_{n=1}^{N} y(m, n) e^{-i(m\lambda + n\mu)} \right|^2. \tag{8.14}$$

Here, the 2-D periodogram function is evaluated at the 2-D Fourier frequencies, namely at $(\pi k / M, \pi j / N), k = 0, \ldots, M, j = 0, \ldots, N$. Clearly the 2-D periodogram function (8.14) is a natural generalization of 1-D periodogram function. Zhang and Mandrekar [28] and Kundu and Nandi [13] used the periodogram function (8.14) to estimate the frequencies of model (8.1). For $p = 1$, Kundu and Nandi [13] proposed the ALSEs as follows:

$$(\widetilde{\lambda}, \widetilde{\mu}) = \arg \max I(\lambda, \mu), \tag{8.15}$$

where the maximization is performed for $0 \leq \lambda \leq \pi$ and $0 \leq \mu \leq \pi$, and the estimates of A and B are obtained as follows:

$$\widetilde{A} = \frac{2}{MN} \sum_{m=1}^{M} \sum_{m=1}^{N} y(m, n) \cos(m\widetilde{\lambda} + n\widetilde{\mu}), \tag{8.16}$$

$$\widetilde{B} = \frac{2}{MN} \sum_{m=1}^{M} \sum_{m=1}^{N} y(m, n) \sin(m\widetilde{\lambda} + n\widetilde{\mu}). \tag{8.17}$$

It has been shown that under Assumption 8.1, the asymptotic distribution of $(\widetilde{A}, \widetilde{B}, \widetilde{\lambda}, \widetilde{\mu})$ is the same as stated in Theorem 8.1. Sequential method as described in the previous section can be applied exactly in the same manner for general p, and the asymptotic distribution of the ALSEs also satisfies Theorem 8.1. Therefore, the LSEs and ALSEs are asymptotically equivalent. This has also been observed in the 1-D sinusoidal model.

8.2.4 Nandi–Prasad–Kundu Algorithm

The ALSEs or the sequential estimators as described in the previous two sections can be obtained by solving a 2-D optimization problem. It is well-known that the least squares surface and the periodogram surface have several local minima and local maxima, respectively. Therefore, the convergence of any optimization algorithm is not guaranteed. Nandi, Prasad, and Kundu [21] proposed a three-step algorithm which produces estimators of the unknown frequencies having the same rate of convergence as the LSEs. We provide the algorithm for $p = 1$. The sequential procedure can be easily used for general p. We use the following notation for describing the algorithm.

$$P_{MN}^1(\lambda, \mu) = \sum_{t=1}^{M} \sum_{s=1}^{N} \left(t - \frac{M}{2} \right) y(t, s) e^{-i(\lambda t + \mu s)} \tag{8.18}$$

$$P_{MN}^2(\lambda, \mu) = \sum_{t=1}^{M} \sum_{s=1}^{N} \left(s - \frac{N}{2} \right) y(t, s) e^{-i(\lambda t + \mu s)} \tag{8.19}$$

$$Q_{MN}(\lambda, \mu) = \sum_{t=1}^{M} \sum_{s=1}^{N} y(t, s) e^{-i(\lambda t + \mu s)} \tag{8.20}$$

$$\widehat{\lambda}^{(r)} = \widehat{\lambda}^{(r-1)} + \frac{12}{M_r^2} \mathrm{Im} \left[\frac{P_{M_r N_r}^1(\widehat{\lambda}^{(r-1)}, \widehat{\mu}^{(0)})}{Q_{M_r N_r}(\widehat{\lambda}^{(r-1)}, \widehat{\mu}^{(0)})} \right], \quad r = 1, 2, \ldots, \tag{8.21}$$

$$\widehat{\mu}^{(r)} = \widehat{\mu}^{(r-1)} + \frac{12}{N_r^2} \mathrm{Im} \left[\frac{P_{M_r N_r}^2(\widehat{\lambda}^{(0)}, \widehat{\mu}^{(r-1)})}{Q_{M_r N_r}(\widehat{\lambda}^{(0)}, \widehat{\mu}^{(r-1)})} \right], \quad r = 1, 2, \ldots. \tag{8.22}$$

Here $\widehat{\lambda}^{(r)}$ and $\widehat{\mu}^{(r)}$ are estimates of λ and μ, respectively and (M_r, N_r) denotes the sample size at the rth step.

Nandi, Prasad, and Kundu [21] suggested to use the following initial estimators of λ and μ. For any fixed $n \in \{1, \dots, N\}$ from the data vector $\{y(1, n), \dots, y(M, n)\}$, obtain $\widehat{\lambda}_n$, the periodogram maximizer over Fourier frequencies. Take

$$\widehat{\lambda}^{(0)} = \frac{1}{N} \sum_{n=1}^{N} \widehat{\lambda}_n. \tag{8.23}$$

Similarly, for fixed $m \in \{1, \dots, M\}$, from the data vector $\{y(m, 1), \dots, y(m, N)\}$, first obtain $\widehat{\mu}_m$, which is the periodogram maximizer over Fourier frequencies, and then consider

$$\widehat{\mu}^{(0)} = \frac{1}{M} \sum_{m=1}^{M} \widehat{\mu}_m. \tag{8.24}$$

It has been shown that $\widehat{\lambda}^{(0)} = O_p(M^{-1}N^{-1/2})$ and $\widehat{\mu}^{(0)} = O_p(M^{-1/2}N^{-1})$. The algorithm can be described as follows.

Algorithm 8.1

- Step 1: Take $r = 1$, choose $M_1 = M^{0.8}$, $N_1 = N$. Compute $\widehat{\lambda}^{(1)}$ from $\widehat{\lambda}^{(0)}$ using (8.21).
- Step 2: Take $r = 2$, choose $M_2 = M^{0.9}$, $N_2 = N$. Compute $\widehat{\lambda}^{(2)}$ from $\widehat{\lambda}^{(1)}$ using (8.21).
- Step 3: Take $r = 3$, choose $M_3 = M$, $N_3 = N$. Compute $\widehat{\lambda}^{(3)}$ from $\widehat{\lambda}^{(2)}$ using (8.21).

Exactly in the same manner, $\widehat{\mu}^{(3)}$ can be obtained from (8.22) interchanging M, M_r, and λ by N, N_r, and μ, respectively. Finally, $\widehat{\lambda}^{(3)}$ and $\widehat{\mu}^{(3)}$ are the proposed estimators of λ and μ, respectively. It has been shown that the proposed estimators have the same asymptotic variances as the corresponding LSEs. The main advantage of the proposed estimators is that they can be obtained in a fixed number of iterations. Extensive simulation results suggest that the proposed algorithm works very well.

8.2.5 Noise Space Decomposition Method

Nandi, Kundu, and Srivastava [20] proposed the noise space decomposition method to estimate the frequencies of the 2-D sinusoidal model (8.1). The proposed method is an extension of the 1-D NSD method which was originally proposed by Kundu and Mitra [11] as described in Sect. 3.4. The NSD method for the 2-D model can be described as follows: From the sth row of the data matrix

$$\mathbf{Y}_N = \begin{bmatrix} y(1,1) & \cdots & y(1,N) \\ \vdots & \ddots & \vdots \\ y(s,1) & \cdots & y(s,N) \\ \vdots & \ddots & \vdots \\ y(M,1) & \cdots & y(M,N) \end{bmatrix}, \tag{8.25}$$

construct a matrix $\mathbf{A}_s$ for any $N - 2p \geq L \geq 2p$ as follows:

$$\mathbf{A}_s = \begin{bmatrix} y(s,1) & \cdots & y(s,L+1) \\ \vdots & \vdots & \vdots \\ y(s,N-L) & \cdots & y(s,N) \end{bmatrix}.$$

Obtain an $(L+1) \times (L+1)$ matrix $\mathbf{B} = \sum_{s=1}^{M} \mathbf{A}_s^T \mathbf{A}_s / ((N-L)M)$. Now using the 1-D NSD method on matrix $\mathbf{B}$, the estimates of $\lambda_1, \ldots, \lambda_p$ can be obtained. Similarly, using the columns of the data matrix $\mathbf{Y}_N$, the estimates of $\mu_1, \ldots, \mu_p$ can be obtained. For details, see Nandi, Kundu, and Srivastava [20]. Finally one needs to estimate the pairs namely $\{(\lambda_k, \mu_k); k = 1, \ldots, p\}$ also. The authors suggested the following two pairing algorithms once the estimates of λ_k and μ_k for $k = 1, \ldots, p$ are obtained. Algorithm 8.2 is based on $p!$ search. It is computationally efficient for small values of p, say $p = 2, 3$ and Algorithm 8.3 is based on p^2-search, so it is efficient for large values of p, that is, when p is greater than 3. Suppose the estimates obtained using the above NSD method are $\{\widehat{\lambda}_{(1)}, \ldots, \widehat{\lambda}_{(p)}\}$ and $\{\widehat{\mu}_{(1)}, \ldots, \widehat{\mu}_{(p)}\}$, then the two algorithms are described as follows.

Algorithm 8.2 Consider all possible $p!$ combination of pairs $\{(\widehat{\lambda}_{(j)}, \widehat{\mu}_{(j)}) : j = 1, \ldots, p\}$ and calculate the sum of the periodogram function for each combination as

$$I_S(\boldsymbol{\lambda}, \boldsymbol{\mu}) = \sum_{k=1}^{p} \frac{1}{MN} \left| \sum_{s=1}^{M} \sum_{t=1}^{N} y(s,t) e^{-i(s\lambda_k + t\mu_k)} \right|^2.$$

Consider that combination as the paired estimates of $\{(\lambda_j, \mu_j) : j = 1, \ldots, p\}$ for which this $I_S(\boldsymbol{\lambda}, \boldsymbol{\mu})$ is maximum.

Algorithm 8.3 Compute $I(\lambda, \mu)$ as defined in (8.14) over $\{(\widehat{\lambda}_{(j)}, \widehat{\mu}_{(k)}), j, k = 1, \ldots, p\}$. Choose the largest p values of $I(\widehat{\lambda}_{(j)}, \widehat{\mu}_{(k)})$ and the corresponding $\{(\widehat{\lambda}_{(k)}, \widehat{\mu}_{(k)}), k = 1, \ldots, p\}$ are the paired estimates of $\{(\lambda_k, \mu_k), k = 1, \ldots, p\}$.

From the extensive experimental results, it is observed that the performance of these estimators are better than the ALSEs, and compare reasonably well with the LSEs. It has also been observed along the same line as the 1-D NSD method that under the assumptions of i.i.d. errors, the frequency estimators obtained by the 2-D NSD method are strongly consistent, but the asymptotic distribution of these estimators has not yet been established.

8.3 2-D Model: Estimating the Number of Components

The estimation of frequencies of the 2-D sinusoidal model has received considerable attention in the signal processing literature. Unfortunately not that much attention has been paid in estimating the number of components, namely p of model (8.1). It may be mentioned that p can be estimated by observing the number of peaks of the 2-D periodogram function, as mentioned in Kay [6], but that is quite subjective in nature and it may not work properly all the times.

Miao, Wu, and Zhao [18] discussed the estimation of the number of components for an equivalent model. Kliger and Francos [7] derived a maximum a posteriori (MAP) model order selection criterion for jointly estimating the parameters of the 2-D sinusoidal model. Their method produces a strongly consistent estimate of the model order p under the assumption of i.i.d. Gaussian random field of the error. Kliger and Francos [8] assumed that the errors are from a stationary random field and again discussed simultaneous estimation of p and other parameters. They first considered the case where the number of sinusoidal components, say q, is underestimated, that is, $q < p$, and the LSEs of the unknown parameters for the q-component model are strongly consistent. In case $q > p$, where the number of sinusoidal components is overestimated, the vector of LSEs contains a $4p$ dimensional sub-vector that converges to the true values of correct parameters of p sinusoidal components. The remaining $q - p$ components are assigned $q - p$ most dominant periodogram peaks of the noise to minimize the residual sum of squares. Finally, they proposed a family of model selection rules based on the idea of BIC/MDL which produces a strongly consistent estimator of the model order in the 2-D sinusoidal model.

Kundu and Nandi [13] proposed a method based on the eigen-decomposition technique and it avoids estimation of the different parameters for different model orders. It only needs the estimation of error variance for different model orders. The method uses the rank of a Vandermond-type matrix and the information theoretic criteria like AIC and MDL. But instead of using any fixed penalty function, a class of penalty functions satisfying some special properties has been used. It is observed that any penalty function from that particular class provides consistent estimates of the unknown parameter p under the assumptions that the errors are i.i.d. random variables. Further, an estimate of the probability of wrong detection for any particular penalty function has been obtained using the matrix perturbation technique. Once an estimate of the probability of wrong detection has been obtained, that penalty function from the class of penalty functions for which the estimated probability of wrong detection is minimum is used to estimate p. The main feature of this method is that the penalty function depends on the observed data, and it has been observed by extensive numerical studies that the data-dependent penalty function works very well in estimating p.

8.4   Multidimensional Model

The M-dimensional sinusoidal model is given by

$$y(\mathbf{n}) = \sum_{k=1}^{p} A_k \cos(n_1\omega_{1k} + n_2\omega_{2k} + \cdots + n_M\omega_{Mk} + \phi_k) + X(\mathbf{n})$$

$$\text{for} \quad n_1 = 1, \ldots, N_1, n_2 = 1, \ldots, N_2, \ldots, n_M = 1, \ldots, N_M.$$

$$(8.26)$$

Here, $\mathbf{n} = (n_1, n_2, \ldots, n_M)$ is an M-tuple and $y(\mathbf{n})$ is the noise corrupted observed signal. For $k = 1, \ldots, p$, A_k is the unknown real-valued amplitude; $\phi_k \in (0, \pi)$ is the unknown phase; $\omega_{1k} \in (-\pi, \pi), \omega_{2k}, \ldots, \omega_{Mk} \in (0, \pi)$ are the frequencies; $X(\mathbf{n})$ is an M-dimensional sequence of random variables with mean zero and finite variance σ^2; and p is the total number of M-dimensional sinusoidal components. Model (8.26) is a similar model in M-dimension as the 1-D model (3.1) and the 2-D model (8.1). It can be seen by expanding the cosine term and renaming the parameters.

In order to estimate the unknown parameters $A_k, \phi_k, \omega_{1k}, \omega_{2k}, \ldots, \omega_{Mk}, k = 1, \ldots, p$, write $\boldsymbol{\theta} = (\boldsymbol{\theta}_1, \ldots, \boldsymbol{\theta}_p)$, $\boldsymbol{\theta}_k = (A_k, \phi_k, \omega_{1k}, \ldots, \omega_{Mk})$. Then, the least squares method minimizes the following residual sum of squares:

$$Q(\boldsymbol{\theta}) = \sum_{n_1}^{N_1} \cdots \sum_{n_M}^{N_M} \left(y(\mathbf{n}) - \sum_{k=1}^{p} A_k \cos(n_1\omega_{1k} + n_2\omega_{2k} + \cdots + n_M\omega_{Mk} + \phi_k) \right)^2$$

$$(8.27)$$

with respect to the unknown parameters. Now, use the following notation to discuss the asymptotic properties of LSEs of $\boldsymbol{\theta}$:

$$N = N_1 \ldots N_M, \quad N_{(1)} = \min(N_1, \ldots, N_M), \quad N_{(M)} = \max(N_1, \ldots, N_M).$$

Kundu [9] provided the strong consistency and asymptotic normality properties of the LSEs under the assumption that $\{X(\mathbf{n})\}$ is an i.i.d. sequence of M-array real-valued random variables. When $A_1, \ldots, A_p$ are arbitrary real numbers not any one of them identically equal to zero and $\omega_{jk}, j = 1, \ldots, M, k = 1, \ldots, p$ are all distinct, Kundu [9] proved that the LSE $\widehat{\boldsymbol{\theta}}$ of $\boldsymbol{\theta}$ is a strongly consistent estimator as $N_{(M)} \to \infty$. This implies that strong consistency of the LSE can be obtained even if all N_ks are small except one.

The limiting distribution of the LSEs $\widehat{\boldsymbol{\theta}}_1, \ldots, \widehat{\boldsymbol{\theta}}_p$ of $\boldsymbol{\theta}_1, \ldots, \boldsymbol{\theta}_p$ was obtained by Kundu [9] as $N_{(1)} \to \infty$. Define a diagonal matrix $\mathbf{D_M}$ of order $(M + 2)$ as

$$\mathbf{D_M} = \text{diag}\{N^{\frac{1}{2}}, N^{\frac{1}{2}}, N_1 N^{\frac{1}{2}}, \ldots, N_M N^{\frac{1}{2}}\}.$$

Then, as $N_{(1)} \to \infty$, $(\widehat{\boldsymbol{\theta}}_k - \boldsymbol{\theta}_k)\mathbf{D_M} \xrightarrow{d} \mathcal{N}_{M+2}(\mathbf{0}, 2\sigma^2 \boldsymbol{\Sigma}_{kM}^{-1})$ where

$$\Sigma_{kM} = \begin{bmatrix} 1 & 0 & 0 & \cdots & \cdots & 0 \\ 0 & A_k^2 & \frac{1}{2}A_k^2 & & \cdots & \frac{1}{2}A_k^2 \\ 0 & \frac{1}{2}A_k^2 & \frac{1}{3}A_k^2 & \frac{1}{4}A_k^2 & \cdots & \frac{1}{4}A_k^2 \\ 0 & \frac{1}{2}A_k^2 & \frac{1}{4}A_k^2 & \frac{1}{3}A_k^2 & \cdots & \frac{1}{4}A_k^2 \\ \vdots & \vdots & \vdots & & \ddots & \vdots \\ 0 & \frac{1}{2}A_k^2 & \frac{1}{4}A_k^2 & \frac{1}{4}A_k^2 & \cdots & \frac{1}{3}A_k^2 \end{bmatrix} \qquad \Sigma_{kM}^{-1} = \begin{bmatrix} 1 & 0 & 0 & 0 & \cdots & 0 \\ 0 & a_k & b_k & b_k & \cdots & b_k \\ 0 & b_k & c_k & 0 & \cdots & 0 \\ 0 & b_k & 0 & c_k & \cdots & 0 \\ \vdots & \vdots & \vdots & \vdots & \ddots & \vdots \\ 0 & b_k & 0 & 0 & \cdots & c_k \end{bmatrix}$$

$$a_k = \frac{3(M-1)+4}{A_k^2}, \quad b_k = -\frac{6}{A_k^2}, \quad c_k = \frac{1}{A_k^2}\frac{36(M-1)+48}{3(M-1)+4},$$

and $\widehat{\boldsymbol{\theta}}_j$ and $\widehat{\boldsymbol{\theta}}_k$ for $j \neq k$ are asymptotically independently distributed. Observe that the asymptotic variances of $\widehat{\omega}_{1k}, \ldots, \widehat{\omega}_{Mk}$ are all same and are inversely proportional to A_k^2. The asymptotic covariance between $\widehat{\omega}_{ik}$ and $\widehat{\omega}_{jk}$ is zero and as M increases, the variances of both $\widehat{\omega}_{ik}$ and $\widehat{\omega}_{jk}$ converge to $\frac{12}{A_k^2}$.

The periodogram estimator may be considered similar to the 2-D periodogram estimator defined in Sect. 8.2.3. M-dimensional periodogram function may be defined as the 2-D periodogram function, extended from the 1-D periodogram function. It can be proved along the same line as 1-D and 2-D cases that periodogram estimator and LSE are asymptotically equivalent and, therefore, will have the same asymptotic distribution. The results of i.i.d. error can be extended for moving average type errors as well as for stationary linear process in M dimension.

8.5 Conclusions

In this chapter, we have considered the 2-D sinusoidal model and discussed different estimators and their properties. In 2-D case also, the LSEs are the most natural choice and they are the most efficient estimators as well. Unfortunately, finding the LSEs is a difficult problem. Due to this reason, several other estimators which may not be as efficient as the LSEs, but easier to compute, have been suggested in the literature, see, for example, Clark and Sharf [4], Li, Razavilar, and Ray [14], Li, Stoica, and Zheng [15], Sandgren, Stoica, and Frigo [27], Sacchini, Steedy, and Moses [26], and see the references cited therein. Prasad and Kundu [22] proposed the 3-D sinusoidal model which can be used quite effectively to model colored textures. They have established the strong consistency and asymptotic normality properties of the LSEs under the assumptions of stationary errors. No attempt has been made to compute the LSEs efficiently or to find some other estimators which can be obtained more conveniently than the LSEs. Several forms of generalization in case of real as well as complex-valued model was considered by Kundu [9]; a special case has been discussed in Sect. 8.4. It seems most of the 2-D results should be possible to extend to 3-D and higher dimensional cases. Further work is needed along that direction.

References

1. Barbieri, M. M., & Barone, P. (1992). A two-dimensional Prony's algorithm for spectral estimation. *IEEE Transactions on Signal Processing, 40*, 2747–2756.
2. Cabrera, S.D., & Bose, N.K. (1993). Prony's method for two-dimensional complex exponential modeling. In S.G. Tzafestas, Marcel, & Dekker (Eds.) *Applied control theory*, New York, pp. 401 – 411 (Chapter 15).
3. Chun, J., & Bose, N. K. (1995). Parameter estimation via signal selectively of signal subspaces (PESS) and its applications. *Digital Signal Processing, 5*, 58–76.
4. Clark, M. P., & Sharf, L. L. (1994). Two-dimensional modal analysis based on maximum likelihood. *IEEE Transactions on Signal Processing, 42*, 1443–1452.
5. Hua, Y. (1992). Estimating two-dimensional frequencies by matrix enhancement and matrix Pencil. *IEEE Transactions on Signal Processing, 40*, 2267–2280.
6. Kay, S. M. (1998). *Fundamentals of statistical signal processing, Vol II - Detection theory.* New York: Prentice Hall.
7. Kliger, M., & Francos, J. M. (2005). MAP model order selection rule for 2-D sinusoids in white noise. *IEEE Transactions on Signal Processing, 53*(7), 2563–2575.
8. Kliger, M., & Francos, J. M. (2013). Strongly consistent model order selection for estimating 2-D sinusoids in colored noise. *IEEE Transactions on Information Theory, 59*(7), 4408–4422.
9. Kundu, D. (2004). A note on the asymptotic properties of the least squares estimators of the multidimensional exponential signals. *Sankhya, 66*(3), 528–535.
10. Kundu, D., & Gupta, R. D. (1998). Asymptotic properties of the least squares estimators of a two dimensional model. *Metrika, 48*, 83–97.
11. Kundu, D., & Mitra, A. (1995). Consistent method of estimating the superimposed exponential signals. *Scandinavian Journal of Statistics, 22*, 73–82.
12. Kundu, D., & Mitra, A. (1996). Asymptotic properties of the least squares estimates of 2-D exponential signals. *Multidimensional Systems and Signal Processing, 7*, 135–150.
13. Kundu, D., & Nandi, S. (2003). Determination of discrete spectrum in a random field. *Statistica Neerlandica, 57*, 258–283.
14. Li, Y., Razavilar, J., & Ray, L. K. J. (1998). A high-resolution technique for multidimensional NMR spectroscopy. *IEEE Transactions on Biomedical Engineering, 45*, 78–86.
15. Li, J., Stoica, P., & Zheng, D. (1996). An efficient algorithm for two-dimensional frequency estimation. *Multidimensional Systems and Signal Processing, 7*, 151–178.
16. Malliavin, P. (1994a). Sur la norme d'une matrice circulate Gaussienne, C.R. Acad. Sc. Paris t, 319 Serie, I, pp. 45–49.
17. Malliavin, P. (1994b). Estimation d'un signal Lorentzien. C.R. Acad. Sc. Paris t, 319. Serie, I, 991–997.
18. Miao, B. Q., Wu, Y., & Zhao, L. C. (1998). On strong consistency of a 2-dimensional frequency estimation algorithm. *Statistica Sinica, 8*, 559–570.
19. Mitra, A., & Stoica, P. (2002). The asymptotic Cramer-Rao bound for 2-D superimposed exponential signals. *Multidimensional Systems and Signal Processing, 13*, 317–331.
20. Nandi, S., Kundu, D., & Srivastava, R. K. (2013). Noise space decomposition method for two-dimensional sinusoidal model. *Computational Statistics and Data Analysis, 58*, 147–161.
21. Nandi, S., Prasad, A., & Kundu, D. (2010). An efficient and fast algorithm for estimating the parameters of two-dimensional sinusoidal signals. *Journal of Statistical Planning and Inference, 140*, 153–168.
22. Prasad, A., & Kundu, D. (2009). Modeling and estimation of symmetric color textures. *Sankhya, Ser B, 71*, 30–54.
23. Prasad, A., Kundu, D., & Mitra, A. (2012). Sequential estimation of two dimensional sinusoidal models. *Journal of Probability and Statistics, 10*(2), 161–178.
24. Rao, C. R., Zhao, L. C., & Zhou, B. (1994). Maximum likelihood estimation of 2-D superimposed exponential signals. *IEEE Transactions on Signal Processing, 42*, 1795–1802.
25. Richards, F. S. G. (1961). A method of maximum likelihood estimation. *Journal of the Royal Statistical Society, B*, 469–475.

26. Sacchini, J. J., Steedy, W. M., & Moses, R. L. (1993). Two-dimensional Prony modeling and parameter estimation. *IEEE Transactions on Signal Processing, 41*, 3127–3136.
27. Sandgren, N., Stoica, P., & Frigo, F. J. (2006). Area-selective signal parameter estimation for two-dimensional MR spectroscopy data. *Journal of Magnetic Resonance, 183*, 50–59.
28. Zhang, H., & Mandrekar, V. (2001). Estimation of hidden frequencies for 2D stationary processes. *Journal of Time Series Analysis, 22*, 613–629.

Chapter 9
Chirp Signal Model

9.1 Introduction

A real-valued one-dimensional (1-D) chirp signal in additive noise can be written mathematically as follows:

$$y(t) = A\cos(\alpha t + \beta t^2) + B\sin(\alpha t + \beta t^2) + X(t); \quad t = 1, \ldots, n. \quad (9.1)$$

Here $y(t)$ is a real-valued signal observed at $t = 1, \ldots, n$; A, B are amplitudes; α and β are the frequency and the frequency rate, respectively. The additive noise $X(t)$ has mean zero, the explicit structure of the error process $\{X(t)\}$ will be discussed later on. This model arises in many applications of signal processing; one of the most important being the radar problem. For instance, consider a radar illuminating a target. Thus, the transmitted signal will undergo a phase shift induced by the distance and relative motion between the target and the receiver. Assuming this motion to be continuous and differentiable, the phase shift can be adequately modeled as $\phi(t) = a_0 + a_1 t + a_2 t^2$, where the parameters a_1 and a_2 are either related to speed and acceleration or range and speed, depending on what the radar is intended for, and on the kind of waveform transmitted, see for example Rihaczek [50] (p. 56–65).

Chirp signals are encountered in many different engineering applications, particularly in radar, active sonar, and passive sonar systems. The problem of parameter estimation of chirp signals has received a considerable amount of attention mainly in the engineering literature; see for example, Abatzoglou [1], Djurić and Kay [8], Saha and Kay [53], Gini, Montanari, and Verrazzani [19], Besson, Giannakis, and Gini [4], Volcker and Otterstern [56], Lu et al. [40], Wang and Yang [58], Guo et al. [23], Yaron, Alon, and Israel [62], Yang, Liu, and Jiang [63], Fourier et al. [14], Gu et al. [22] and the references cited therein. A limited amount of work can be found in the statistical literature, interested readers are referred to the following; Nandi and Kundu [43], Kundu and Nandi [31], Robertson [51], Robertson, Gray, and Woodward [52], Lahiri, Kundu, and Mitra [34, 36, 37], Kim et al. [29], Mazumder [41], and Grover, Kundu, and Mitra [20, 21].

© Springer Nature Singapore Pte Ltd. 2020
S. Nandi and D. Kundu, *Statistical Signal Processing*,
https://doi.org/10.1007/978-981-15-6280-8_9

An alternative formulation of model (9.1), is also available mainly in the engineering literature, and that is as follows:

$$y(t) = Ae^{i(\alpha t + \beta t^2)} + X(t); \quad t = 1, \ldots, n. \tag{9.2}$$

Here $y(t)$ is the complex-valued signal; the amplitude A and the error component $X(t)$ are also complex-valued; α, β are same as defined before. Although all physical signals are real-valued, it might be advantageous from an analytical, a notational or an algorithmic point of view to work with signals in their analytic form which is complex-valued, see, for example, Gabor [18]. For a real-valued continuous signal, its analytic form can be easily obtained using the Hilbert transformation. Hence, it is equivalent to work either with model (9.1) or (9.2). In this chapter, we concentrate on real-valued chirp model only.

Observe that the chirp model (9.1), is a generalization of the well-known sinusoidal frequency ;

$$y(t) = A \cos(\alpha t) + B \sin(\alpha t) + X(t); \quad t = 1, \ldots, n, \tag{9.3}$$

where, A, B, α, and $X(t)$ are same as in (9.1). In sinusoidal frequency model; (9.3), the frequency α is constant over time and does not change like chirp signal model (9.1). In chirp signal model, frequency changes linearly over time. A more general form of (9.3) can be written as

$$y(t) = \sum_{k=1}^{p} [A_k \cos(\alpha_k t) + B_k \sin(\alpha_k t)] + X(t); \quad t = 1, \ldots, n. \tag{9.4}$$

It is clear that model (9.3), can be written as a special case of model (9.1). Moreover, model (9.1), can be treated as a frequency modulated sinusoidal model of (9.3), also.

In this chapter, we introduce the 1-D chirp model (9.1), and the more general multicomponent chirp model as defined below;

$$y(t) = \sum_{k=1}^{p} [A_k \cos(\alpha_k t + \beta_k t^2) + B_k \sin(\alpha_k t + \beta_k t^2)] + X(t); \quad t = 1, \ldots, n,$$
$$\tag{9.5}$$

and some related models. We discuss different issues related to these models, mainly concentrate on the estimation of the unknown parameters and establishing their properties. One major difficulty associated with chirp signal is the estimation of the unknown parameters. Although model (9.1), can be seen as a nonlinear regression model, the model does not satisfy the sufficient conditions of Jennrich [27], or Wu [60], required for establishing the consistency and asymptotic normality of the least squares estimators (LSEs) of the unknown parameters of a standard nonlinear regression model. Therefore, the consistency and the asymptotic normality of the LSEs or the maximum likelihood estimators (MLEs) are not immediate. Hence, similar to the sinusoidal model, in this case also it needs to be established independently. Moreover, the chirp signal model is a highly nonlinear model in its parameters, hence finding the MLEs or the LSEs becomes a nontrivial problem. An extensive amount of work has been done to compute different efficient estimators, and in deriving their properties under different error assumptions. We provide different classical and Bayesian estimation procedures available till date and discuss their properties.

The rest of the chapter is organized as follows. In Sect. 9.2, we provide details of one-dimensional chirp model. Two-dimensional chirp and two-dimensional polynomial phase models are presented in Sect. 9.3 and Sect. 9.4, respectively. Finally, we point out some related models which are discussed in details in Chap. 11, and draw conclusion in Sect. 9.5.

9.2   One-Dimensional Chirp Model

In this section, first we discuss different issues involved with single chirp model, and then the multiple chirp model will be discussed.

9.2.1   Single Chirp Model

The one-dimensional single chirp model can be written as follows:

$$y(t) = A^0 \cos(\alpha^0 t + \beta^0 t^2) + B^0 \sin(\alpha^0 t + \beta^0 t^2) + X(t); \quad t = 1, \ldots, n. \quad (9.6)$$

Here $y(t)$ is the real-valued signal as mentioned before, and it is observed at $t = 1, \ldots, n$. A^0, B^0 are real-valued amplitudes; α^0, β^0 are the frequency and the frequency rate, respectively. The problem is to estimate the unknown parameters namely A^0, B^0, α^0, and β^0, based on the observed sample. Different methods have been proposed in the literature. We provide different methods of estimation and discuss theoretical properties of these estimators. First, it is assumed that the errors $X(t)$s are i.i.d. normal random variables with mean 0 and variance σ^2. A more general form of the error random variable $X(t)$ will be considered later on.

9.2.1.1   Maximum Likelihood Estimators

We need to obtain the log-likelihood function of the observed data to compute the MLEs. The log-likelihood function without the additive constant can be written as

$$l(\boldsymbol{\Theta}) = -\frac{n}{2} \ln \sigma^2 - \frac{1}{2\sigma^2} \sum_{t=1}^{n} \left(y(t) - A \cos(\alpha t + \beta t^2) - B \sin(\alpha t + \beta t^2) \right)^2,$$

$$(9.7)$$

where $\boldsymbol{\Theta} = (A, B, \alpha, \beta, \sigma^2)^T$. Hence, the MLEs of A^0, B^0, α^0, β^0, and σ^2, say $\widehat{A}$, $\widehat{B}$, $\widehat{\alpha}$, $\widehat{\beta}$, and $\widehat{\sigma}^2$, respectively, can be obtained by maximizing $l(\boldsymbol{\Theta})$, with respect to the unknown parameters. It is immediate that $\widehat{A}$, $\widehat{B}$, $\widehat{\alpha}$, and $\widehat{\beta}$ can be obtained by minimizing $Q(\boldsymbol{\Gamma})$, where, $\boldsymbol{\Gamma} = (A, B, \alpha, \beta)^T$ and

$$Q(\boldsymbol{\Gamma}) = \sum_{t=1}^{n} \left(y(t) - A\cos(\alpha t + \beta t^2) - B\sin(\alpha t + \beta t^2) \right)^2, \qquad (9.8)$$

with respect to the unknown parameters. Therefore, as it is expected under normality assumption, $\widehat{A}$, $\widehat{B}$, $\widehat{\alpha}$, and $\widehat{\beta}$ are the LSEs of the corresponding parameters also. Once, $\widehat{A}$, $\widehat{B}$, $\widehat{\alpha}$, and $\widehat{\beta}$ are obtained, $\widehat{\sigma}^2$ can be obtained as

$$\widehat{\sigma}^2 = \frac{1}{n} \sum_{t=1}^{n} \left(y(t) - \widehat{A}\cos(\widehat{\alpha} t + \widehat{\beta} t^2) - \widehat{B}\sin(\widehat{\alpha} t + \widehat{\beta} t^2) \right)^2. \qquad (9.9)$$

In the following, we first provide the procedure how to obtain $\widehat{A}$, $\widehat{B}$, $\widehat{\alpha}$, and $\widehat{\beta}$, and then we discuss about their asymptotic properties. Observe that, $Q(\boldsymbol{\Gamma})$ can be written as

$$Q(\boldsymbol{\Gamma}) = [\mathbf{Y} - \mathbf{W}(\alpha, \beta)\boldsymbol{\theta}]^T [\mathbf{Y} - \mathbf{W}(\alpha, \beta)\boldsymbol{\theta}], \qquad (9.10)$$

using the notation $\mathbf{Y} = (y(1), \ldots, y(n))^T$ as the data vector, $\boldsymbol{\theta} = (A, B)^T$, the linear parameter vector and

$$\mathbf{W}(\alpha, \beta) = \begin{bmatrix} \cos(\alpha + \beta) & \sin(\alpha + \beta) \\ \cos(2\alpha + 4\beta) & \sin(2\alpha + 4\beta) \\ \vdots & \vdots \\ \cos(n\alpha + n^2\beta) & \sin(n\alpha + n^2\beta) \end{bmatrix}, \qquad (9.11)$$

the matrix with nonlinear parameters. From (9.10), it is immediate that for fixed (α, β), the MLE of $\boldsymbol{\theta}^0$, say $\widehat{\boldsymbol{\theta}}(\alpha, \beta)$, can be obtained as

$$\widehat{\boldsymbol{\theta}}(\alpha, \beta) = \left[\mathbf{W}^T(\alpha, \beta)\mathbf{W}(\alpha, \beta) \right]^{-1} \mathbf{W}^T(\alpha, \beta)\mathbf{Y}, \qquad (9.12)$$

and the MLEs of α and β can be obtained as

$$(\widehat{\alpha}, \widehat{\beta}) = \arg\min_{\alpha, \beta} Q(\widehat{A}(\alpha, \beta), \widehat{B}(\alpha, \beta), \alpha, \beta). \qquad (9.13)$$

Replacing $\boldsymbol{\theta}$ by $\widehat{\boldsymbol{\theta}}(\alpha, \beta)$ in (9.10), it can be easily seen that

$$(\widehat{\alpha}, \widehat{\beta}) = \arg\max_{\alpha, \beta} Z(\alpha, \beta), \qquad (9.14)$$

where

$$Z(\alpha, \beta) = \mathbf{Y}^T \mathbf{W}(\alpha, \beta)[\mathbf{W}^T(\alpha, \beta)\mathbf{W}(\alpha, \beta)]^{-1}\mathbf{W}^T(\alpha, \beta)\mathbf{Y} = \mathbf{Y}^T \mathbf{P}_{\mathbf{W}(\alpha,\beta)}\mathbf{Y}. \quad (9.15)$$

Here, $\mathbf{P}_{\mathbf{W}(\alpha,\beta)}$ is the projection matrix on the column space of $\mathbf{W}(\alpha, \beta)$. Then, the MLE of θ^0 can be obtained as $\widehat{\theta} = \widehat{\theta}(\widehat{\alpha}, \widehat{\beta})$. Clearly, $\widehat{\alpha}$ and $\widehat{\beta}$ cannot be obtained analytically. Different numerical methods may be used to compute $\widehat{\alpha}$ and $\widehat{\beta}$. Saha and Kay [53], proposed to use the method of Pincus [49], in computing $\widehat{\alpha}$ and $\widehat{\beta}$, and they can be described as follows. Using the main theorem of Pincus [49], it follows that

$$\widehat{\alpha} = \lim_{c \to \infty} \frac{\int_0^\pi \int_0^\pi \alpha \exp(cZ(\alpha, \beta))d\beta d\alpha}{\int_0^\pi \int_0^\pi \exp(cZ(\alpha, \beta))d\beta d\alpha} \quad \text{and} \quad \widehat{\beta} = \lim_{c \to \infty} \frac{\int_0^\pi \int_0^\pi \beta \exp(cZ(\alpha, \beta))d\beta d\alpha}{\int_0^\pi \int_0^\pi \exp(cZ(\alpha, \beta))d\beta d\alpha}. \quad (9.16)$$

Therefore, one needs to compute two-dimensional integration to compute the MLEs. Alternatively, importance sampling technique can be used quite effectively in this case to compute the MLEs of α^0 and β^0 using the following algorithm.

The following simple algorithm of the importance sampling method can be used to compute $\widehat{\alpha}$ and $\widehat{\beta}$ as defined in (9.16).

Algorithm 9.1

- Step 1: Generate $\alpha_1, \ldots, \alpha_M$ from uniform$(0, \pi)$ and similarly, generate $\beta_1, \ldots, \beta_M$ from uniform$(0, \pi)$.
- Step 2: Consider a sequence of $\{c_k\}$, such that $c_1 < c_2 < c_3 < \cdots ,$. For fixed $c = c_k$, compute

$$\widehat{\alpha}(c) = \frac{\frac{1}{M}\sum_{i=1}^M \alpha_i \exp(cZ(\alpha_i, \beta_i))}{\frac{1}{M}\sum_{i=1}^M \exp(cZ(\alpha_i, \beta_i))} \quad \text{and} \quad \widehat{\beta}(c) = \frac{\frac{1}{M}\sum_{i=1}^M \beta_i \exp(cZ(\alpha_i, \beta_i))}{\frac{1}{M}\sum_{i=1}^M \exp(cZ(\alpha_i, \beta_i))}.$$

Stop the iteration if the convergence takes place. ∎

It may be noted that one can use the same α_n's and β_n's to compute $\widehat{\alpha}(c_k)$ and $\widehat{\beta}(c_k)$ for each k. Other methods like Newton–Raphson, Gauss–Newton or Downhill simplex methods may be used to compute the MLEs in this case. But one needs very good initial guesses for any iterative process to converge. The $Z(\alpha, \beta)$ surface for $0 < \alpha, \beta < \pi$ has several local maxima, hence any iterative process without very good initial guesses often converges to a local maximum rather than a global maximum.

Now we discuss the properties of the MLEs of the unknown parameters. Model (9.6), can be seen as a typical nonlinear regression model with an additive error, where the errors are i.i.d. normal random variables. For different nonlinear regression models, see, for example, Seber and Wild [54]. Jennrich [27] and Wu [60], developed several sufficient conditions under which the MLEs of the unknown parameters in a nonlinear regression model with additive Gaussian errors are consistent and asymptotically normally distributed. Unfortunately, the sufficient conditions proposed by Jennrich [27] or Wu [60], do not hold here. Therefore, the consistency and the asymptotic normality of the MLEs are not immediate.

Kundu and Nandi [31], developed the consistency and asymptotic normality results of the MLEs under certain regularity conditions and they will be presented below.

Theorem 9.1 *If there exists a K, such that* $0 < |A^0| + |B^0| < K$, $0 < \alpha^0$, $\beta^0 < \pi$, *and* $\sigma^2 > 0$, *then* $\widehat{\Theta} = (\widehat{A}, \widehat{B}, \widehat{\alpha}, \widehat{\beta}, \widehat{\sigma}^2)^T$ *is a strongly consistent estimate of* $\Theta^0 = (A^0, B^0, \alpha^0, \beta^0, \sigma^2)^T$. ∎

Along with the consistency results, the asymptotic normality properties of the MLEs of A^0, B^0, α^0, and β^0 have been obtained by Kundu and Nandi [37], but the elements of the asymptotic variance covariance matrix look quite complicated. Lahiri [32], (see also Lahiri, Kundu, and Mitra [37]) establish the following result using a number theoretic result which simplifies the entries of the asymptotic variance covariance matrix.

Result 9.1 *If* $(\xi, \eta) \in (0, \pi) \times (0, \pi)$, *then except for countable number of points, the following results are true*

$$\lim_{n \to \infty} \frac{1}{n} \sum_{t=1}^{n} \cos(\xi t + \eta t^2) = \lim_{n \to \infty} \frac{1}{n} \sum_{t=1}^{n} \sin(\xi t + \eta t^2) = 0,$$

$$\lim_{n \to \infty} \frac{1}{n^{k+1}} \sum_{t=1}^{n} t^k \cos^2(\xi t + \eta t^2) = \lim_{n \to \infty} \frac{1}{n^{k+1}} \sum_{t=1}^{n} t^k \sin^2(\xi t + \eta t^2) = \frac{1}{2(k+1)},$$

$$\lim_{n \to \infty} \frac{1}{n^{k+1}} \sum_{t=1}^{n} t^k \sin(\xi t + \eta t^2) \cos(\xi t + \eta t^2) = 0,$$

where $k = 0, 1, 2, \ldots$. ∎

Interestingly, it is observed that the rate of convergence of the nonlinear parameters are quite different than the linear ones. The results are presented in the following theorem, and for the proofs of these theorems, see Kundu and Nandi [31], and Lahiri [32].

Theorem 9.2 *Under the same assumptions as in Theorem 9.1,*

$$\left(n^{1/2}(\widehat{A} - A^0), n^{1/2}(\widehat{B} - B^0), n^{3/2}(\widehat{\alpha} - \alpha^0), n^{5/2}(\widehat{\beta} - \beta^0)\right)^T \xrightarrow{d} \mathcal{N}_4(\mathbf{0}, 2\sigma^2 \boldsymbol{\Sigma}),$$
(9.17)

here $\xrightarrow{d}$ *means convergence in distribution, and* $\boldsymbol{\Sigma}$ *is given by*

$$\boldsymbol{\Sigma} = \frac{2}{A^{0^2} + B^{0^2}} \begin{bmatrix} \frac{1}{2}(A^{0^2} + 9B^{0^2}) & -4A^0 B^0 & -18B^0 & 15B^0 \\ -4A^0 B^0 & \frac{1}{2}(9A^{0^2} + B^{0^2}) & 18A^0 & -15A^0 \\ -18B^0 & 18A^0 & 96 & -90 \\ 15B^0 & -15A^0 & -90 & 90 \end{bmatrix}.$$

∎

Theorem 9.1 establishes the consistency of the MLEs, whereas Theorem 9.2 states the asymptotic distribution along with the rate of convergence of the MLEs. It is interesting to observe that the rates of convergence of the MLEs of the linear parameters are significantly different than the corresponding nonlinear parameters. The nonlinear parameters are estimated more efficiently than the linear parameters for a given sample size. Moreover, as $A^{0^2} + B^{0^2}$ decreases, the asymptotic variances of the MLEs increase.

Theorem 9.1 provides the consistency results of the MLEs under the boundedness assumptions on the linear parameters. The following open problem might be of interest.

Open Problem 9.1 Develop the consistency properties of the MLEs without any boundedness assumptions on the linear parameters. ∎

Theorem 9.2 provides the asymptotic distributions of the MLEs. The asymptotic distribution may be used to construct approximate confidence intervals of the unknown parameters, and also to develop testing of hypotheses problems. So far the construction of confidence intervals or testing of hypotheses problems have not been considered in the literature. The following open problem has some practical applications.

Open Problem 9.2 Construct different bootstrap confidence intervals like percentile bootstrap, biased corrected bootstrap, etc. and compare them with the approximate confidence intervals based on asymptotic distributions of the MLEs. ∎

From Theorem 9.2, it is observed that the rate of convergence of the MLEs of the linear parameters is $O_p(n^{-1/2})$, whereas the rates of convergence of the MLEs for the frequency and frequency rate are, $O_p(n^{-3/2})$ and $O_p(n^{-5/2})$, respectively. Hence, the MLEs for the frequency and frequency rate converge faster than the linear parameters. These results are quite different than the usual rate of convergence of the MLEs for a general nonlinear regression model, see for example, Seber and Wild [54]. Although we have provided the results for the MLEs when the errors are normally distributed, it is observed that the consistency and the asymptotic normality results are valid for the general LSEs when the errors are i.i.d. random variables with mean zero and finite variance σ^2. Now we discuss the properties of the LSEs for more general errors.

9.2.1.2 Least Squares Estimators

In the last section, we have discussed about the estimation of the unknown parameters when the errors are normally distributed. It may be mentioned that when the errors are normally distributed, then the LSEs and the MLEs are the same. In this section, we discuss about the estimation of the unknown parameters when the errors may not be normally distributed. The following assumption on $\{X(t)\}$ has been made.

Assumption 9.1 Suppose

$$X(t) = \sum_{j=-\infty}^{\infty} a(j)e(t-j), \tag{9.18}$$

where $\{e(t)\}$ is a sequence of i.i.d. random variables with mean zero and finite fourth moment, and

$$\sum_{j=-\infty}^{\infty} |a(j)| < \infty. \tag{9.19}$$

Assumption 9.1 is a standard assumption for a stationary linear process and similar to Assumption 3.2. Any finite dimensional stationary AR, MA or ARMA process can be represented as (9.18), when $a(j)$s are absolutely summable, that is, satisfy (9.19).

Kundu and Nandi [31], first discussed the estimation of the unknown parameters of model (9.1), when $\{X(t)\}$ satisfies Assumption 9.1. Clearly, the MLEs are not possible to obtain as the exact distribution of the error process is not known. Kundu and Nandi [31], considered the LSEs of the unknown parameters and they can be obtained by minimizing $Q(\Gamma)$ as defined in (9.8), with respect to Γ. Hence, the same computational procedure which has been used to compute the MLEs can be used here also. But the consistency and the asymptotic normality of the LSEs need to be developed independently. Kundu and Nandi [31], developed the following two results similar to Theorems 9.1 and 9.2.

Theorem 9.3 *If there exists a K, such that* $0 < |A^0| + |B^0| < K$, $0 < \alpha^0, \beta^0 < \pi$, $\sigma^2 > 0$ *and* $\{X(t)\}$ *satisfies Assumption 9.1, then the LSE* $\widehat{\Gamma} = (\widehat{A}, \widehat{B}, \widehat{\alpha}, \widehat{\beta})^T$ *is a strongly consistent estimate of* $\Gamma^0 = (A^0, B^0, \alpha^0, \beta^0)^T$. ∎

Theorem 9.4 *Under the same assumptions as in Theorem 9.3,*

$$\left(n^{1/2}(\widehat{A} - A^0), n^{1/2}(\widehat{B} - B^0), n^{3/2}(\widehat{\alpha} - \alpha^0), n^{5/2}(\widehat{\beta} - \beta^0)\right)^T \xrightarrow{d} \mathcal{N}_4(\mathbf{0}, 2c\sigma^2 \Sigma), \tag{9.20}$$

here Σ *is same as defined in Theorem 9.2, and* $c = \sum_{j=-\infty}^{\infty} a^2(j)$. ∎

Note that Theorem 9.4, provides the rates of convergence of the LSEs under the assumption of stationary linear process on the error. This asymptotic distribution stated in Theorem 9.4, can also be used to construct asymptotic confidence intervals of the unknown parameters and in developing testing of hypotheses provided one can obtain a good estimate of $c\sigma^2$. In this respect, the method proposed by Nandi and Kundu [44], may be used. The following open problems will be of interest.

Open Problem 9.3 Construct confidence intervals based on the asymptotic distributions of the LSEs and develop different bootstrap confidence intervals also. ∎

Open Problem 9.4 Develop different testing of hypotheses based on the LSEs associated with the chirp model parameters. ∎

In the next subsection, we discuss the properties of the least absolute deviation (LAD) estimators, which are more robust than the LSEs, if the errors are heavy-tailed or there are outliers in the data.

9.2.1.3 Least Absolute Deviation Estimators

In this section, we discuss about LAD estimators of A^0, B^0, α^0, and β^0. The LAD estimators of the unknown parameters can be obtained by minimizing

$$R(\boldsymbol{\Gamma}) = \sum_{t=1}^{n} |y(t) - A\cos(\alpha t + \beta t^2) - B\sin(\alpha t + \beta t^2)|. \qquad (9.21)$$

Here $\boldsymbol{\Gamma}$ is same as defined before. With the abuse of notations, we denote the LAD estimators of A^0, B^0, α^0, and β^0 by $\widehat{A}$, $\widehat{B}$, $\widehat{\alpha}$, and $\widehat{\beta}$, respectively. They can be obtained as

$$(\widehat{A}, \widehat{B}, \widehat{\alpha}, \widehat{\beta}) = \arg\min R(\boldsymbol{\Gamma}). \qquad (9.22)$$

The minimization of $R(\boldsymbol{\Gamma})$ with respect to the unknown parameters is a challenging problem. Even when α and β are known, unlike the LSEs of A and B, the LAD estimators of A and B cannot be obtained in closed form. For known α and β, the LAD estimators of A and B can be obtained by solving a $2n$ dimensional linear programming problem, which is quite intensive computationally, see, for example, Kennedy and Gentle [28]. Hence, finding the LAD estimators of $\boldsymbol{\Gamma}$ efficiently, is a challenging problem. Assume that $-K \le A, B \le K$, for some $K > 0$, then grid search method may be used to compute the LAD estimators in this case. But clearly, it is also a very computationally intensive method. The importance sampling technique as it was used to find the MLEs can be used here also, and the LAD estimators can be obtained as follows:

$$\widehat{A} = \lim_{c \to \infty} \frac{\int_{-K}^{K} \int_{-K}^{K} \int_{0}^{\pi} \int_{0}^{\pi} A \exp(cR(\boldsymbol{\Gamma})) d\beta d\alpha dA dB}{\int_{-K}^{K} \int_{-K}^{K} \int_{0}^{\pi} \int_{0}^{\pi} \exp(cR(\boldsymbol{\Gamma})) d\beta d\alpha dA dB},$$

$$\widehat{B} = \lim_{c \to \infty} \frac{\int_{-K}^{K} \int_{-K}^{K} \int_{0}^{\pi} \int_{0}^{\pi} B \exp(cR(\boldsymbol{\Gamma})) d\beta d\alpha dA dB}{\int_{-K}^{K} \int_{-K}^{K} \int_{0}^{\pi} \int_{0}^{\pi} \exp(cR(\boldsymbol{\Gamma})) d\beta d\alpha dA dB},$$

$$\widehat{\alpha} = \lim_{c \to \infty} \frac{\int_{-K}^{K} \int_{-K}^{K} \int_{0}^{\pi} \int_{0}^{\pi} \alpha \exp(cR(\boldsymbol{\Gamma})) d\beta d\alpha dA dB}{\int_{-K}^{K} \int_{-K}^{K} \int_{0}^{\pi} \int_{0}^{\pi} \exp(cR(\boldsymbol{\Gamma})) d\beta d\alpha dA dB},$$

$$\widehat{\beta} = \lim_{c \to \infty} \frac{\int_{-K}^{K} \int_{-K}^{K} \int_{0}^{\pi} \int_{0}^{\pi} \beta \exp(cR(\boldsymbol{\Gamma})) d\beta d\alpha dA dB}{\int_{-K}^{K} \int_{-K}^{K} \int_{0}^{\pi} \int_{0}^{\pi} \exp(cR(\boldsymbol{\Gamma})) d\beta d\alpha dA dB}.$$

Therefore, in this case also, using similar algorithm as Algorithm 9.1, one can obtain the LAD estimators of the unknown parameters. It will be interesting to see performance of the above estimators based on simulations. More work is needed in that direction. Efficient technique of computing the LAD estimators is not available till date, but Lahiri, Kundu, and Mitra [36], established the asymptotic properties of the LAD estimators under some regularity conditions. The results can be stated as follows, the proofs are available in Lahiri, Kundu, and Mitra [36] and Lahiri [32].

Assumption 9.2 The error random variable $\{X(t)\}$ is a sequence of i.i.d. random variables with mean zero, variance $\sigma^2 > 0$, and it has a probability density function (PDF) $f(x)$. The PDF $f(x)$ is symmetric and differentiable in $(0, \varepsilon)$ and $(-\varepsilon, 0)$, for some $\varepsilon > 0$, and $f(0) > 0$.

Theorem 9.5 *If there exists a K, such that $0 < |A^0| + |B^0| < K$, $0 < \alpha^0, \beta^0 < \pi$, $\sigma^2 > 0$, and $\{X(t)\}$ satisfies Assumption 9.2, then $\widehat{\boldsymbol{\Gamma}} = (\widehat{A}, \widehat{B}, \widehat{\alpha}, \widehat{\beta})^T$ is a strongly consistent estimate of $\boldsymbol{\Gamma}^0$.* ∎

Theorem 9.6 *Under the same assumptions as in Theorem 9.5*

$$\left(n^{1/2}(\widehat{A} - A^0), n^{1/2}(\widehat{B} - B^0), n^{3/2}(\widehat{\alpha} - \alpha^0), n^{5/2}(\widehat{\beta} - \beta^0)\right)^T \xrightarrow{d} \mathcal{N}_4(0, \frac{1}{f^2(0)}\boldsymbol{\Sigma}),$$

(9.23)

here $\boldsymbol{\Sigma}$ is same as defined in (9.17). ∎

Therefore, it is observed that the LSEs and LAD estimators both provide consistent estimators of the unknown parameters, and they have the same rate of convergence. Similarly, as in Theorem 9.2, Theorem 9.6 also can be used to construct approximate confidence intervals of the unknown parameters and to develop different testing of hypotheses provided one can obtain a good estimate of $f^2(0)$. The kernel density estimator of the estimated error can be used for numerical experiments involving confidence intervals. It will be important to see the performances of the approximate confidence intervals in terms of the coverage percentages and confidence lengths. The following problems related to LAD estimators will be of interest.

Open Problem 9.5 Develop an efficient algorithm to compute the LAD estimators for the chirp parameters under different error assumptions. ∎

Open Problem 9.6 Construct confidence intervals of the unknown parameters of the chirp model and develop the testing of hypotheses based on the LAD estimators. ∎

9.2.1.4 Finite Step Efficient Algorithm

In previous sections of this chapter, it has been observed that the MLEs, LSEs, and LAD estimators provide consistent estimators of the frequency and frequency rate, but all of them need to be computed by using some optimization technique or the importance sampling method described before. Any optimization method involving nonlinear models needs to be solved using some iterative procedure, and any iterative procedure has its own problem of convergence. In this section, we present an algorithm where the frequency and frequency rate estimators can be obtained by using an iterative procedure. The interesting point of this algorithm is that it converges in a fixed number of iterations, and at the same time both the frequency and frequency rate estimators attain the same rate of convergence as the MLEs or LAD estimators. The main idea of this algorithm came from a finite step efficient algorithm in case of sinusoidal signals, proposed by Bai et al. [3]. It is observed that if we start the initial guesses of α^0 and β^0 with convergence rates $O_p(n^{-1})$ and $O_p(n^{-2})$, respectively, then after four iterations, the algorithm produces an estimate of α^0 with convergence rate $O_p(n^{-3/2})$, and an estimate of β^0 with convergence rate $O_p(n^{-5/2})$. Before providing the algorithm in details first we show how to improve the estimators of α^0 and β^0. Then, the exact computational algorithm for practical implementation will be provided. If $\widetilde{\alpha}$ is an estimator of α^0, such that for $\delta_1 > 0$, $\widetilde{\alpha} - \alpha^0 = O_p(n^{-1-\delta_1})$, and $\widetilde{\beta}$ is an estimator of β^0, such that for $\delta_2 > 0$, $\widetilde{\beta} - \beta^0 = O_p(n^{-2-\delta_2})$, then the improved estimators of α^0 and β^0, can be obtained as

$$\widetilde{\widetilde{\alpha}} = \widetilde{\alpha} + \frac{48}{n^2} Im\left(\frac{P_n^\alpha}{Q_n}\right) \tag{9.24}$$

$$\widetilde{\widetilde{\beta}} = \widetilde{\beta} + \frac{45}{n^4} Im\left(\frac{P_n^\beta}{Q_n}\right), \tag{9.25}$$

respectively, where

$$P_n^\alpha = \sum_{t=1}^{n} y(t)\left(t - \frac{n}{2}\right) e^{-i(\widetilde{\alpha}t + \widetilde{\beta}t^2)},$$

$$P_n^\beta = \sum_{t=1}^{n} y(t)\left(t^2 - \frac{n^2}{3}\right) e^{-i(\widetilde{\alpha}t + \widetilde{\beta}t^2)},$$

$$Q_n^\alpha = \sum_{t=1}^{n} y(t) e^{-i(\widetilde{\alpha}n + \widetilde{\beta}n^2)}.$$

Here, if C is a complex number, then $Im(C)$ means the imaginary part of C.

The following two theorems provide the justification for the improved estimators, and the proofs of these two theorems can be obtained in Lahiri, Kundu, and Mitra [34].

Theorem 9.7 *If* $\widetilde{\alpha} - \alpha^0 = O_p(n^{-1-\delta_1})$ *for* $\delta_1 > 0$, *then*

$$(a)\ (\widetilde{\widetilde{\alpha}} - \alpha^0) = O_p(n^{-1-2\delta_1}) \qquad if\ \delta_1 \leq 1/4,$$
$$(b)\ n^{3/2}(\widetilde{\widetilde{\alpha}} - \alpha^0) \xrightarrow{d} \mathcal{N}(0, \sigma_1^2)\ if\ \delta_1 > 1/4,$$

where $\sigma_1^2 = \dfrac{384\sigma^2}{A^{0^2} + B^{0^2}}$.   ∎

Theorem 9.8 *If* $\widetilde{\beta} - \beta^0 = O_p(n^{-2-\delta_2})$ *for* $\delta_2 > 0$, *then*

$$(a)\ (\widetilde{\widetilde{\beta}} - \beta^0) = O_p(n^{-2-2\delta_2}) \qquad if\ \delta_2 \le 1/4,$$
$$(b)\ n^{5/2}(\widetilde{\widetilde{\beta}} - \beta^0) \xrightarrow{d} \mathcal{N}(0, \sigma_2^2)\ if\ \delta_1 > 1/4,$$

where $\sigma_2^2 = \dfrac{360\sigma^2}{A^{0^2} + B^{0^2}}$.   ∎

Implementation: Now we show that starting from initial guesses $\widetilde{\alpha}, \widetilde{\beta}$ with convergence rates $\widetilde{\alpha} - \alpha^0 = O_p(n^{-1})$ and $\widetilde{\beta} - \beta^0 = O_p(n^{-2})$, respectively, how the above procedure can be used to obtain efficient estimators. We note that finding initial guesses with the above convergence rates is not difficult. Idea is similar to Nandi and Kundu algorithm discussed in Sect. 3.16. It can be obtained by finding the minimum of $Q(\widehat{A}(\alpha, \beta), \widehat{B}(\alpha, \beta), \alpha, \beta)$ over the grid $\left(\dfrac{\pi j}{n}, \dfrac{\pi k}{n^2} \right); j = 1, \ldots, n$ and $k = 1, \ldots, n^2$. The main idea is not to use the whole sample at the beginning, as it was originally suggested by Bai et al. [3]. A part of the sample is used at the beginning, and gradually proceed towards the complete sample. With varying sample size, more and more data points are used with the increasing number of iteration. The algorithm can be described as follows. Denote the estimates of α^0 and β^0 obtained at the jth iteration as $\widetilde{\alpha}^{(j)}$ and $\widetilde{\beta}^{(j)}$, respectively.

Algorithm 9.2

- Step   1:   Choose $n_1 = n^{8/9}$.   Therefore, $\widetilde{\alpha}^{(0)} - \alpha^0 = O_p(n^{-1}) = O_p(n_1^{-1-1/8})$ and $\widetilde{\beta}^{(0)} - \beta^0 = O_p(n^{-2}) = O_p(n_1^{-2-1/4})$. Perform steps (9.24) and (9.25). Therefore, after 1-st iteration, we have

$$\widetilde{\alpha}^{(1)} - \alpha^0 = O_p(n_1^{-1-1/4}) = O_p(n^{-10/9}) \ \text{and} \ \widetilde{\beta}^{(1)} - \beta^0 = O_p(n_1^{-2-1/2}) = O_p(n^{-20/9}).$$

- Step 2: Choose $n_2 = n^{80/81}$. Therefore, $\widetilde{\alpha}^{(1)} - \alpha^0 = O_p(n_2^{-1-1/8})$ and $\widetilde{\beta}^{(1)} - \beta^0 = O_p(n_2^{-2-1/4})$. Perform steps (9.24) and (9.25). Therefore, after 2-nd iteration, we have

$$\widetilde{\alpha}^{(2)} - \alpha^0 = O_p(n_2^{-1-1/4}) = O_p(n^{-100/81}) \ \text{and} \ \widetilde{\beta}^{(2)} - \beta^0 = O_p(n_2^{-2-1/2}) = O_p(n^{-200/81}).$$

- Step 3: Choose $n_3 = n$. Therefore, $\widetilde{\alpha}^{(2)} - \alpha^0 = O_p(n_3^{-1-19/81})$ and $\widetilde{\beta}^{(2)} - \beta^0 = O_p(n_3^{-2-38/81})$. Perform steps (9.24) and (9.25). Therefore, after 3-rd iteration, we have

$$\widetilde{\alpha}^{(3)} - \alpha^0 = O_p(n^{-1-38/81}) \ \text{and} \ \widetilde{\beta}^{(3)} - \beta^0 = O_p(n^{-2-76/81}).$$

- Step 4: Choose $n_4 = n$ and perform steps (9.24) and (9.25). Now we obtain the required convergence rates, i.e.

$$\tilde{\alpha}^{(4)} - \alpha^0 = O_p(n^{-3/2}) \quad \text{and} \quad \tilde{\beta}^{(4)} - \beta^0 = O_p(n^{-5/2}).$$

∎

The above algorithm can be used quite efficiently to compute estimators in four steps which are equivalent to the LSEs. It may be mentioned that the fraction of the sample sizes which have been used in each step is not unique. It is possible to obtain equivalent estimators with different choices of subsamples. Although they are asymptotically equivalent, the finite sample performances might be different. The fraction $\frac{8}{9}$ or $\frac{80}{81}$ in Steps 1 or 2 is not unique and several other choices are possible. It might be interesting to compare the finite sample performances of the different estimators by extensive Monte Carlo simulations.

9.2.1.5 Approximate Least Squares Estimators

Recently, Grover, Kundu, and Mitra [20] proposed a periodogram like estimators of the unknown parameters of a chirp signal. Consider the periodogram like function defined for the chirp signal as follows:

$$I(\alpha, \beta) = \frac{2}{n} \left\{ \left(\sum_{t=1}^{n} y(t) \cos(\alpha t + \beta t^2) \right)^2 + \left(\sum_{t=1}^{n} y(t) \sin(\alpha t + \beta t^2) \right)^2 \right\}.$$

(9.26)

Recall that the periodogram function for the sinusoidal signal (9.3), is defined as follows:

$$I(\alpha) = \frac{2}{n} \left\{ \left(\sum_{t=1}^{n} y(t) \cos(\alpha t) \right)^2 + \left(\sum_{t=1}^{n} y(t) \sin(\alpha t) \right)^2 \right\}.$$

(9.27)

The periodogram estimator or the approximate least squares estimator (ALSE) of α can be obtained by maximizing $I(\alpha)$ as defined in (9.27), over the range $(0, \pi)$. If $y(t)$ satisfies the sinusoidal model assumption (9.3), then the ALSE of α is consistent and asymptotically equivalent to the LSE of α^0 of model (9.3). Therefore, $I(\alpha, \beta)$ in (9.26), is a natural generalization of $I(\alpha)$ in (9.27), to compute the estimators of the frequency and the frequency rate of the one component chirp model (9.1).

Grover, Kundu, and Mitra [20], proposed the approximate maximum likelihood estimators (AMLEs) of α and β as

$$(\tilde{\alpha}, \tilde{\beta}) = \arg \max_{\alpha, \beta} I(\alpha, \beta),$$

(9.28)

under the normality assumption on the error process $\{X(t)\}$. Once $\widetilde{\alpha}$ and $\widetilde{\beta}$ are obtained, $\widetilde{A}$ and $\widetilde{B}$ can be calculated as

$$\widetilde{A} = \frac{2}{n} \sum_{t=1}^{n} \cos(\widetilde{\alpha}t + \widetilde{\beta}t^2), \quad \widetilde{B} = \frac{2}{n} \sum_{t=1}^{n} \sin(\widetilde{\alpha}t + \widetilde{\beta}t^2). \tag{9.29}$$

A^0 and B^0 can also be estimated using $\widehat{\theta}(\widetilde{\alpha}, \widetilde{\beta})$, where $\widehat{\theta}(\alpha, \beta)$ is given in (9.12). In order to compute the AMLEs one can use Algorithm 9.1, by replacing $Z(\alpha, \beta)$ with $I(\alpha, \beta)$. Grover, Kundu, and Mitra [20], established the asymptotic properties of the ALSE of A^0, B^0, α^0, and β^0. It is observed that the ALSEs also have the consistency and asymptotic normality properties as the LSEs but in a slightly weaker condition. The following results can be found in Grover, Kundu, and Mitra [20].

Theorem 9.9 *If* $0 < |A^0| + |B^0|$, $0 < \alpha^0, \beta^0 < \pi$, *and* $\{X(t)\}$ *satisfies Assumption 9.1, then* $\widetilde{\Gamma} = (\widetilde{A}, \widetilde{B}, \widetilde{\alpha}, \widetilde{\beta})^T$ *is a strongly consistent estimate of* Γ^0. ∎

Theorem 9.10 *Under the same assumptions as in Theorem 9.9*

$$\left(n^{1/2}(\widetilde{A} - A^0), n^{1/2}(\widetilde{B} - B^0), n^{3/2}(\widetilde{\alpha} - \alpha^0), n^{5/2}(\widetilde{\beta} - \beta^0)\right)^T \xrightarrow{d} \mathcal{N}_4(0, 2\sigma^2 c \Sigma). \tag{9.30}$$
here c and Σ are same as defined in Theorem 9.4. ∎

Comparing Theorems 9.3 and 9.9, it may be observed that in Theorem 9.9, the boundedness condition on the linear parameters has been removed. By extensive simulation experiments, it has been observed that computationally the ALSEs have a slight advantage over the LSEs, although the mean squared errors (MSEs) of the ALSEs are slightly more than the corresponding LSEs. Again, as in case of LSEs, Theorem 9.10 can also be used for the construction of confidence intervals of the unknown parameters, and for different testing of hypotheses provided one can obtain a good estimate of $\sigma^2 c$. The following problem will be of interest.

Open Problem 9.7 Construct confidence intervals of the unknown parameters based on LSEs and ALSEs, and compare them in terms of average lengths of the confidence intervals and their respective coverage percentages. ∎

9.2.1.6 Bayes Estimates

Recently, Mazumder [41], considered model (9.1), when $X(t)$s are i.i.d. normal random variables and provided a Bayesian solution. The author provided the Bayes estimates and the associated credible intervals of the unknown parameters. The following transformations on the model parameters have been made.

$$A^0 = r^0 \cos(\theta^0), \quad B^0 = r^0 \sin(\theta^0), \quad r^0 \in (0, K], \quad \theta^0 \in [0, 2\pi], \quad \alpha, \beta \in (0, \pi).$$

The following assumptions on prior distributions have been made on the above unknown parameters.

$$r \sim \text{uniform}(0, K)$$
$$\theta \sim \text{uniform}(0, 2\pi)$$
$$\alpha \sim \text{von Misses}(a_0, a_1)$$
$$\beta \sim \text{von Misses}(b_0, b_1)$$
$$\sigma^2 \sim \text{inverse gamma}(c_0, c_1).$$

It may be mentioned that r and θ have noninformative priors, σ^2 has a conjugate prior. In this case, 2α and 2β are circular random variables. That is why, von Misses distribution, the natural analog of the normal distribution in circular data has been considered. We denote the prior densities of r, θ, α, β, and σ^2 as $[r]$, $[\theta]$, $[\alpha]$, $[\beta]$, and $[\sigma^2]$, respectively, and Y is the data vector as defined in Sect. 9.2.1.1. If it is assumed that the priors are independently distributed, then the joint posterior density function of r, θ, α, β, and σ^2 can be obtained as

$$[r, \theta, \alpha, \beta, \sigma^2 | Y] \propto [r][\theta][\alpha][\beta][\sigma^2][Y | r, \theta, \alpha, \beta, \sigma^2].$$

Now to compute the Bayes estimates of the unknown parameters using the Gibbs sampling technique, one needs to compute the conditional distribution of each parameter given all the parameters, known as the full conditional distribution, denoted by $[\cdot | \ldots]$, and they are given by

$$[r | \ldots] \propto [r][Y | r, \theta, \alpha, \beta, \sigma^2]$$

$$[\theta | \ldots] \propto [\theta][Y | r, \theta, \alpha, \beta, \sigma^2]$$

$$[\alpha | \ldots] \propto [\alpha][Y | r, \theta, \alpha, \beta, \sigma^2]$$

$$[\beta | \ldots] \propto [\beta][Y | r, \theta, \alpha, \beta, \sigma^2]$$

$$[\sigma^2 | \ldots] \propto [\sigma^2][Y | r, \theta, \alpha, \beta, \sigma^2].$$

The closed form expression of the full conditionals cannot be obtained. Mazumder [41], proposed to use the random walk Markov Chain Monte Carlo (MCMC) technique to update these parameters. The author has used this method to predict future observation also.

9.2.1.7 Testing of Hypothesis

Recently, Dhar, Kundu, and Das [7], considered the following testing of the hypothesis problem for a single chirp model. They have considered model (9.1), and it is assumed that $X(t)$s are i.i.d. random variables with mean zero and finite variance σ^2. Some more assumptions are needed in developing the properties of the test statistics, and those will be explicitly mentioned later. We have used the vector $\boldsymbol{\Gamma}^0 = (A^0, B^0, \alpha^0, \beta^0)$, then we want to test the following hypothesis

$$H_0 : \boldsymbol{\Gamma} = \boldsymbol{\Gamma}^0 \quad versus \quad H_1 : \boldsymbol{\Gamma} \neq \boldsymbol{\Gamma}^0. \tag{9.31}$$

This is a typical testing of the hypothesis problem, and it mainly tests whether the data are coming from a specific chirp model or not.

Four different tests to test the hypothesis (9.31), have been proposed by Dhar, Kundu, and Das [7], based on the following test statistics:

$$T_{n,1} = ||\mathbf{D}^{-1}(\widehat{\boldsymbol{\Gamma}}_{n,LSE} - \boldsymbol{\Gamma}^0)||_2^2, \tag{9.32}$$

$$T_{n,2} = ||\mathbf{D}^{-1}(\widehat{\boldsymbol{\Gamma}}_{n,LAD} - \boldsymbol{\Gamma}^0)||_2^2, \tag{9.33}$$

$$T_{n,3} = ||\mathbf{D}^{-1}(\widehat{\boldsymbol{\Gamma}}_{n,LSE} - \boldsymbol{\Gamma}^0)||_1^2, \tag{9.34}$$

$$T_{n,4} = ||\mathbf{D}^{-1}(\widehat{\boldsymbol{\Gamma}}_{n,LAD} - \boldsymbol{\Gamma}^0)||_1^2. \tag{9.35}$$

Here, $\widehat{\boldsymbol{\Gamma}}_{n,LSE}$ and $\widehat{\boldsymbol{\Gamma}}_{n,LAD}$ denote the LSE and LAD estimate of $\boldsymbol{\Gamma}^0$, as discussed in Sects. 9.2.1.2 and 9.2.1.3, respectively. The 4×4 diagonal matrix $\mathbf{D}$ is as follows:

$$\mathbf{D} = diag\{n^{-1/2}, n^{-1/2}, n^{-3/2}, n^{-5/2}\}.$$

Further, $|| \cdot ||_2$ and $|| \cdot ||_1$ denote the Euclidean and L_1 norms, respectively. Note that all these test statistics are based on some normalized values of the distances between the estimates and the parameter value under the null hypothesis. Here, Euclidean and L_1 distances have been chosen, but any other distance function can also be considered. In all these cases, clearly, the null hypothesis will be rejected if the values of the test statistics are large.

Now to choose the critical values of the test statistics, the following assumptions are made. Other than the assumptions required on the parameter values defined in Theorem 9.3, the following assumptions are also required.

Assumption 9.3 The i.i.d. error random variables have the positive density function $f(\cdot)$ with finite second moment.

Assumption 9.4 Let F_t be the distribution function of $y(t)$ with the probability density function $f_t(y, \boldsymbol{\Gamma})$, which is twice continuously differentiable with respect to $\boldsymbol{\Gamma}$. It is assumed that $E\left[\dfrac{\partial}{\partial \Gamma_i} f_t(y, \boldsymbol{\Gamma})\right]_{\boldsymbol{\Gamma} = \boldsymbol{\Gamma}^0}^{2+\delta} < \infty$, for some $\delta > 0$ and $E\left[\dfrac{\partial^2}{\partial \Gamma_i \partial \Gamma_j} f_t(y, \boldsymbol{\Gamma})\right]_{\boldsymbol{\Gamma} = \boldsymbol{\Gamma}^0}^{2} < \infty$, for all $t = 1, 2, \ldots, n$. Here Γ_i and Γ_j, for $1 \leq i, j \leq 4$, are the ith and jth component of $\boldsymbol{\Gamma}$.

Note that Assumptions 9.3 and 9.4, are not quite unnatural. Assumption 9.3, holds for most of the well-known probability density functions, e.g. normal, Laplace, and Cauchy probability density functions. The smoothness assumptions in Assumption 9.4 is required to prove the asymptotic normality of the test statistics under contiguous alternatives. Such assumptions are quite common in general across the asymptotic statistics.

Now we are in a position to provide the asymptotic properties of the above test statistics. We use the following notation. Suppose $A = (A_1, A_2, A_3, A_4)^T$ is a four-dimensional Gaussian random vector with mean vector $\mathbf{0}$ and the dispersion matrix $\Sigma_1 = 2\sigma^2 \Sigma$, where Σ is same as defined in Theorem 9.2. Further, let $B = (B_1, B_2, B_3, B_4)^T$ be also a four-dimensional Gaussian random vector with mean vector $\mathbf{0}$ and dispersion matrix $\Sigma_2 = \frac{1}{\{f(M)\}^2} \Sigma$. Here, M denotes the median of the distribution function associated with the density function $f(\cdot)$. Then we have the following results.

Theorem 9.11 *Let $c_{1\eta}$ be $(1 - \eta)$th quantile of the distribution of $\sum_{i=1}^{4} \lambda_i Z_i^2$, where λ_is are the eigenvalues of Σ_1, as defined above, and Z_is are independent standard normal random variables. If the assumption required on the parameter values defined in Theorem 9.3 and Assumption 9.3 hold true, then the test based on $T_{n,1}$ will have asymptotic size $= \eta$, when $T_{n,1} \geq c_{1\eta}$. Moreover, under the same set of assumptions $P_{H_1}[T_{n,1} \geq c_{1\eta}] \to 1$, as $n \to \infty$, i.e. the test based on $T_{n,1}$ will be a consistent test.* ∎

Theorem 9.12 *Let $c_{2\eta}$ be $(1 - \eta)$th quantile of the distribution of $\sum_{i=1}^{4} \lambda_i^* Z_i^2$, where λ_i^*s are the eigenvalues of Σ_2, as defined above, and Z_is are independent standard normal random variables. Under the same assumptions as Theorem 9.11, the test based on $T_{n,2}$ will have asymptotic size $= \eta$, when $T_{n,2} \geq c_{2\eta}$. Moreover, under the same set of assumptions $P_{H_1}[T_{n,2} \geq c_{2\eta}] \to 1$, as $n \to \infty$, i.e. the test based on $T_{n,2}$ will be a consistent test.* ∎

Theorem 9.13 *Let $c_{3\eta}$ be $(1 - \eta)$th quantile of the distribution of $\left\{ \sum_{i=1}^{4} |A_i| \right\}^2$, where A_is are same as defined above. Under the same assumptions as Theorem 9.11, the test based on $T_{n,3}$ will have asymptotic size $= \eta$, when $T_{n,3} \geq c_{3\eta}$. Moreover, under the same set of assumptions $P_{H_1}[T_{n,3} \geq c_{3\eta}] \to 1$, as $n \to \infty$, i.e. the test based on $T_{n,3}$ will be a consistent test.* ∎

Theorem 9.14 *Let $c_{4\eta}$ be $(1 - \eta)$th quantile of the distribution of $\left\{ \sum_{i=1}^{4} |B_i| \right\}^2$, where B_is are same as defined above. Under the same assumptions as Theorem 9.11, the test based on $T_{n,4}$ will have asymptotic size $= \eta$, when $T_{n,4} \geq c_{4\eta}$. Moreover, under the same set of assumptions $P_{H_1}[T_{n,4} \geq c_{4\eta}] \to 1$, as $n \to \infty$, i.e. the test based on $T_{n,4}$ will be a consistent test.* ∎

The proofs of the Theorems 9.11–9.14, can be found in Dhar, Kundu, and Das [7]. Extensive simulations have been performed by the authors to assess the finite sample performances of the four tests. It is observed that all the four tests are able to maintain the level of significance. In terms of the power, it is observed that $T_{n,1}$ and $T_{n,3}$ (based on least squares methodology) perform well when the data are obtained from a light-tailed distribution like the normal distribution. On the other hand, the tests based on $T_{n,2}$ and $T_{n,4}$ (based on least absolute deviation methodology) perform better than $T_{n,1}$ and $T_{n,3}$ for heavy tailed distribution like Laplace and t distribution with 5 degrees of freedom. Overall, one can prefer the least absolute deviation methodologies when the data are likely to have influential observations/ outliers.

The authors further obtained the asymptotic distributions of the tests based on local alternatives. The following form of the local alternatives has been considered.

$$H_0 : \boldsymbol{\Gamma} = \boldsymbol{\Gamma}^0 \quad vs. \quad H_{1,n} : \boldsymbol{\Gamma} = \boldsymbol{\Gamma}_n = \boldsymbol{\Gamma}^0 + \boldsymbol{\delta}_n, \tag{9.36}$$

where $\boldsymbol{\delta}_n = \left(\dfrac{\delta_1}{n^{1/2}}, \dfrac{\delta_2}{n^{1/2}}, \dfrac{\delta_3}{n^{3/2}}, \dfrac{\delta_4}{n^{5/2}} \right)$. If the assumption required on the parameter values defined in Theorem 9.3, and Assumptions 9.3 and 9.4 hold true, then the authors obtained the asymptotic distributions of $T_{n,1}$, $T_{n,2}$, $T_{n,3}$, and $T_{n,4}$ under the alternative hypothesis. The exact forms of the asymptotic distributions are quite involved and they are not presented here. Interested readers are referred to the original article of Dhar, Kundu, and Das [7] for details.

So far we have discussed the one component chirp model, now we consider the multicomponent chirp model.

9.2.2 Multicomponent Chirp Model

The multicomponent chirp model can be written as follows:

$$y(t) = \sum_{j=1}^{p} \{ A_j^0 \cos(\alpha_j^0 t + \beta_j^0 t^2) + B_j^0 \sin(\alpha_j^0 t + \beta_j^0 t^2) \} + X(t). \tag{9.37}$$

Here $y(t)$ is the real-valued signal as mentioned before, and it is observed at $t = 1, \ldots, n$. For $j = 1, \ldots, p$, A_j^0, B_j^0 are real-valued amplitudes, α_j^0, β_j^0 are the frequency and the frequency rate, respectively. The problem is to estimate the unknown parameters, A_j^0, B_j^0, α_j^0, β_j^0, for $j = 1, \ldots, p$ and the number of components p, based on the observed sample. Different methods have been proposed in the literature. In this subsection, we provide different methods and discuss their theoretical properties. Throughout this section, it is assumed that p is known in advance. As in the case of single chirp model, it is first assumed that the errors $X(t)$s are i.i.d. normal random variables with mean 0 and variance σ^2. More general forms of errors will be considered later on. Before progressing further, write $\boldsymbol{\alpha}^0 = (\alpha_1^0, \ldots, \alpha_p^0)^T$, $\boldsymbol{\beta}^0 = (\beta_1^0, \ldots, \beta_p^0)^T$, $\boldsymbol{A} = (A_1^0, \ldots, A_p^0)^T$, $\boldsymbol{B} = (B_1^0, \ldots, B_p^0)^T$ and $\boldsymbol{\Gamma}_j = (A_j, B_j, \alpha_j, \beta_j)^T$ for $j = 1, \ldots, p$.

9.2.3 Maximum Likelihood Estimators

Proceeding as before, the log-likelihood function without the additive constant can be written as

$$
l(\boldsymbol{\Gamma}_1, \ldots, \boldsymbol{\Gamma}_p, \sigma^2) = -\frac{n}{2}\ln\sigma^2 - \frac{1}{2\sigma^2}\sum_{t=1}^{n}\left(y(t) - \sum_{j=1}^{p}\left\{A_j\cos(\alpha_j t + \beta_j t^2) + B_j\sin(\alpha_j t + \beta_j t^2)\right\}\right)^2.
$$
(9.38)

The MLEs of the unknown parameters can be obtained by maximizing (9.38), with respect to the unknown parameters $\boldsymbol{\Gamma}_1, \ldots, \boldsymbol{\Gamma}_p$ and σ^2. As in the case of single chirp model, the MLEs of $\boldsymbol{\Gamma}_j = (A_j, B_j, \alpha_j, \beta_j)^T$, for $j = 1, \ldots, p$, can be obtained by minimizing $Q(\boldsymbol{\Gamma}_1, \ldots, \boldsymbol{\Gamma}_p)$, where

$$
Q(\boldsymbol{\Gamma}_1, \ldots, \boldsymbol{\Gamma}_p) = \sum_{t=1}^{n}\left(y(t) - \sum_{j=1}^{p}\left\{A_j\cos(\alpha_j t + \beta_j t^2) + B_j\sin(\alpha_j t + \beta_j t^2)\right\}\right)^2,
$$

with respect to the unknown parameters. If the MLEs of $\boldsymbol{\Gamma}_j$ is denoted by $\widehat{\boldsymbol{\Gamma}}_j$, for $j = 1, \ldots, p$, the MLE of σ^2 can be obtained as before

$$
\widehat{\sigma}^2 = \frac{1}{n}\sum_{t=1}^{n}\left(y(t) - \sum_{j=1}^{p}\left\{\widehat{A}_j\cos(\widehat{\alpha}_j t + \widehat{\beta}_j t^2) + \widehat{B}_j\sin(\widehat{\alpha}_j t + \widehat{\beta}_j t^2)\right\}\right)^2.
$$

It may be observed that $Q(\boldsymbol{\Gamma}_1, \ldots, \boldsymbol{\Gamma}_p)$ can be written as follows:

$$
Q(\boldsymbol{\Gamma}_1, \ldots, \boldsymbol{\Gamma}_p) = \left[\mathbf{Y} - \sum_{j=1}^{p}\mathbf{W}(\alpha_j, \beta_j)\boldsymbol{\theta}_j\right]^T\left[\mathbf{Y} - \sum_{j=1}^{p}\mathbf{W}(\alpha_j, \beta_j)\boldsymbol{\theta}_j\right], \quad (9.39)
$$

where $\mathbf{Y} = (y(1), \ldots, y(n))^T$ is the $n \times 1$ data vector, the $n \times 2$ matrix $\mathbf{W}(\alpha, \beta)$ is same as defined in (9.11) and $\boldsymbol{\theta}_j = (A_j, B_j)^T$ is a 2×1 vector for $j = 1, \ldots, p$. The MLEs of the unknown parameters can be obtained by minimizing (9.39), with respect to the unknown parameters. Now define the $n \times 2p$ matrix $\widetilde{\mathbf{W}}(\boldsymbol{\alpha}, \boldsymbol{\beta})$ as

$$
\widetilde{\mathbf{W}}(\boldsymbol{\alpha}, \boldsymbol{\beta}) = [\mathbf{W}(\alpha_1, \beta_1) : \cdots : \mathbf{W}(\alpha_p, \beta_p)],
$$

then, for fixed α and β, the MLEs of $\theta_1, \ldots, \theta_p$, the linear parameter vectors, can be obtained as

$$[\widehat{\boldsymbol{\theta}}_1^T(\alpha_1, \beta_1) : \cdots : \widehat{\boldsymbol{\theta}}_p^T(\alpha_p, \beta_p)]^T = [\widetilde{\mathbf{W}}^T(\boldsymbol{\alpha}, \boldsymbol{\beta})\widetilde{\mathbf{W}}(\boldsymbol{\alpha}, \boldsymbol{\beta})]^{-1} \widetilde{\mathbf{W}}^T(\boldsymbol{\alpha}, \boldsymbol{\beta})\mathbf{Y}.$$

We note that because $\widetilde{\mathbf{W}}^T(\boldsymbol{\alpha}, \boldsymbol{\beta})\widetilde{\mathbf{W}}(\boldsymbol{\alpha}, \boldsymbol{\beta})$ is a diagonal matrix for large n, $\widehat{\boldsymbol{\theta}}_j = \widehat{\boldsymbol{\theta}}_j(\alpha_j, \beta_j)$, the MLE of θ_j, $j = 1, \ldots, p$ can also be expressed as

$$\widehat{\boldsymbol{\theta}}_j(\alpha_j, \beta_j) = [\mathbf{W}^T(\alpha_j, \beta_j)\mathbf{W}(\alpha_j, \beta_j)]^{-1} \mathbf{W}^T(\alpha_j, \beta_j)\mathbf{Y}.$$

Using similar techniques as in Sect. 9.2.1.1, the MLEs of α and β can be obtained as the argument maximum of

$$\mathbf{Y}^T \widetilde{\mathbf{W}}(\boldsymbol{\alpha}, \boldsymbol{\beta})[\widetilde{\mathbf{W}}^T(\boldsymbol{\alpha}, \boldsymbol{\beta})\widetilde{\mathbf{W}}(\boldsymbol{\alpha}, \boldsymbol{\beta})]^{-1}\widetilde{\mathbf{W}}^T(\boldsymbol{\alpha}, \boldsymbol{\beta})\mathbf{Y}. \tag{9.40}$$

The criterion function, given in (9.40), is a highly nonlinear function of α and β, therefore, the MLEs of α and β cannot be obtained in closed form. Saha and Kay [53], suggested to use the method of Pincus [49], to maximize (9.40). Alternatively, different other methods can also be used to maximize (9.40).

It may be easily observed that the MLEs obtained above are the same as the LSEs. Kundu and Nandi [31], first derived the consistency and the asymptotic normality properties of the LSEs when the error random variables follow Assumption 9.1. It has been shown that as the sample size $n \to \infty$, the LSEs are strongly consistent. Kundu and Nandi [31], obtained the following consistency and asymptotic normality results.

Theorem 9.15 *Suppose there exists a K, such that for $j = 1, \ldots, p$, $0 < |A_j^0| + |B_j^0| < K, 0 < \alpha_j^0, \beta_j^0 < \pi$, α_j^0 are distinct, similarly β_j^0 are also distinct and $\sigma^2 > 0$. If $\{X(t)\}$ satisfies Assumption 9.1, then $\widehat{\boldsymbol{\Gamma}}_j = (\widehat{A}_j, \widehat{B}_j, \widehat{\alpha}_j, \widehat{\beta}_j)^T$ is a strongly consistent estimate of $\boldsymbol{\Gamma}_j^0 = (A_j^0, B_j^0, \alpha_j^0, \beta_j^0)^T$, for $j = 1, \ldots, p$.* ∎

Theorem 9.16 *Under the same assumptions as in Theorem 9.15, for $j = 1, \ldots, p$,*

$$\left(n^{1/2}(\widehat{A}_j - A_j^0), n^{1/2}(\widehat{B}_j - B^0), n^{3/2}(\widehat{\alpha}_j - \alpha_j^0), n^{5/2}(\widehat{\beta}_j - \beta_j^0)\right)^T \xrightarrow{d} \mathcal{N}_4(0, 2c\sigma^2 \boldsymbol{\Sigma}_j), \tag{9.41}$$

where $\boldsymbol{\Sigma}_j$ can be obtained from the matrix $\boldsymbol{\Sigma}$ defined in Theorem 9.6, by replacing A^0 and B^0 with A_j^0 and B_j^0, respectively, and c is same as defined in Theorem 9.4. Moreover, $\widehat{\boldsymbol{\Gamma}}_j$ and $\widehat{\boldsymbol{\Gamma}}_k$, for $j \neq k$ are asymptotically independently distributed. ∎

The above consistency results, the asymptotic normality properties of $\widehat{\boldsymbol{\Gamma}}_j$ have been obtained by Kundu and Nandi [31], but the form of the asymptotic variance covariance matrix is quite complicated. Lahiri, Kundu, and Mitra [37], simplified the entries of the asymptotic variance covariance matrix using Result 9.1.

9.2.4 Sequential Estimation Procedures

It is known that the MLEs are the most efficient estimators, but to compute the MLEs one needs to solve a $2p$ dimensional optimization problem. Hence, for moderate or large p it can be a computationally challenging problem. At the same time, due to the highly nonlinear nature of the likelihood (least squares) surface, any numerical algorithm often converges to a local maximum (minimum) rather than the global maximum (minimum) unless the initial guesses are very close to the true maximum (minimum).

In order to avoid this problem, Lin and Djurić [38], see also Lahiri, Kundu, and Mitra [37], proposed a numerically efficient sequential estimation technique that can produce estimators which are asymptotically equivalent to the LSEs, but it involves solving only p separate two-dimensional optimization problems. Therefore, computationally it is quite easy to implement, and even for large p it can be used quite conveniently. The following algorithm may be used to obtain the sequential estimators of the unknown parameters of the multicomponent chirp model (9.37).

Algorithm 9.3

- Step 1: First maximize $Z(\alpha, \beta)$, given in (9.15), with respect to (α, β) and take the estimates of α and β as the argument maximum of $Z(\alpha, \beta)$. Obtain the estimates of associated A and B by using the separable least squares method of Richards [48]. Denote the estimates of $\alpha_1^0, \beta_1^0, A_1^0, B_1^0$ as $\widehat{\alpha}_1, \widehat{\beta}_1, \widehat{A}_1$, and $\widehat{B}_1$, respectively.
- Step 2: Now to compute the estimates of $\alpha_2^0, \beta_2^0, A_2^0, B_2^0$, take out the effect of the first component from the signal. Consider the new data vector as

$$\mathbf{Y}^1 = \mathbf{Y} - \mathbf{W}(\widehat{\alpha}_1, \widehat{\beta}_1) \begin{bmatrix} \widehat{A}_1 \\ \widehat{B}_1 \end{bmatrix},$$

  where $\mathbf{W}$ matrix is same as defined in (9.11).
- Step 3: Repeat Step 1 by replacing $\mathbf{Y}$ with $\mathbf{Y}^1$ and obtain the estimates of α_2^0, β_2^0, A_2^0, B_2^0 as $\widehat{\alpha}_2, \widehat{\beta}_2, \widehat{A}_2$, and $\widehat{B}_2$, respectively.
- Step 4: Continue the process p times and obtain the estimates sequentially.

We note that the above algorithm reduces the computational time significantly. A natural question is: how efficient are these sequential estimators? Lahiri, Kundu, and Mitra [37], established that the sequential estimators are strongly consistent and they have the same asymptotic distribution as the LSEs if the following famous number theoretic conjecture, see Montgomery [42], holds. Extensive simulation results indicate that sequential estimators behave very similarly as the LSEs in terms of MSEs and biases.

Conjecture *If $\xi, \eta, \xi', \eta' \in (0, \pi)$, then except for countable number of points*

$$\lim_{n \to \infty} \frac{1}{n^{k+1/2}} \sum_{t=1}^{n} n^k \sin(\xi t + \eta t^2) \cos(\xi' t + \eta' t^2) = 0; \quad t = 0, 1, 2.$$

In addition if $\eta \neq \eta'$, then

$$\lim_{n\to\infty} \frac{1}{n^{k+1/2}} \sum_{t=1}^{n} n^k \cos(\xi t + \eta t^2)\cos(\xi' t + \eta' t^2) = 0; \quad t = 0, 1, 2.$$

and

$$\lim_{n\to\infty} \frac{1}{n^{k+1/2}} \sum_{t=1}^{n} n^k \sin(\xi t + \eta t^2)\sin(\xi' t + \eta' t^2) = 0; \quad t = 0, 1, 2.$$

In the same paper, it has also been established that if the sequential procedure is continued beyond p steps, then the corresponding amplitude estimates converge to zero almost surely. Therefore, the sequential procedure in a way provides an estimator of the number of components p which is usually unknown in practice.

Recently, Grover, Kundu, and Mitra [20], provided another sequential estimator which is based on the ALSEs as proposed in Sect. 9.2.1.5. At each step instead of using the LSEs the authors proposed to use the ALSEs. It has been shown by Grover, Kundu, and Mitra [20] that the proposed estimators are strongly consistent and have the same asymptotic distribution as the LSEs provided that the above Conjecture holds. Similar to LSEs, in case of overestimation of p, it has been established that if the sequential procedure is continued beyond p steps, then the corresponding amplitude estimates converge to zero almost surely. It has been further observed based on extensive simulation results that the computational time is significantly smaller if ALSEs are used instead of LSEs studied by Lahiri, Kundu, and Mitra [37], although the performances are very similar in nature. It should be mentioned here that all the methods proposed for one component chirp model can be used sequentially for the multicomponent chirp model also.

In case of multicomponent chirp model (9.37), LAD estimation method has not been used. Therefore, we have the following open problem

Open Problem 9.8 Develop the properties of the LAD estimators for the multicomponent chirp model.

9.2.5 Heavy-Tailed Error

So far we have discussed the estimation of the unknown parameters of the chirp model when the errors have second and higher order moments. Recently, Nandi, Kundu, and Grover [45], considered the estimation of one component and multicomponent chirp models in presence of heavy-tailed error. It is assumed that the error random variables $X(t)$s are i.i.d. random variables with mean zero but it may not have finite variance. The following explicit assumptions are made in this case by Nandi, Kundu, and Grover [45].

Assumption 9.5 The sequence of error random variables $\{X(t)\}$ is a sequence of i.i.d. random variables with mean zero and $E|X(t)|^{1+\delta} < \infty$, for some $0 < \delta < 1$.

Assumption 9.6 The error random variables $\{X(t)\}$ is a sequence of i.i.d. random variables with mean zero and distributed as a symmetric α-stable distribution with the scale parameter $\gamma > 0$, where $1 + \delta < \alpha < 2$, $0 < \delta < 1$. It implies that the characteristic function of $X(t)$ is of the following form:

$$E[e^{itX(t)}] = e^{-\gamma^\alpha |t|^\alpha}.$$

Nandi, Kundu, and Grover [45], consider the LSEs and the ALSEs of the unknown parameters of the one component and multicomponent chirp models. In the case of one component chirp model (9.6), the following results have been obtained. The details can be found in Nandi, Kundu, and Grover [45].

Theorem 9.17 *If there exists K, such that $0 < |A^0|^2 + |B^0|^2 < K$, $0 < \alpha^0$, $\beta^0 < \pi$, $\{X(t)\}$ satisfies Assumption 9.5, then $\widehat{\Gamma} = (\widehat{A}, \widehat{B}, \widehat{\alpha}, \widehat{\beta})^T$, the LSE of Γ^0, is strongly consistent.* ∎

Theorem 9.18 *Under the same assumptions as Theorem 9.17, and further if $\{X(t)\}$ satisfies Assumption 9.6, then*

$$(n^{\frac{\alpha-1}{\alpha}}(\widehat{A} - A^0), n^{\frac{\alpha-1}{\alpha}}(\widehat{B} - B^0), n^{\frac{2\alpha-1}{\alpha}}(\widehat{\alpha} - \alpha^0), n^{\frac{3\alpha-1}{\alpha}}(\widehat{\beta} - \beta^0))^T$$

converges to a symmetric α-stable distribution of dimension four. ∎

It has been shown that the LSEs and ALSEs are asymptotically equivalent. Moreover, in case of multiple chirp model (9.37), the sequential procedure can be adopted, and their asymptotic properties along the same line as Theorems 9.17 and 9.18, have been obtained. It may be mentioned that in the presence of heavy-tailed error, the LSEs or the ALSEs may not be robust. In that case M-estimators which are more robust, may be considered. Developing both the theoretical properties and numerically efficient algorithm will be of interest. More work is needed in this direction.

9.2.6 Polynomial Phase Chirp Model

The real-valued single-component polynomial phase chirp (PPC) model of order q can be written as follows:

$$y(t) = A^0 \cos(\alpha_1^0 t + \alpha_2^0 t^2 + \ldots + \alpha_q^0 t^q) + B^0 \sin(\alpha_1^0 t + \alpha_2^0 t^2 + \ldots + \alpha_q^0 t^q) + X(t). \tag{9.42}$$

Here $y(t)$ is the real-valued signal observed at $t = 1, \ldots, n$; A^0 and B^0 are amplitudes; $\alpha_m^0 \in (0, \pi)$ for $m = 1, \ldots, q$ are parameters of the polynomial phase. Here α_1^0 is the frequency and change in frequency is governed by $\alpha_2^0, \ldots, \alpha_q^0$. The error component $\{X(t)\}$ is from a stationary linear process and satisfies Assumption 9.1. The problem remains the same, that is, estimate the unknown parameters namely A^0, B^0, $\alpha_1^0, \ldots, \alpha_q^0$ given a sample of size n. The corresponding complex model was originally proposed by Djurić and Kay [8], and Nandi and Kundu [43] established the consistency and asymptotic normality properties of the LSEs of the unknown parameters for the complex PPC model. Along similar lines, the following consistency and asymptotic normality results of LSEs of the unknown parameters of model (9.42) can be easily established.

Theorem 9.19 *Suppose there exists a K, such that for $j = 1, \ldots, p$, $0 < |A^0| + |B^0| < K$, $0 < \alpha_1^0, \ldots, \alpha_k^0 < \pi$ and $\sigma^2 > 0$. If $\{X(t)\}$ satisfies Assumption 9.1, then $\widehat{\boldsymbol{\Gamma}} = (\widehat{A}, \widehat{B}, \widehat{\alpha}_1, \ldots, \widehat{\alpha}_q)^T$ is a strongly consistent estimate of $\boldsymbol{\Gamma}^0 = (A^0, B^0, \alpha_1^0, \ldots, \alpha_q^0)^T$.* ∎

Theorem 9.20 *Under the same assumptions as in Theorem 9.19,*

$$\left(n^{1/2}(\widehat{A} - A^0), n^{1/2}(\widehat{B} - B^0), n^{3/2}(\widehat{\alpha}_1 - \alpha_1^0), \ldots, n^{(2q+1)/2}(\widehat{\alpha}_q - \alpha_q^0) \right)^T$$

$$\xrightarrow{d} \mathcal{N}_{q+2}(\mathbf{0}, 2c\sigma^2 \boldsymbol{\Sigma}),$$

where

$$\boldsymbol{\Sigma}^{-1} = \begin{bmatrix} 1 & 0 & \frac{B^0}{2} & \frac{B^0}{3} & \cdots & \frac{B^0}{q+1} \\ 0 & 1 & -\frac{A^0}{2} & -\frac{A^0}{3} & \cdots & -\frac{A^0}{q+1} \\ \frac{B^0}{2} & -\frac{A^0}{2} & \frac{A^{02}+B^{02}}{3} & \frac{A^{02}+B^{02}}{4} & \cdots & \frac{A^{02}+B^{02}}{q+2} \\ \frac{B^0}{3} & -\frac{A^0}{3} & \frac{A^{02}+B^{02}}{4} & \frac{A^{02}+B^{02}}{5} & \cdots & \frac{A^{02}+B^{02}}{q+3} \\ \vdots & \vdots & \vdots & \vdots & \vdots & \vdots \\ \frac{B^0}{q+1} & -\frac{A^0}{q+1} & \frac{A^{02}+B^{02}}{q+2} & \frac{A^{02}+B^{02}}{q+3} & \cdots & \frac{A^{02}+B^{02}}{2q+1} \end{bmatrix}.$$

Here σ^2 and c are same as defined in Theorem 9.4. ∎

The theoretical properties of the LSEs of the parameters of model (9.42) can be obtained, but at the same time, the computation of the LSEs is a challenging problem. Not much attention has been paid in this area. Moreover, the asymptotic distribution of the LSEs, as provided in Theorem 9.20, can be used in constructing confidence intervals and also in addressing different testing of hypotheses issues. The following problems will be of interest.

Open Problem 9.9 Develop numerically efficient algorithm to compute the LSEs of the unknown parameters of model (9.42). ∎

Open Problem 9.10 Construct different confidence intervals of the unknown parameters of model (9.42), under different error assumptions, and compare their performances in terms of average confidence lengths and coverage percentages. ∎

Open Problem 9.11 Consider multicomponent polynomial phase signal and develop the theoretical properties of LSEs. ∎

Some of the other methods which have been proposed in the literature are very specific to the complex chirp model. They are based on complex data and its conjugate, see, for example, Ikram, Abed-Meraim, and Hua [25], Besson, Giannakis, and Gini [5], Xinghao, Ran, and Siyong [61], Zhang et al. [64], Liu and Yu [39], Wang, Su, and Chen [57], and the references cited therein. There are some related models which have been used in place of the chirp model. We will be discussing those in detail in Chap. 11. Now we consider the two-dimensional chirp models in the next section.

9.3 Two-Dimensional Chirp Model

A multicomponent two-dimensional (2-D) chirp model can be expressed as follows:

$$y(m, n) = \sum_{k=1}^{p} (A_k^0 \cos(\alpha_k^0 m + \beta_k^0 m^2 + \gamma_k^0 n + \delta_k^0 n^2) +$$

$$B_k^0 \sin(\alpha_k^0 m + \beta_k^0 m^2 + \gamma_k^0 n + \delta_k^0 n^2)) + X(m, n);$$
$$m = 1, \ldots, M, n = 1, \ldots, N. \qquad (9.43)$$

Here $y(m, n)$ is the observed value, $A_k^0 s$, $B_k^0 s$ are the amplitudes, $\alpha_k^0 s$, $\gamma_k^0 s$ are the frequencies, and $\beta_k^0 s$, $\delta_k^0 s$ are the frequency rates. The error random variables $X(m, n)$s have mean zero and they may have some dependence structure. The details will be mentioned later.

The above model (9.43) is a natural generalization of the 1-D chirp model. Model (9.43) and some of its variants have been used quite extensively in modeling and analyzing magnetic resonance imaging (MRI), optical imaging and different texture imaging. It has been used quite extensively in modeling black and white "gray" images, and to analyze fingerprint images data. This model has wide applications in modeling Synthetic Aperture Radar (SAR) data, and in particular, Interferometric SAR data. See, for example, Pelag and Porat [47], Hedley and Rosenfeld [24], Friedlander and Francos [17], Francos and Friedlander [15, 16], Cao, Wang, and Wang [6], Zhang and Liu [65], Zhang, Wang, and Cao [66], and the references cited therein.

In this section also first we discuss different estimation procedures and their properties for single-component 2-D chirp model, and then we consider the multiple 2-D chirp model.

9.3.1 Single 2-D Chirp Model

The single 2-D chirp model can be written as follows:

$$y(m, n) = A^0 \cos(\alpha^0 m + \beta^0 m^2 + \gamma^0 n + \delta^0 n^2) + B^0 \sin(\alpha^0 m + \beta^0 m^2 + \gamma^0 n + \delta^0 n^2)$$
$$+ X(m, n); \quad m = 1, \ldots, M, n = 1, \ldots, N. \qquad (9.44)$$

Here all the quantities are same as defined for model (9.43). The problem remains the same as 1-D chirp model, i.e. based on the observed data $\{y(m, n)\}$, one needs to estimate the unknown parameters under a suitable error assumption. First assume that $X(m, n)$s are i.i.d. Gaussian random variables with mean zero and variance σ^2, for $m = 1, \ldots, M$ and $n = 1, \ldots, N$. More general error assumptions will be considered in subsequent sections.

9.3.1.1 Maximum Likelihood and Least Squares Estimators

The log-likelihood function of the observed data without the additive constant can be written as

$$l(\boldsymbol{\Theta}) = -\frac{MN}{2} \ln \sigma^2 - \frac{1}{2\sigma^2} \sum_{m=1}^{M} \sum_{n=1}^{N} \left(y(m, n) - A\cos(\alpha m + \beta m^2 + \gamma n + \delta n^2) - \right.$$
$$\left. B\sin(\alpha m + \beta m^2 + \gamma n + \delta n^2) \right)^2.$$

Here, $\boldsymbol{\Theta} = (A, B, \alpha, \beta, \gamma, \delta, \sigma^2)^T$. The MLE of the unknown parameter vector $\boldsymbol{\Gamma}^0 = (A^0, B^0, \alpha^0, \beta^0, \gamma^0, \delta^0)^T$ can be obtained, similarly as 1-D chirp model, as the argument minimum of

$$Q(\boldsymbol{\Gamma}) = \sum_{m=1}^{M} \sum_{n=1}^{N} \left(y(m, n) - A\cos(\alpha m + \beta m^2 + \gamma n + \delta n^2) \right.$$
$$\left. - B\sin(\alpha m + \beta m^2 + \gamma n + \delta n^2) \right)^2. \quad (9.45)$$

If $\widehat{\boldsymbol{\Gamma}} = (\widehat{A}, \widehat{B}, \widehat{\alpha}, \widehat{\beta}, \widehat{\gamma}, \widehat{\delta})^T$, is the MLE of $\boldsymbol{\Gamma}^0$, which minimizes (9.45), then the MLEs of $\widehat{\sigma}^2$ can be obtained as

$$\widehat{\sigma}^2 = \frac{1}{MN} \sum_{m=1}^{M} \sum_{n=1}^{N} \left(y(m, n) - \widehat{A}\cos(\widehat{\alpha}m + \widehat{\beta}m^2 + \widehat{\gamma}n + \widehat{\delta}n^2) \right.$$
$$\left. - \widehat{B}\sin(\widehat{\alpha}m + \widehat{\beta}m^2 + \widehat{\gamma}n + \widehat{\delta}n^2) \right)^2.$$

As expected, the MLEs cannot be obtained in explicit forms. One needs to use some numerical techniques to compute the MLEs. Newton–Raphson, Gauss–Newton, Genetic algorithm or simulated annealing method may be used for this purpose. Alternatively, the method suggested by Saha and Kay [53], as described in Sect. 9.2.1.1, may be used to find the MLEs of the unknown parameters. The details are not provided here.

Lahiri [32], see also Lahiri and Kundu [33], in this respect, established the asymptotic properties of the MLEs. Under a fairly general set of conditions, it has been shown that the MLEs are strongly consistent and they are asymptotically normally distributed. The results are provided in details below.

Theorem 9.21 *If there exists a K, such that* $0 < |A^0| + |B^0| < K$, $0 < \alpha^0, \beta^0$, $\gamma^0, \delta^0 < \pi$, *and* $\sigma^2 > 0$, *then* $\widehat{\Theta} = (\widehat{A}, \widehat{B}, \widehat{\alpha}, \widehat{\beta}, \widehat{\gamma}, \widehat{\delta}, \widehat{\sigma}^2)^T$ *is a strongly consistent estimate of* $\Theta^0 = (A^0, B^0, \alpha^0, \beta^0, \gamma^0, \delta^0, \sigma^2)^T$. ∎

Theorem 9.22 *Under the same assumptions as in Theorem 9.21, if we denote* **D** *as a* 6×6 *diagonal matrix as*

$$\mathbf{D} = diag\left\{ M^{1/2}N^{1/2}, M^{1/2}N^{1/2}, M^{3/2}N^{1/2}, M^{5/2}N^{1/2}, M^{1/2}N^{3/2}, M^{1/2}N^{5/2} \right\},$$

then

$$\mathbf{D}(\widehat{A} - A^0, \widehat{B} - B^0, \widehat{\alpha} - \alpha^0, \widehat{\beta} - \beta^0, \widehat{\gamma} - \gamma^0, \widehat{\delta} - \delta^0)^T \xrightarrow{d} \mathcal{N}_6(\mathbf{0}, 2\sigma^2\mathbf{\Sigma}),$$

where

$$
\mathbf{\Sigma}^{-1} =
\begin{bmatrix}
1 & 0 & \frac{B^0}{2} & \frac{B^0}{3} & \frac{B^0}{2} & \frac{B^0}{3} \\
0 & 1 & -\frac{A^0}{2} & -\frac{A^0}{3} & -\frac{A^0}{2} & -\frac{A^0}{3} \\
\frac{B^0}{2} & -\frac{A^0}{2} & \frac{A^{02}+B^{02}}{3} & \frac{A^{02}+B^{02}}{4} & \frac{A^{02}+B^{02}}{4} & \frac{A^{02}+B^{02}}{4.5} \\
\frac{B^0}{3} & -\frac{A^0}{3} & \frac{A^{02}+B^{02}}{4} & \frac{A^{02}+B^{02}}{5} & \frac{A^{02}+B^{02}}{4.5} & \frac{A^{02}+B^{02}}{5} \\
\frac{B^0}{2} & -\frac{A^0}{2} & \frac{A^{02}+B^{02}}{4} & \frac{A^{02}+B^{02}}{4.5} & \frac{A^{02}+B^{02}}{3} & \frac{A^{02}+B^{02}}{4} \\
\frac{B^0}{3} & -\frac{A^0}{3} & \frac{A^{02}+B^{02}}{4.5} & \frac{A^{02}+B^{02}}{5} & \frac{A^{02}+B^{02}}{4} & \frac{A^{02}+B^{02}}{5}
\end{bmatrix}.
$$

∎

Note that the asymptotic distribution of the MLEs can be used to construct confidence intervals of the unknown parameters. They can be used for testing of hypothesis problem also.

Since the LSEs are MLEs under the assumption of i.i.d. Gaussian errors, the above MLEs are the LSEs also. In fact, the LSEs can be obtained by minimizing (9.45) for a more general set of error assumptions. Let us make the following assumption on the error component $X(m, n)$.

Assumption 9.7 The error component $X(m, n)$ has the following form:

$$X(m, n) = \sum_{j=-\infty}^{\infty} \sum_{k=-\infty}^{\infty} a(j, k) e(m - j, n - k),$$

with

$$\sum_{j=-\infty}^{\infty} \sum_{k=-\infty}^{\infty} |a(j, k)| < \infty,$$

where $\{e(m, n)\}$ is a double array sequence of i.i.d. random variables with mean zero, variance σ^2, and with finite fourth moment.

Observe that Assumption 9.7 is a natural generalization of Assumption 9.1 from 1-D to 2-D. The additive error $X(m, n)$ which satisfies Assumption 9.7, is known as the 2-D linear process. Lahiri [32] established that under Assumption 9.7, the LSE of $\boldsymbol{\Theta}^0$ is strongly consistent under the same assumption as in Theorem 9.21. Moreover, the LSEs are asymptotically normally distributed as provided in the following theorem. Let us denote $\widehat{\boldsymbol{\Gamma}}$ as the LSE of $\boldsymbol{\Gamma}^0$.

Theorem 9.23 *Under the same assumptions as in Theorem 9.21 and Assumption 9.7*

$$\mathbf{D}(\widehat{\boldsymbol{\Gamma}} - \boldsymbol{\Gamma}^0) \xrightarrow{d} N_6(\mathbf{0}, 2c\sigma^2\boldsymbol{\Sigma}),$$

where matrices $\mathbf{D}$ *and* $\boldsymbol{\Sigma}$ *are same as defined in Theorem 9.22 and*

$$c = \sum_{j=-\infty}^{\infty} \sum_{k=-\infty}^{\infty} a^2(j, k).$$

∎

9.3.1.2   Approximate Least Squares Estimators

Along the same line as the 1-D chirp signal model, Grover, Kundu, and Mitra [21], considered a 2-D periodogram type function as follows:

$$I(\alpha, \beta, \gamma, \delta) = \frac{2}{MN} \left\{ \left(\sum_{m=1}^{M} \sum_{n=1}^{N} y(m, n) \cos(\alpha m + \beta m^2 + \gamma n + \delta n^2) \right)^2 + \left(\sum_{m=1}^{M} \sum_{n=1}^{N} y(m, n) \sin(\alpha m + \beta m^2 + \gamma n + \delta n^2) \right)^2 \right\}. \quad (9.46)$$

The main idea about the above 2-D periodogram type function has been obtained from the periodogram estimator in case of the 2-D sinusoidal model, discussed in Sect. 8.2.3. It has been observed by Kundu and Nandi [30], that the 2-D periodogram type estimators for a 2-D sinusoidal model are consistent and asymptotically equivalent to the corresponding LSEs. The 2-D periodogram type estimators or the ALSEs of the 2-D chirp model can be obtained by the argument maximum of $I(\alpha, \beta, \gamma, \delta)$ given in (9.46) over the range $(0, \pi) \times (0, \pi) \times (0, \pi) \times (0, \pi)$. The explicit solutions of the argument maximum of $I(\alpha, \beta, \gamma, \delta)$ cannot be obtained analytically. Numerical methods are required to compute the ALSEs. Extensive simulation experiments have been performed by Grover, Kundu, and Mitra [21], and it has been observed that the Downhill Simplex method, discussed in Sect. 2.4.2, performs quite well to compute the ALSEs, provided the initial guesses are quite close to the true values. It has been observed in simulation studies that although both LSEs and ALSEs involve solving 4-D optimization problem, computational time of the ALSEs is significantly lower than the LSEs. It has been established that the LSEs and ALSEs are asymptotically equivalent. Therefore, the ALSEs are strongly consistent and the asymptotic distribution of the ALSEs is same as the LSEs.

Since the computation of the LSEs or the ALSEs is a challenging problem, several other computationally efficient methods are available in the literature. But in most of the cases, either the asymptotic properties are unknown or they may not have the same efficiency as the LSEs or ALSEs. Next, we provide estimators which have the same asymptotic efficiency as the LSEs or ALSEs and at the same time they can be computed more efficiently than the LSEs or the ALSEs.

9.3.1.3 2-D Finite Step Efficient Algorithm

Lahiri, Kundu, and Mitra [35], extended the 1-D efficient algorithm as has been described in Sect. 9.2.1.4, for 2-D single chirp signal model. The idea is quite similar. It is observed that if we start with the initial guesses of α^0 and γ^0 having convergence rates $O_p(M^{-1}N^{-1/2})$ and $O_p(N^{-1}M^{-1/2})$, respectively, and β^0 and δ^0 having convergence rates $O_p(M^{-2}N^{-1/2})$ and $O_p(N^{-2}M^{-1/2})$, respectively, then after four iterations the algorithm produces estimates of α^0 and γ^0 having convergence rates $O_p(M^{-3/2}N^{-1/2})$ and $O_p(N^{-3/2}M^{-1/2})$, respectively, and β^0 and δ^0 having convergence rates $O_p(M^{-5/2}N^{-1/2})$ and $O_p(N^{-5/2}M^{-1/2})$, respectively. Therefore, the efficient algorithm produces estimates which have the same rates of convergence as the LSEs or the ALSEs. Moreover, it is guaranteed that the algorithm stops after four iterations.

Before providing the algorithm in details, we introduce the following notation and some preliminary results similarly as in Sect. 9.2.1.4. If $\widetilde{\alpha}$ is an estimator of α^0 such that $\widetilde{\alpha} - \alpha^0 = O_p(M^{-1-\lambda_{11}}N^{-\lambda_{12}})$, for some $0 < \lambda_{11}, \lambda_{12} \leq 1/2$, and $\widetilde{\beta}$ is an estimator of β^0 such that $\widetilde{\beta} - \beta^0 = O_p(M^{-2-\lambda_{21}}N^{-\lambda_{22}})$, for some $0 < \lambda_{21}, \lambda_{22} \leq 1/2$, then an improved estimator of α^0 can be obtained as

$$\widetilde{\widetilde{\alpha}} = \widetilde{\alpha} + \frac{48}{M^2}\text{Im}\left(\frac{P_{MN}^\alpha}{Q_{MN}^{\alpha,\beta}}\right), \tag{9.47}$$

with

$$P_{MN}^\alpha - \sum_{n=1}^{N}\sum_{m=1}^{M} y(m,n)\left(m - \frac{M}{2}\right)e^{-i(\widetilde{\alpha}m + \widetilde{\beta}m^2)} \tag{9.48}$$

$$Q_{MN}^{\alpha,\beta} = \sum_{n=1}^{N}\sum_{m=1}^{M} y(m,n)e^{-i(\widetilde{\alpha}m + \widetilde{\beta}m^2)}. \tag{9.49}$$

Similarly, an improved estimator of β^0 can be obtained as

$$\widetilde{\widetilde{\beta}} = \widetilde{\beta} + \frac{45}{M^4}\text{Im}\left(\frac{P_{MN}^{\beta}}{Q_{MN}^{\alpha,\beta}}\right), \tag{9.50}$$

with

$$P_{MN}^{\beta} = \sum_{n=1}^{N}\sum_{m=1}^{M} y(m,n)\left(m^2 - \frac{M^2}{3}\right) e^{-i(\widetilde{\alpha}m + \widetilde{\beta}m^2)} \tag{9.51}$$

and $Q_{MN}^{\alpha,\beta}$ is same as defined above in (9.49).

The following two results provide the justification for the improved estimators, whose proofs can be obtained in [35].

Theorem 9.24 *If* $\widetilde{\alpha} - \alpha^0 = O_p(M^{-1-\lambda_{11}}N^{-\lambda_{12}})$ *for* $\lambda_{11}, \lambda_{12} > 0$, *then*

  (a) $(\widetilde{\widetilde{\alpha}} - \alpha^0) = O_p(M^{-1-2\lambda_{11}}N^{-\lambda_{12}})$ *if* $\lambda_{11} \leq 1/4$,
  (b) $M^{3/2}N^{1/2}(\widetilde{\widetilde{\alpha}} - \alpha^0) \xrightarrow{d} \mathcal{N}(0, \sigma_1^2)$ *if* $\lambda_{11} > 1/4, \lambda_{12} = 1/2$,

where $\sigma_1^2 = \dfrac{384c\sigma^2}{A^{0^2} + B^{0^2}}$, *the asymptotic variance of the LSE of* α^0, *and c is same as defined in Theorem 9.23.*   ∎

Theorem 9.25 *If* $\widetilde{\beta} - \beta^0 = O_p(M^{-2-\lambda_{21}}N^{-\lambda_{22}})$ *for* $\lambda_{21}, \lambda_{22} > 0$, *then*

  (a) $(\widetilde{\widetilde{\beta}} - \beta^0) = O_p(M^{-2-\lambda_{21}}N^{-\lambda_{22}})$ *if* $\lambda_{21} \leq 1/4$,
  (b) $M^{5/2}N^{1/2}(\widetilde{\widetilde{\beta}} - \beta^0) \xrightarrow{d} \mathcal{N}(0, \sigma_2^2)$ *if* $\lambda_{21} > 1/4, \lambda_{22} = 1/2$,

where $\sigma_2^2 = \dfrac{360c\sigma^2}{A^{0^2} + B^{0^2}}$, *the asymptotic variance of the LSE of* β^0, *and c is same as defined in the previous theorem.*   ∎

In order to find estimators of γ^0 and δ^0, interchange the roles of M and N. If $\widetilde{\gamma}$ is an estimator of γ^0 such that $\widetilde{\gamma} - \gamma^0 = O_p(N^{-1-\kappa_{11}}N^{-\kappa_{12}})$, for some $0 < \kappa_{11}, \kappa_{12} \leq 1/2$, and $\widetilde{\delta}$ is an estimator of δ^0 such that $\widetilde{\delta} - \delta^0 = O_p(M^{-2-\kappa_{21}}N^{-\kappa_{22}})$, for some $0 < \kappa_{21}, \kappa_{22} \leq 1/2$, then an improved estimator of γ^0 can be obtained as

$$\widetilde{\widetilde{\gamma}} = \widetilde{\gamma} + \frac{48}{N^2}\text{Im}\left(\frac{P_{MN}^{\gamma}}{Q_{MN}^{\gamma,\delta}}\right), \tag{9.52}$$

with

$$P_{MN}^{\gamma} = \sum_{n=1}^{N} \sum_{m=1}^{M} y(m,n)\left(n - \frac{N}{2}\right) e^{-i(\tilde{\gamma}n + \tilde{\delta}n^2)} \tag{9.53}$$

$$Q_{MN}^{\gamma,\delta} = \sum_{n=1}^{N} \sum_{m=1}^{M} y(m,n) e^{-i(\tilde{\gamma}n + \tilde{\delta}n^2)}, \tag{9.54}$$

and an improved estimator of δ^0 can be obtained as

$$\tilde{\tilde{\delta}} = \tilde{\delta} + \frac{45}{N^4} \mathrm{Im}\left(\frac{P_{MN}^{\delta}}{Q_{MN}^{\gamma,\delta}}\right), \tag{9.55}$$

with

$$P_{MN}^{\delta} = \sum_{n=1}^{N} \sum_{m=1}^{M} y(m,n)\left(n^2 - \frac{N^2}{3}\right) e^{-i(\tilde{\gamma}n + \tilde{\delta}n^2)} \tag{9.56}$$

and $Q_{MN}^{\gamma,\delta}$ is same as defined in (9.54).

In this case, the following two results provide the justification for the improved estimators and the proofs can be obtained in Lahiri, Kundu, and Mitra [35].

Theorem 9.26 *If* $\tilde{\gamma} - \gamma^0 = O_p(N^{-1-\kappa_{11}} M^{-\kappa_{12}})$ *for* $\kappa_{11}, \kappa_{12} > 0$, *then*

$$(a)(\tilde{\tilde{\gamma}} - \gamma^0) = O_p(N^{-1-2\kappa_{11}} M^{-\kappa_{12}}) \quad \text{if } \kappa_{11} \leq 1/4,$$
$$(b)N^{3/2}M^{1/2}(\tilde{\tilde{\gamma}} - \gamma^0) \xrightarrow{d} \mathcal{N}(0, \sigma_1^2) \text{ if } \kappa_{11} > 1/4, \kappa_{12} = 1/2.$$

Here σ_1^2 *and* c *are same as defined in Theorem 9.24.* ∎

Theorem 9.27 *If* $\tilde{\delta} - \delta^0 = O_p(N^{-2-\kappa_{21}} M^{-\kappa_{22}})$ *for* $\kappa_{21}, \kappa_{22} > 0$, *then*

$$(a)(\tilde{\tilde{\delta}} - \delta^0) = O_p(N^{-2-\kappa_{21}} M^{-\kappa_{22}}) \quad \text{if } \kappa_{21} \leq 1/4,$$
$$(b)N^{5/2}M^{1/2}(\tilde{\tilde{\delta}} - \delta^0) \xrightarrow{d} \mathcal{N}(0, \sigma_2^2) \text{ if } \kappa_{21} > 1/4, \kappa_{22} = 1/2.$$

Here σ_2^2 *and* c *are same as defined in Theorem 9.25.* ∎

In the following, we show that starting from the initial guesses $\tilde{\alpha}, \tilde{\beta}$, with convergence rates $\tilde{\alpha} - \alpha^0 = O_p(M^{-1}N^{-1/2})$ and $\tilde{\beta} - \beta^0 = O_p(M^{-2}N^{-1/2})$, respectively, how the above results can be used to obtain efficient estimators, which have the same rate of convergence as the LSEs.

Similar to the idea of finite steps efficient algorithm for 1-D chirp model, discussed in Sect. 9.2.1.4, the idea here is not to use the whole sample at the beginning. We use part of the sample at the beginning and gradually proceed towards the complete sample. The algorithm is described below. We denote the estimates of α^0 and β^0 obtained at the jth iteration as $\tilde{\alpha}^{(j)}$ and $\tilde{\beta}^{(j)}$, respectively.

Algorithm 9.4

- Step 1: Choose $M_1 = M^{8/9}$, $N_1 = N$. Therefore,

$$\tilde{\alpha}^{(0)} - \alpha^0 = O_p(M^{-1}N^{-1/2}) = O_p(M_1^{-1-1/8}N_1^{-1/2}) \quad \text{and}$$
$$\tilde{\beta}^{(0)} - \beta^0 = O_p(M^{-2}N^{-1/2}) = O_p(M_1^{-2-1/4}N_1^{-1/2}).$$

Perform steps (9.47) and (9.50). Therefore, after 1-st iteration, we have

$$\tilde{\alpha}^{(1)} - \alpha^0 = O_p(M_1^{-1-1/4}N_1^{-1/2}) = O_p(M^{-10/9}N^{-1/2}) \quad \text{and}$$
$$\tilde{\beta}^{(1)} - \beta^0 = O_p(M_1^{-2-1/2}N_1^{-1/2}) = O_p(M^{-20/9}N^{-1/2}).$$

- Step 2: Choose $M_2 = M^{80/81}$, $N_1 = N$. Therefore,

$$\tilde{\alpha}^{(1)} - \alpha^0 = O_p(M_2^{-1-1/8}N_2^{-1/2}) \quad \text{and}$$
$$\tilde{\beta}^{(1)} - \beta^0 = O_p(M_2^{-2-1/4}N_2^{-1/2}).$$

Perform steps (9.47) and (9.50). Therefore, after 2-nd iteration, we have

$$\tilde{\alpha}^{(2)} - \alpha^0 = O_p(M_2^{-1-1/4}N_2^{-1/2}) = O_p(M^{-100/81}N^{-1/2}) \quad \text{and}$$
$$\tilde{\beta}^{(2)} - \beta^0 = O_p(M_2^{-2-1/2}N_2^{-1/2}) = O_p(M^{-200/81}N^{-1/2}).$$

- Step 3: Choose $M_3 = M$, $N_3 = N$. Therefore,

$$\tilde{\alpha}^{(2)} - \alpha^0 = O_p(M_3^{-1-19/81}N_3^{-1/2}) \quad \text{and}$$
$$\tilde{\beta}^{(2)} - \beta^0 = O_p(M_2^{-2-38/81}N_3^{-1/2}).$$

Again, performing steps (9.47) and (9.50), after 3-rd iteration, we have

$$\tilde{\alpha}^{(3)} - \alpha^0 = O_p(M^{-1-38/81}N^{-1/2}) \quad \text{and}$$
$$\tilde{\beta}^{(3)} - \beta^0 = O_p(M^{-2-76/81}N^{-1/2}).$$

- Step 4: Choose $M_4 = M$, $N_4 = N$, and after performing steps (9.47) and (9.50) we obtain the required convergence rates, i.e.

$$\tilde{\alpha}^{(4)} - \alpha^0 = O_p(M^{-3/2}N^{-1/2}) \quad \text{and}$$
$$\tilde{\beta}^{(4)} - \beta^0 = O_p(M^{-5/2}N^{-1/2}).$$

Similarly, interchanging the role of M and N, we can get the algorithm corresponding to γ^0 and δ^0. Extensive simulation experiments have been carried out by Lahiri [32], and it is observed that the performance of the finite step algorithm is quite satisfactory in terms of MSEs and biases. In initial steps, the part of the sample is selected in such a way that the dependence structure is maintained in the subsample. The MSEs and biases of the finite step algorithm are very similar to the corresponding performance of the LSEs. Therefore, the finite step algorithm can be used quite efficiently in practice. Now we will introduce more general 2-D polynomial phase model and discuss its applications, estimation procedures, and their properties.

9.4 Two-Dimensional Polynomial Phase Signal Model

In previous sections, we have discussed about 1-D and 2-D chirp models in details. But 2-D polynomial phase signal model also has received significant amount of attention in the signal processing literature. Francos and Friedlander [15] first introduced the most general 2-D polynomial (of degree r) phase signal model, and it can be described as follows:

$$
y(m, n) = A^0 \cos\left(\sum_{p=1}^{r}\sum_{j=0}^{p} \alpha^0(j, p-j)m^j n^{p-j}\right) +
$$

$$
B^0 \sin\left(\sum_{p=1}^{r}\sum_{j=0}^{p} \alpha^0(j, p-j)m^j n^{p-j}\right) + X(m, n);
$$

$$
m = 1, \ldots, M; \quad n = 1, \ldots, N, \tag{9.57}
$$

here $X(m, n)$ is the additive error with mean 0, A^0 and B^0 are non-zero amplitudes, and for $j = 0, \ldots, p$, for $p = 1, \ldots, r$, $\alpha^0(j, p-j)$'s are distinct frequency rates of order $(j, p-j)$, respectively, and lie strictly between 0 and π. Here $\alpha^0(0, 1)$ and $\alpha^0(1, 0)$ are called frequencies. The explicit assumptions on the error $X(m, n)$ will be provided later.

Different specific forms of model (9.57), have been used quite extensively in the literature. Friedlander and Francos [17] used 2-D polynomial phase signal model to analyze fingerprint type data, and Djurović, Wang, and Ioana [12], used a specific 2-D cubic phase signal model due to its applications in modeling SAR data, and in particular, Interferometric SAR data. Further, the 2-D polynomial phase signal model also has been used in modeling and analyzing MRI data, optical imaging, and different texture imaging. For some of the other specific applications of this model, one may refer to Djukanović and Djurović [9], Ikram and Zhou [26], Wang and Zhou [59], Tichavsky and Handel [55], Amar, Leshem, and van der Veen [2], and see the references cited therein.

Different estimation procedures have been suggested in the literature based on the assumption that the error components are i.i.d. random variables with mean zero and finite variance. For example, Farquharson, O'Shea, and Ledwich [13] provided a computationally efficient estimation procedures of the unknown parameters of a polynomial phase signals. Djurović and Stanković [11] proposed the quasi maximum likelihood estimators based on the normality assumption of the error random variables. Recently, Djurović, Simeunović, and Wang [10], considered an efficient estimation method of the polynomial phase signal parameters by using a cubic phase function. Some of the refinement of the parameter estimation of the polynomial phase signals can be obtained in O'Shea [46].

Interestingly, a significant amount of work has been done in developing different estimation procedures to compute parameter estimates of the polynomial phase signal, but not much work has been done in developing the properties of the estimators. In most of the cases, the MSEs or the variances of the estimates are compared with the corresponding Cramer-Rao lower bound, without establishing formally that the asymptotic variances of the maximum likelihood estimators attain the lower bound in case the errors are i.i.d. normal random variables. Recently, Lahiri, and Kundu [33], established formally the asymptotic properties of the LSEs of parameters in model (9.57), under a fairly general assumptions on the error random variables. From the results of Lahiri and Kundu [33], it can be easily obtained that when the errors are i.i.d. normally distributed, the asymptotic variance of the MLEs attain the Cramer-Rao lower bound.

Lahiri and Kundu [33] established the consistency and the asymptotic normality properties of the LSEs of the parameters of model (9.57) under a general error assumption. It is assumed that the errors $X(m, n)$s follow Assumption 9.7 as defined in Sect. 9.3.1.1, and the parameter vector satisfies Assumption 9.8, given below. For establishing the consistency and asymptotic normality of the LSEs, the following Assumption is required for further development.

Assumption 9.8 Denote the true parameters by $\boldsymbol{\Gamma}^0 = (A^0, B^0, \alpha^0(j, p - j), j = 0, \ldots, p, p = 1, \ldots, r)^T$ and the parameter space by $\boldsymbol{\Theta} = [-K, K] \times [-K, K] \times [0, \pi]^{r(r+3)/2}$. Here $K > 0$ is an arbitrary constant and $[0, \pi]^{r(r+3)/2}$ denotes $r(r + 3)/2$ fold of $[0, \pi]$. It is assumed that $\boldsymbol{\Gamma}^0$ is an interior point of $\boldsymbol{\Theta}$.

Lahiri and Kundu [33] established the following results under the assumption of 2-D stationary error process. The details can be obtained in that paper.

Theorem 9.28 *If Assumptions 9.7 and 9.8 are satisfied, then $\widehat{\boldsymbol{\Gamma}}$, the LSE of $\boldsymbol{\Gamma}^0$, is a strongly consistent estimator of $\boldsymbol{\Gamma}^0$.* ∎

Theorem 9.29 *Under Assumptions 9.7 and 9.8, $\mathbf{D}(\widehat{\boldsymbol{\Gamma}} - \boldsymbol{\Gamma}^0) \to \mathcal{N}_d(0, 2c\sigma^2 \boldsymbol{\Sigma}^{-1})$ where the matrix $\mathbf{D}$ is a $(2 + \frac{r(r+3)}{2}) \times (2 + \frac{r(r+3)}{2})$ diagonal matrix of the form*

$$\mathbf{D} = diag\left(M^{\frac{1}{2}}N^{\frac{1}{2}}, M^{\frac{1}{2}}N^{\frac{1}{2}}, M^{j+\frac{1}{2}}N^{(p-j)+\frac{1}{2}}, j = 0, \ldots, p, p = 1, \ldots, r\right),$$

and

$$\boldsymbol{\Sigma} = \begin{bmatrix} 1 & 0 & \mathbf{V}_1 \\ 0 & 1 & \mathbf{V}_2 \\ \mathbf{V}_1^T & \mathbf{V}_2^T & \mathbf{W} \end{bmatrix} \tag{9.58}$$

Here $\mathbf{V}_1 = \left(\frac{B^0}{(j+1)(p-j+1)}, \ j = 0, \dots, p, \ p = 1, \dots, r\right)$, $\mathbf{V}_2 = \left(-\frac{A^0}{(j+1)(p-j+1)}, \ j = 0, \dots, p, \ p = 1, \dots, r\right)$, *are vectors of order* $\frac{r(r+3)}{2}$, $W = \left(\frac{A^{0^2}+B^{0^2}}{(j+k+1)(p+q-j-k+1)}, \ j = 0, \dots, p, \ p = 1, \dots, r, \ k = 0, \dots, q, \ q = 1, \dots, r\right)$, *is a matrix of order* $\frac{r(r+3)}{2} \times \frac{r(r+3)}{2}$, $c = \sum_{j=-\infty}^{\infty} \sum_{k=-\infty}^{\infty} a(j,k)^2$, *and* $d = 2 + \frac{r(r+3)}{2}$, ∎

Theoretical properties of the LSEs have been established by Lahiri and Kundu [33], but finding the LSEs is a computationally challenging problem. No work has been done along this line. It is a very important practical problems. Further work is needed along that direction.

Open Problem 9.12 Develop an efficient algorithm to estimate the unknown parameters of 2-D polynomial phase signal model.

9.5 Conclusions

In this chapter, we have provided a brief review of 1-D and 2-D chirp and some related models which have received a considerable amount of attention in recent years in the signal processing literature. These models have been used in analyzing different real-life signals or images quite efficiently. It is observed that several sophisticated statistical and computational techniques are needed to analyze these models and in developing estimation procedures. We have provided several open problems for future research. Some of the related models of chirp signal model, like random amplitude chirp model, harmonic chirp, and the chirp like model are discussed in Chaps. 10 and 11. These models have received some attention in recent years. It is observed that the chirp like model can conveniently be used in place of multicomponent chirp model in practice, and is quite useful for analyzing real-life signals.

References

1. Abatzoglou, T. (1986). Fast maximum likelihood joint estimation of frequency and frequency rate. *IEEE Transactions on Aerospace and Electronic Systems, 22,* 708–715.
2. Amar, A., Leshem, A., & van der Veen, A. J. (2010). A low complexity blind estimator of narrow-band polynomial phase signals. *IEEE Transactions on Signal Processing, 58,* 4674–4683.
3. Bai, Z. D., Rao, C. R., Chow, M., & Kundu, D. (2003). An efficient algorithm for estimating the parameters of superimposed exponential signals. *Journal of Statistical Planning and Inference, 110,* 23–34.
4. Benson, O., Ghogho, M., & Swami, A. (1999). Parameter estimation for random amplitude chirp signals. *IEEE Transactions on Signal Processing, 47,* 3208–3219.

5. Besson, O., Giannakis, G. B., & Gini, F. (1999). Improved estimation of hyperbolic frequency modulated chirp signals. *IEEE Transactions on Signal Processing.*, *47*, 1384–1388.
6. Cao, F.,Wang, S., & Wang, F. (2006). Cross-spectral method based on 2-D cross polynomial transform for 2-D chirp signal parameter estimation. In *ICSP2006 Proceedings*. https://doi.org/10.1109/ICOSP.2006.344475.
7. Dhar, S. S., Kundu, D., & Das, U. (2019). On testing parameters of chirp signal model. *IEEE Transactions on Signal Processing*, *67*, 4291–4301.
8. Djurić, P. M., & Kay, S. M. (1990). Parameter estimation of chirp signals. *IEEE Transactions on Acoustics, Speech and Signal Processing*, *38*, 2118–2126.
9. Djukanović, S., & Djurović, I. (2012). Aliasing detection and resolving in the estimation of polynomial-phase signal parameters. *Signal Processing*, *92*, 235–239.
10. Djurović, I., Simeunović, M., & Wang, P. (2017). Cubic phase function: A simple solution for polynomial phase signal analysis. *Signal Processing*, *135*, 48–66.
11. Djurović, I., & Stanković, L. J. (2014). Quasi maximum likelihood estimators of polynomial phase signals. *IET Signal Processing*, *13*, 347–359.
12. Djurović, I., Wang, P., & Ioana, C. (2010). Parameter estimation of 2-D cubic phase signal function using genetic algorithm. *Signal Processing*, *90*, 2698–2707.
13. Farquharson, M., O'Shea, P., & Ledwich, G. (2005). A computationally efficient technique for estimating the parameters phase signals from noisy observations. *IEEE Transactions on Signal Processing*, *53*, 3337–3342.
14. Fourier, D., Auger, F., Czarnecki, K., & Meignen, S. (2017). Chirp rate and instantaneous frequency estimation: application to recursive vertical synchrosqueezing. *IEEE Signal Processing Letters*, *24*, 1724–1728.
15. Francos, J. M., & Friedlander, B. (1998). Two-dimensional polynomial phase signals: Parameter estimation and bounds. *Multidimensional Systems and Signal Processing*, *9*, 173–205.
16. Francos, J. M., & Friedlander, B. (1999). Parameter estimation of 2-D random amplitude polynomial phase signals. *IEEE Transactions on Signal Processing*, *47*, 1795–1810.
17. Friedlander, B., & Francos, J. M. (1996). An estimation algorithm for 2-D polynomial phase signals. *IEEE Transactions on Image Processing*, *5*, 1084–1087.
18. Gabor, D. (1946). Theory of communication. Part 1: The analysis of information. *Journal of the Institution of Electrical Engineers - Part III: Radio and Communication Engineering*, *93*, 429–441.
19. Gini, F., Montanari, M., & Verrazzani, L. (2000). Estimation of chirp signals in compound Gaussian clutter: A cyclostationary approach. *IEEE Transactions on Acoustics, Speech and Signal Processing*, *48*, 1029–1039.
20. Grover, R., Kundu, D., & Mitra, A. (2018). On approximate least squares estimators of parameters of one-dimensional chirp signal. *Statistics*, *52*, 1060–1085.
21. Grover, R., Kundu, D., & Mitra, A. (2018). Asymptotic of approximate least squares estimators of parameters of two-dimensional chirp signal. *Journal of Multivariate Analysis*, *168*, 211–220.
22. Gu, T., Liai, G., Li, Y., Guo, Y., & Huang, Y. (2020). Parameter estimation of multicomponent LFM signals based on GAPCK. *Digital Signal Processing*, *100*. Article ID. 102683.
23. Guo, J., Zou, H., Yang, X., & Liu, G. (2011). Parameter estimation of multicomponent chirp signals via sparse representation. *IEEE Transactions on Aerospace and Electronic Systems*, *47*, 2261–2268.
24. Hedley, M., & Rosenfeld, D. (1992). A new two-dimensional phase unwrapping algorithm for MRI images. *Magnetic Resonance in Medicine*, *24*, 177–181.
25. Ikram, M. Z., Abed-Meraim, K., & Hua, Y. (1998). Estimating the parameters of chirp signals: An iterative aproach. *IEEE Transactions on Signal Processing*, *46*, 3436–3441.
26. Ikram, M. Z., & Zhou, G. T. (2001). Estimation of multicomponent phase signals of mixed orders. *Signal Processing*, *81*, 2293–2308.
27. Jennrich, R. I. (1969). Asymptotic properties of the nonlinear least squares estimators. *Annals of Mathematical Statistics*, *40*, 633–643.
28. Kennedy, W. J, Jr., & Gentle, J. E. (1980). *Statistical Computing*. New York: Marcel Dekker Inc.

29. Kim, G., Lee, J., Kim, Y., & Oh, H.-S. (2015). Sparse Bayesian representation in time-frequency domain. *Journal of Statistical Planning and Inference, 166*, 126–137.
30. Kundu, D., & Nandi, S. (2003). Determination of discrete spectrum in a random field. *Statistica Neerlandica, 57*, 258–283.
31. Kundu, D., & Nandi, S. (2008). Parameter estimation of chirp signals in presence of stationary Christensen, noise. *Statistica Sinica, 18*, 187–201.
32. Lahiri, A. (2011). *Estimators of parameters of chirp signals and their properties*. Ph.D. Dissertation, Department of Mathematics and Statistics, Indian Institute of Technology, Kanpur, India.
33. Lahiri, A., & Kundu, D. (2017). On parameter estimation of two-dimensional polynomial phase signal model. *Statistica Sinica, 27*, 1779–1792.
34. Lahiri, A., Kundu, D., & Mitra, A. (2012). Efficient algorithm for estimating the parameters of chirp signal. *Journal of Multivariate Analysis, 108*, 15–27.
35. Lahiri, A., Kundu, D., & Mitra, A. (2013). Efficient algorithm for estimating the parameters of two dimensional chirp signal. *Sankhya Series B, 75*, 65–89.
36. Lahiri, A., Kundu, D., & Mitra, A. (2014). On least absolute deviation estimator of one dimensional chirp model. *Statistics, 48*, 405–420.
37. Lahiri, A., Kundu, D., & Mitra, A. (2015). Estimating the parameters of multiple chirp signals. *Journal of Multivariate Analysis, 139*, 189–205.
38. Lin, C-C., & Djurić, P. M. (2000). Estimation of chirp signals by MCMC. *ICASSP-1998, 1*, 265–268.
39. Liu, X., & Yu, H. (2013). Time-domain joint parameter estimation of chirp signal based on SVR. *Mathematical Problems in Engineering*. Article ID: 952743.
40. Lu, Y., Demirli, R., Cardoso, G., & Saniie, J. (2006). A successive parameter estimation algorithm for chirplet signal decomposition. *IEEE Transactions on Ultrasonic, Ferroelectrics and Frequency Control, 53*, 2121–2131.
41. Mazumder, S. (2017). Single-step and multiple-step forecasting in one-dimensional chirp signal using MCMC-based Bayesian analysis. *Communications in Statistics - Simulation and Computation, 46*, 2529–2547.
42. Montgomery, H. L. (1990). *Ten lectures on the interface between analytic number theory and harmonic analysis* (vol. 196). Providence: American Mathematical Society.
43. Nandi, S., & Kundu, D. (2004). Asymptotic properties of the least squares estimators of the parameters of the chirp signals. *Annals of the Institute of Statistical Mathematics, 56*, 529–544.
44. Nandi, S., & Kundu, D. (2006). Analyzing non-stationary signals using a cluster type model. *Journal of Statistical Planning and Inference, 136*, 3871–3903.
45. Nandi, S., Kundu, D., & Grover, R. (2019). Estimation of parameters of multiple chirp signal in presence of heavy tailed errors.
46. O'Shea, P. (2010). On refining polynomial phase signal parameter estimates. *IEEE Transactions on Aerospace, Electronic Syatems, 4*, 978–987.
47. Pelag, S., & Porat, B. (1991). Estimation and classification of polynomial phase signals. *IEEE Transactions on Information Theory, 37*, 422–430.
48. Richards, F. S. G. (1961). A method of maximum likelihood estimation. *Journal of Royal Statistical Society Series B, 23*, 469–475.
49. Pincus, M. (1968). A closed form solution of certain programming problems. *Operation Research, 16*, 690–694.
50. Rihaczek, A. W. (1969). *Principles of high resolution radar*. New York: McGraw-Hill.
51. Robertson, S. D. (2008). Generalization and application of the linear chirp. Ph.D. Dissertation, Southern Methodist University, Department of Statistical Science.
52. Robertson, S. D., Gray, H. L., & Woodward, W. A. (2010). The generalized linear chirp process. *Journal of Statistical Planning and Inference, 140*, 3676–3687.
53. Saha, S., & Kay, S. M. (2002). Maximum likelihood parameter estimation of superimposed chirps using Monte Carlo importance sampling. *IEEE Transactions on Signal Processing, 50*, 224–230.
54. Seber, G. A. F., & Wild, C. J. (1989). *Nonlinear regression*. New York: Wiley.

55. Ticahvsky, P., & Handel, P. (1999). Multicomponent polynomial phase signal analysis using a tracking algorithm. *IEEE Transactions on Signal Processing, 47,* 1390–1395.
56. Volcker, B., & Ottersten, B. (2001). Chirp parameter estimation from a sample covariance matrix. *IEEE Transactions on Signal Processing, 49,* 603–612.
57. Wang, J. Z., Su, S. Y., & Chen, Z. (2015). Parameter estimation of chirp signal under low SNR. *Science China: Information Sciences, 58.* Article ID 020307.
58. Wang, P., & Yang, J. (2006). Multicomponent chirp signals analysis using product cubic phase function. *Digital Signal Processing, 16,* 654–669.
59. Wang, Y., & Zhou, Y. G. T. (1998). On the use of high-order ambiguity function for multi-component polynomial phase signals. *Signal Processing, 5,* 283–296.
60. Wu, C. F. J. (1981). Asymptotic theory of the nonlinear least squares estimation. *Annals of Statistics, 9,* 501–513.
61. Xinghao, Z., Ran, T., & Siyong, Z. (2003). A novel sequential estimation algorithm for chirp signal parameters. In *IEEE Conference in Neural Networks and & Signal Processing,* Nanjing, China (pp. 628–631). Retrieved Dec 14–17, 2003
62. Yaron, D., Alon, A., & Israel, C. (2015). Joint model order selection and parameter estimation of chirps with harmonic components. *IEEE Transactions on Signal Processing, 63,* 1765–1778.
63. Yang, P., Liu, Z., & Jiang, W.-L. (2015). Parameter estimation of multicomponent chirp signals based on discrete chirp Fourier transform and population Monte Carlo. *Signal, Image and Video Processing, 9,* 1137–1149.
64. Zhang, H., Liu, H., Shen, S., Zhang, Y., & Wang, X. (2013). Parameter estimation of chirp signals based on fractional Fourier transform. *The Journal of China Universities of Posts and Telecommunications, 20*(Suppl. 2), 95–100.
65. Zhang, H., & Liu, Q. (2006). Estimation of instantaneous frequency rate for multicomponent polynomial phase signals. In *ICSP2006 Proceedings* (pp. 498–502). https://doi.org/10.1109/ICOSP.2006.344448.
66. Zhang, K., Wang, S., & Cao, F. (2008). Product cubic phase function algorithm for estimating the instantaneous frequency rate of multicomponent two-dimensional chirp signals. In *2008 Congress on Image and Signal Processing.* https://doi.org/10.1109/CISP.2008.352.

Chapter 10
Random Amplitude Sinusoidal and Chirp Model

10.1 Introduction

In this monograph, we have considered sinusoidal frequency model and many of its variants in one and higher dimensions. In all these models considered so far, amplitudes are assumed to be unknown constants. In this chapter, we allow the amplitudes to be random or some deterministic function of the index. Such random amplitude sinusoidal and chirp models generalize most of the models considered previously. A different set of assumptions will be required. The random amplitude can be thought of as a multiplicative error. Therefore, in this chapter, we are going to consider models having sinusoidal components observed in the presence of additive, as well as multiplicative error.

Some of the models discussed in this chapter are complex-valued. It should be mentioned that although all physical signals are real-valued, it might be advantageous from an analytical point of view to work with analytic form, that is, with complex-valued models, see Gabor [9]. The corresponding analytic form of a real-valued continuous signal can be obtained by using the Hilbert transform. Therefore, it is sometime quite natural and useful to work with complex-valued model and derive the theory for complex models and use them to the corresponding real models. In most of the signal processing literature, it is observed that the complex-valued models are being used for analytical derivation or algorithmic development, although they can be used for analyzing any real-valued signal, see, for example, Stoica and Moses [17].

In many applications, it is more appropriate to consider a multiplicative noise sinusoidal model than the usual additive noise model, for example, when the vehicle speed is to be estimated from an on-board Doppler radar. This results in a sinusoidal model with slowly varying amplitudes due to the effects of reflections and trajectories from different points of the vehicle. This phenomenon is somewhat related to fading and has been observed in many situations. Besson and Castanie [2] introduced a multiplicative sinusoidal model called ARCOS (AutoRegressive amplitude modulated COSinusoid) in order to describe this phenomenon. The model is described

© Springer Nature Singapore Pte Ltd. 2020
S. Nandi and D. Kundu, *Statistical Signal Processing*,
https://doi.org/10.1007/978-981-15-6280-8_10

as

$$y(t) = z(t)\cos(t\omega^0 + \phi), \tag{10.1}$$

where $z(t)$ is an Autoregressive process of order p (AR(p)), defined as

$$z(t) = -\sum_{k=1}^{p} \phi_k z(t-k) + \varepsilon(t), \tag{10.2}$$

$\{\varepsilon(t)\}$ is a sequence of i.i.d. random variables with mean zero and finite variance σ_ε^2. The frequency ω^0 is referred to as a Doppler frequency. The characteristic equation of the AR(p) process $z(t)$ is

$$\Phi(x) = 0, \quad \Phi(x) = \sum_{k=0}^{p} \phi_k x^{-k}, \quad \phi_0 = 1.$$

The AR(p) process is stationary provided the roots of the above polynomial equation are inside the unit circle. Besson and Castanie [2] and Besson [1] considered model (10.1) which describes a sinusoidal signal embedded in low-pass multiplicative noise, such that the zeros of $\Phi(x)$ are much smaller than ω^0.

A more general model than the ARCOS model was proposed by Besson and Stoica [5], where both multiplicative, as well as additive errors are present. The model is a complex-valued model and is given by

$$y(t) = A\alpha(t)e^{i\omega^0 t} + e(t), \tag{10.3}$$

where A is a complex-valued amplitude, $\alpha(t)$ is a real-valued time-varying envelope, ω^0 is the frequency and $\{e(t)\}$ is a sequence of complex-valued i.i.d. circular Gaussian random variables with zero mean and finite variance σ_e^2. The parameter of interest is ω^0 and time-varying amplitude $A\alpha(t)$ is considered as nuisance parameter. No specific structure was imposed on the envelope $\{\alpha(t)\}$ by Besson and Stoica [5]. The random amplitude sinusoidal model has received considerable attention in Signal Processing literature, for example, see Francos and Friedlander [8], Besson and Stoica [4], Zhou and Ginnakis [18], etc. If $\alpha(t)$ is identically equal to one, model (10.3), is nothing but the single-component complex-valued sinusoidal model. Ghogho, Nandi, and Swami [10], studied the maximum likelihood estimation for random amplitude phase-modulated signals. Ciblata et al. [6] considered model (10.3), and its equivalent real model and addressed the problem of harmonic retrieval.

In order to estimate the parameter of interest ω^0, present in model (10.3), one can consider the maximization of the periodogram, given a sample of size n. In case of constant amplitude, that is, $\alpha(t) = 1$, the nonlinear least squares approach is equivalent to maximizing the periodogram function and denote the estimate of ω^0 as $\widehat{\omega}_c$ in this case. Therefore,

$$\widehat{\omega}_c = \arg\max_\omega \frac{1}{n} \left| \sum_{t=1}^n y(t)e^{-i\omega t} \right|^2. \tag{10.4}$$

Besson and Stoica [5] proposed an approach that maximizes another form of peri-odogram function and is given by

$$\widehat{\omega}_r = \arg\max_\omega \frac{1}{n} \left| \sum_{t=1}^n y^2(t)e^{-i2\omega t} \right|^2 = \arg\max_\omega I_R(\omega), \quad (say) \tag{10.5}$$

when $\alpha(t)$ is time-varying and unknown. If we write $y^2(t) = x(t)$ and $2\omega = \beta$, then $I_R(\omega)$ is nothing but the periodogram function of $\{x(t)\}$. Also $I_R(\omega)$ can be viewed as the periodogram of the data points squared and interlaced with zeros, $\{y^2(1), 0, y^2(2), 0, \ldots\}$. As an alternative procedure, the maximum peak of $I_R(\omega)$ estimates the twice of the frequency ω^0, and hence ω^0 can be estimated by the argument maximum of $I_R(\omega)$ divided by two.

In this chapter, we consider the complex-valued random amplitude chirp signal in detail in Sect. 10.2, the single-component as well as multicomponent models. Some of the results related to these models are proved at the end of this chapter in appendices. A more general real-valued model is introduced in Sect. 10.3. Some related issues are discussed in Sect. 10.4. Random amplitude sinusoidal model is not considered separately as the chirp model is nothing but the sinusoidal model when the frequency rate is equal to zero. Therefore, the estimation methods and results discussed in subsequent sections are also applicable to random amplitude sinusoidal model.

10.2 Random Amplitude Chirp Model: Complex-Valued Model

A form of complex-valued chirp signal with random time-varying amplitude studied by Nandi and Kundu [16], is given by

$$y(t) = \alpha(t)e^{i(\theta_1^0 t + \theta_2^0 t^2)} + e(t), \quad t = 1, \ldots, n, \tag{10.6}$$

where θ_1^0 and θ_2^0 are frequency and chirp rate similar to the constant amplitude chirp model as defined in (9.1). The amplitude $\{\alpha(t)\}$ is a sequence of real-valued i.i.d. random variables with nonzero mean μ_α and variance σ_α^2. The additive error $\{e(t)\}$ is a sequence of complex-valued i.i.d. random variables with mean zero and variance σ^2. Write $e(t) = e_R(t) + ie_I(t)$, then $\{e_R(t)\}$ and $\{e_I(t)\}$ are i.i.d. $(0, \frac{\sigma^2}{2})$, have finite fourth moment γ and are independently distributed. It is also assumed that the multiplicative error $\{\alpha(t)\}$ is independent of $\{e(t)\}$ and the fourth moment of $\{\alpha(t)\}$ exists.

Model (10.6) is known as the random amplitude chirp signal model and with slight variation, it was first introduced by Besson, Ghogho, and Swami [3]. They originally considered the following model;

$$y(t) = \alpha(t) e^{i(\theta_0^0 + \theta_1^0 t + \theta_2^0 t^2)} + e(t), \quad t = 1, \dots, n, \tag{10.7}$$

where $\alpha(t)$, $e(t)$, θ_1^0 and θ_2^0 are same as in (10.6), but in model (10.7), there is a phase term θ_0^0, which has not been considered in model (10.6). Observe that when the mean of the multiplicative error, μ_α, is unknown and a phase term θ_0^0, which is also unknown, is present, then they are not identifiable. Therefore, if both are present one has to be known. Hence, a separate phase term has not been considered by Nandi and Kundu [16], and it is basically included in $\alpha(t)$ without loss of any generality. Therefore, (10.6) and (10.7), are equivalent models.

Nandi and Kundu [16] studied the theoretical properties of the estimator proposed by Besson, Ghogho, and Swami [3]. The frequency θ_1^0 and the frequency rate θ_2^0 are estimated by maximizing a periodogram like function similar to $I_R(\omega)$, defined in (10.5) and is given by

$$Q(\boldsymbol{\theta}) = \frac{1}{n} \left| \sum_{t=1}^{n} y^2(t) e^{-i2(\theta_1 t + \theta_2 t^2)} \right|^2, \tag{10.8}$$

where $\boldsymbol{\theta} = (\theta_1, \theta_2)$. Nandi and Kundu [16] proved the strong consistency of $\widehat{\boldsymbol{\theta}} = (\widehat{\theta}_1, \widehat{\theta}_2)$ that maximizes $Q(\boldsymbol{\theta})$ under the assumptions stated after model (10.6), and when (θ_1^0, θ_2^0) is an interior point of its parameter space. Under the same assumptions, the asymptotic distribution of $\widehat{\boldsymbol{\theta}}$ is also derived and is given by

$$\left(n^{\frac{3}{2}} (\widehat{\theta}_1 - \theta_1^0), n^{\frac{5}{2}} (\widehat{\theta}_2 - \theta_2^0) \right) \xrightarrow{d} \mathcal{N}_2(\mathbf{0}, 4(\sigma_\alpha^2 + \mu_\alpha^2)^2 \boldsymbol{\Sigma}^{-1} \boldsymbol{\Gamma} \boldsymbol{\Sigma}^{-1}),$$

where with $C_\alpha = 8(\sigma_\alpha^2 + \mu_\alpha^2)\sigma^2 + \frac{1}{2}\gamma + \frac{1}{8}\sigma^4$,

$$\boldsymbol{\Sigma} = \frac{2(\sigma_\alpha^2 + \mu_\alpha^2)^2}{3} \begin{pmatrix} 1 & 1 \\ 1 & \frac{16}{15} \end{pmatrix}, \quad \boldsymbol{\Gamma} = C_\alpha \begin{bmatrix} \frac{1}{3} & \frac{1}{4} \\ \frac{1}{4} & \frac{1}{5} \end{bmatrix}.$$

The strong consistency of $\widehat{\boldsymbol{\theta}}$ is proved in Appendix A and the asymptotic distribution is derived in Appendix B.

Remark 10.1 The asymptotic variances of $\widehat{\theta}_1$ and $\widehat{\theta}_2$ depend on the mean μ_α and the variance σ_α^2 of the random amplitude and variance σ^2 and fourth moment γ of the additive error. According to the asymptotic distribution, $\widehat{\theta}_1 = O_p(n^{-\frac{3}{2}})$ and $\widehat{\theta}_2 = O_p(n^{-\frac{5}{2}})$ similar to the constant amplitude chirp model discussed in Chap. 9. This implies that for a given sample size, the chirp rate θ_2 can be estimated more accurately than the frequency θ_1.

Remark 10.2 The random amplitude sinusoidal signal model $y(t) = \alpha(t)e^{i\theta_1^0 t} + e(t)$ is a special case of model (10.6). In this case, the effective frequency does not change over time and is constant throughout as the frequency rate $\theta_2^0 = 0$. The unknown frequency can be estimated by maximizing a similar function as $Q(\theta)$, defined in (10.8). The consistency and asymptotic normality of the estimator follow in the same way. The estimation procedure and the theoretical properties for random amplitude sinusoidal model can be deduced from the random amplitude chirp model, hence, it is not separately discussed.

Remark 10.3 Random amplitude generalized chirp is a complex-valued model of the form $y(t) = \alpha(t)e^{i(\theta_1^0 t + \theta_2^0 t^2 + \cdots + \theta_q^0 t^q)} + e(t)$. This is called as random amplitude polynomial phase signal. In this case, the frequency rate is not linear and change in frequency is governed by the term $\theta_2^0 t^2 + \cdots + \theta_q^0 t^q$. The parameters are estimated using a periodogram like function of the squared signal. Under similar assumptions on $\{\alpha(t)\}$, $\{e(t)\}$ and the true values of the parameter as in model (10.6), the consistency and asymptotic normality can be obtained. Estimation of q, the degree of the polynomial, is another important problem in the case of random amplitude polynomial phase signal.

10.2.1 Multicomponent Random Amplitude Chirp Model

Multicomponent random amplitude chirp model is a general model and captures the presence of multiple frequency and chirp rate pairs instead of a single pair as in model (10.6). The model can be written as

$$y(t) = \sum_{k=1}^{p} \alpha_k(t)e^{i(\theta_{1k}^0 t + \theta_{2k}^0 t^2)} + e(t). \tag{10.9}$$

The sequence of additive errors $\{e(t)\}$ is same as in the case of model (10.6). The multiplicative error corresponding to k-th component $\{\alpha_k(t)\}$ is a sequence of i.i.d. real-valued random variables with mean $\mu_{k\alpha}$, variance $\sigma_{k\alpha}^2$, and finite fourth moment, $\mu_{k\alpha} \neq 0$ and $\sigma_{k\alpha}^2 > 0$, $k = 1, \ldots, p$. It is also assumed that $\{\alpha_j(t)\}$ and $\{\alpha_k(t)\}$ for $j \neq k$ are independent and $\{e(t)\}$ is assumed to be independent of $\{\alpha_1(t)\}, \ldots, \{\alpha_p(t)\}$.

The unknown parameters for multicomponent model (10.9) are estimated by maximizing $Q(\theta)$, defined in (10.8) locally. Write $\theta_k = (\theta_{1k}, \theta_{2k})$ and let θ_k^0 be the true value of θ_k. Use the notation N_k as a neighborhood of θ_k^0. Then estimate θ_k as

$$\widehat{\theta}_k = \arg \max_{(\theta_1, \theta_2) \in N_k} \frac{1}{n} \left| \sum_{t=1}^{n} y^2(t)e^{-i2(\theta_1 t + \theta_2 t^2)} \right|^2,$$

where $y(t)$ is given in (10.9).

If $\{(\theta_{11}^0, \theta_{21}^0), (\theta_{12}^0, \theta_{22}^0), \ldots, (\theta_{1p}^0, \theta_{2p}^0)\}$ is an interior points of its parameter space and $(\theta_{1j}^0, \theta_{2j}^0) \neq (\theta_{1k}^0, \theta_{2k}^0)$ for $j \neq k$, $j, k = 1, \ldots, p$, then under the above set-up, $\widehat{\boldsymbol{\theta}}_k$, which maximizes $Q(\boldsymbol{\theta})$ in N_k, is a strongly consistent estimator of θ_k^0.

Under the same set-up, as $n \to \infty$,

$$\left(n^{\frac{3}{2}}(\widehat{\theta}_{1k} - \theta_{1k}^0), n^{\frac{5}{2}}(\widehat{\theta}_{2k} - \theta_{2k}^0)\right) \xrightarrow{d} \mathcal{N}_2(\mathbf{0}, 4(\sigma_{k\alpha}^2 + \mu_{k\alpha}^2)^2 \boldsymbol{\Sigma}_k^{-1} \boldsymbol{\Gamma}_k \boldsymbol{\Sigma}_k^{-1}),$$

where $C_{k\alpha} = 8(\sigma_{k\alpha}^2 + \mu_{k\alpha}^2)\sigma^2 + \frac{1}{2}\gamma + \frac{1}{8}\sigma^4$,

$$\boldsymbol{\Sigma}_k = \frac{2(\sigma_{k\alpha}^2 + \mu_{k\alpha}^2)^2}{3} \begin{pmatrix} 1 & 1 \\ 1 & \frac{16}{15} \end{pmatrix}, \quad \boldsymbol{\Gamma}_k = C_{k\alpha} \begin{bmatrix} \frac{1}{3} & \frac{1}{4} \\ \frac{1}{4} & \frac{1}{5} \end{bmatrix}.$$

Additionally, it has been observed that $\left(n^{\frac{3}{2}}(\widehat{\theta}_{1k} - \theta_{1k}^0), n^{\frac{5}{2}}(\widehat{\theta}_{2k} - \theta_{2k}^0)\right)$ and $\left(n^{\frac{3}{2}}(\widehat{\theta}_{1j} - \theta_{1j}^0), n^{\frac{5}{2}}(\widehat{\theta}_{2j} - \theta_{2j}^0)\right)$ for $k \neq j$ are asymptotically independently distributed.

The consistency of $\widehat{\boldsymbol{\theta}}_k$, $k = 1, \ldots, p$ can be established in the similar way as in the case of single-component model (10.6), given in Appendix A. In this case, $z_R(t)$ and $z_I(t)$ (defined in Appendix A) will be based on p-component $y(t)$, given in (10.9). The proof of the asymptotic distribution of the estimators of the unknown parameters in the case of model (10.9) goes along the same line as the proof, provided in Appendix B, once we have equivalent results of Lemma 10.5 (see Appendix A) for the multicomponent model (10.9). The asymptotic independence of the estimators corresponding to different components can be proved by finding the covariances between them because the estimators are asymptotically normally distributed. Interested readers are referred to Nandi and Kundu [16] for details.

10.2.2 Some Practical Issues

We would like to emphasize that all real-life data sets are real-valued. The complex model has been considered in the literature and in the present section due to analytical reasons. The corresponding analytic (complex) form of a real-valued continuous signal can be obtained by using Hilbert transform. For illustration, analysis of a real data "aaa" (already discussed in Chapter 7) using model (10.9) is discussed here. The data set contains 512 data points sampled at 10 kHz frequency and plotted in Fig. 10.1. Because "aaa" data is a real data set, the corresponding imaginary component is not readily available. To use complex-valued model (10.9), Hilbert transform is applied to the real data set and the corresponding complex data have been obtained. The chirp periodogram function

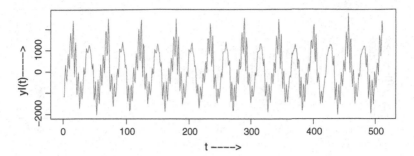

Fig. 10.1 The observed "aaa" signal

$$I_C(\alpha, \beta) = \frac{1}{n} \left| \sum_{t=1}^{n} y(t) e^{-i(\alpha t + \beta t^2)} \right|^2$$

is maximized over a fine grid $\left(\frac{\pi j}{M}, \frac{\pi k}{M^2} \right)$, $j = 1, \ldots, M$, $k = 1, \ldots, M^2$ to find the initial estimates of the frequencies and frequency rates. The number of components is used as nine, same as used by Grover, Kundu, and Mitra [11], for this data set. The unknown parameters θ_{1k} and θ_{2k}, $k = 1, \ldots, 9$ are estimated using the method described in Sect. 10.2.1. In order to find the predicted values of the observed data, it is required to estimate the mean values of random amplitudes corresponding to each component. The real part is estimated as

$$\widehat{y}_R(t) = \sum_{k=1}^{p} \widehat{\mu}_{k\alpha} \cos(\widehat{\theta}_{1k}t + \widehat{\theta}_{2k}t^2),$$

where $\mu_{k\alpha}$, $k = 1 \ldots, p$ are estimated as follows:

$$\widehat{\mu}_{k\alpha} = \frac{1}{n} \sum_{t=1}^{n} \left[y_R(t) \cos(\widehat{\theta}_{1k}t + \widehat{\theta}_{2k}t^2) + y_I(t) \sin(\widehat{\theta}_{1k}t + \widehat{\theta}_{2k}t^2) \right]. \tag{10.10}$$

It can be shown that $\widehat{\mu}_{k\alpha} \longrightarrow \mu_{k\alpha}$ a.s. using a results of the form

$$n(\widehat{\theta}_{1k} - \theta_{1k}^0) \longrightarrow 0, \quad n^2(\widehat{\theta}_{2k} - \theta_{2k}^0) \longrightarrow 0, \quad a.s.$$

for $k = 1, \ldots, p$ as $n \to \infty$. Details have not been discussed here. The predicted values (blue) of "aaa" data are plotted in Fig. 10.2 along with the mean-corrected "aaa" data (red). They match quite well.

This analysis of a real data set illustrates the use of model (10.9). This is also an example where the theory has been developed for a complex-valued model, but that has been used to a real data set.

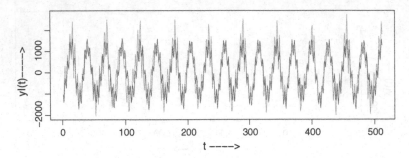

Fig. 10.2 The observed (red) and estimated (blue) "aaa" signals

10.3   Random Amplitude Chirp Model: Real-Valued Model

In this section, we consider a real-valued random amplitude chirp model which is in some way a general version of the complex-valued model discussed in Sect. 10.2. The model is given by

$$y(t) = A^0 \delta(t) \cos(\alpha^0 t + \beta^0 t^2) + B^0 \delta(t) \sin(\alpha^0 t + \beta^0 t^2) + e(t). \qquad (10.11)$$

Here, A^0 and B^0 denote the constant parts of the amplitudes and assumed to be bounded; α^0 and β^0 are frequency and frequency rate and $(\alpha^0, \beta^0) \in (0, \pi)$; $\{\delta(t)\}$ is a sequence of i.i.d. random variables with nonzero mean $\mu_{\delta 0}$ and variance $\sigma_{\delta 0}^2$; the fourth moment of $\{\delta(t)\}$ exists; $\{e(t)\}$ is a sequence of i.i.d random variables with mean zero and variance σ^2; the effective amplitudes $A^0 \delta(t)$ and $B^0 \delta(t)$ are random and not independently distributed. For identifiability issues, it is assumed that $\mu_{\delta 0}$ is known. Therefore, without loss of generality $\mu_{\delta 0}$ is assume to be 1. It is also assumed that $\{\delta(t)\}$ and $\{e(t)\}$ are independently distributed. Then

$$E[y(t)] = A^0 \cos(\alpha^0 t + \beta^0 t^2) + B^0 \sin(\alpha^0 t + \beta^0 t^2).$$

In model (10.6), the real part of $y(t)$ is $\alpha(t) \cos(\alpha^0 t + \beta^0 t^2) + e_R(t)$, whereas in model (10.11), an extra sine term is present such that the corresponding effective amplitudes $A^0 \delta(t)$ and $B^0 \delta(t)$ are not independently distributed.

In order to estimate the unknown parameters A, B, α, and β, we consider minimization of the following function:

$$R(\xi) = \sum_{t=1}^{n} \left[y(t) - A \cos(\alpha t + \beta t^2) - B \sin(\alpha t + \beta t^2) \right]^2,$$

with respect to ξ, where $\xi = (A, B, \alpha, \beta)$. Write $\xi^0 = (A^0, B^0, \alpha^0, \beta^0)$ as the true value of ξ and $\widehat{\xi} = (\widehat{A}, \widehat{B}, \widehat{\alpha}, \widehat{\beta})$ minimizes $R(\xi)$. Then, under the set-up of this section, it can be shown that $\widehat{\xi}$ is a strongly consistent estimator of ξ^0. Also the following asymptotic distribution can be derived:

$$\left(\sqrt{n}(\widehat{A} - A^0), \sqrt{n}(\widehat{B} - B^0), n\sqrt{n}(\widehat{\alpha} - \alpha^0), n^2\sqrt{n}(\widehat{\beta} - \beta^0)\right)$$

$$\xrightarrow{d} \mathcal{N}_4\left(\mathbf{0}, \boldsymbol{\Sigma}^{-1}\boldsymbol{\Gamma}\boldsymbol{\Sigma}^{-1}\right).$$

Here, with $(A^{0^2} + B^{0^2}) = \rho^0$

$$\boldsymbol{\Sigma} = \begin{pmatrix} 1 & 0 & \frac{1}{2}B^0 & \frac{1}{3}B^0 \\ 0 & 1 & -\frac{1}{2}A^0 & -\frac{1}{3}A^0 \\ \frac{1}{2}B^0 & -\frac{1}{2}A^0 & \frac{1}{3}\rho^0 & \frac{1}{4}\rho^0 \\ \frac{1}{3}B^0 & -\frac{1}{3}A^0 & \frac{1}{4}\rho^0 & \frac{1}{5}\rho^0 \end{pmatrix}$$

and $\boldsymbol{\Gamma} = ((\gamma_{ij}))$ is a symmetric matrix, where

$$\gamma_{11} = \frac{1}{2}\sigma_\delta^2(3A^{0^2} + B^{0^2}) + 2\sigma^2, \quad \gamma_{12} = 0, \quad \gamma_{13} = \frac{1}{4}B^0\sigma_\delta^2\rho^0 + B^0\sigma^2,$$

$$\gamma_{14} = \frac{1}{6}B^0\sigma_\delta^2\rho^0 + \frac{2}{3}B^0\sigma^2, \quad \gamma_{22} = \frac{1}{2}\sigma_\delta^2(A^{0^2} + 3B^{0^2}) + 2\sigma^2,$$

$$\gamma_{23} = -\frac{1}{4}A^0\sigma_\delta^2\rho^0 - A^0\sigma^2, \quad \gamma_{24} = -\frac{1}{6}A^0\sigma_\delta^2\rho^0 - \frac{2}{3}A^0\sigma^2,$$

$$\gamma_{33} = \frac{1}{6}\sigma_\delta^2\rho^{0^2} + \frac{2}{3}\sigma^2\rho^0, \quad \gamma_{34} = \frac{1}{8}\sigma_\delta^2\rho^{0^2} + \frac{1}{2}\sigma^2\rho^0,$$

$$\gamma_{44} = \frac{1}{10}\sigma_\delta^2\rho^{0^2} + \frac{2}{5}\sigma^2\rho^0.$$

If we compare $R(\boldsymbol{\xi})$ with the residual sum of squares for constant amplitude single chirp model (see model (9.1) in Chapter 9), then $A\mu_\delta$ and $B\mu_\delta$ correspond to cosine and sine amplitudes, respectively. Therefore, they are linear parameters. The asymptotic distribution states that the rate of convergence of $\widehat{A}$ as well as $\widehat{B}$ is $O_p(n^{-\frac{1}{2}})$, whereas for frequency and frequency rate, they are $O_p(n^{-\frac{3}{2}})$ and $O_p(n^{-\frac{5}{2}})$, respectively. Thus, for a given sample size n, the frequency rate can be estimated more accurately than the frequency and the frequency can be estimated more accurately than A and B, a similar feature observed in constant amplitude chirp model.

The multicomponent model in line of (10.11) can be given as

$$y(t) = \sum_{k=1}^{p}\left[A_k^0\delta_k(t)\cos(\alpha_k^0 t + \beta_k^0 t^2) + B_k^0\delta_k(t)\sin(\alpha_k^0 t + \beta_k^0 t^2)\right] + e(t), \quad (10.12)$$

where for $k = 1, \ldots, p$, A_k^0 and B_k^0 are constant parts of amplitudes; $\{\delta_k(t)\}$ is sequence of i.i.d. random variables with mean $\mu_{k\delta 0} \neq 0$ and variance $\sigma_{k\delta 0}^2$; $\{e(t)\}$ is a sequence i.i.d. random variables with mean zero and variance σ^2. It is assumed that $\{\delta_k(t)\}$ and $\{\delta_j(t)\}$, for $k \neq j$ are independently distributed and $\{e(t)\}$ is independently distributed of $\{\delta_k(t)\}$, $k = 1, \ldots, p$. Similar to single-component model (10.11), $\mu_{k\delta 0}, k = 1, \ldots, p$ are assumed to be known and without loss of generality, all are assumed to be 1.

The unknown parameters, present in model (10.12), can be estimated by minimizing a similar function as $R(\boldsymbol{\xi})$ defined in the case of multicomponent model. Write $\boldsymbol{\xi}_k = (A_k, B_k, \alpha_k, \beta_k)$, $\boldsymbol{\eta}_p = (\boldsymbol{\xi}_1, \ldots, \boldsymbol{\xi}_p)$ and let $\boldsymbol{\xi}_k^0$ be the true value of $\boldsymbol{\xi}_k$ and so is $\boldsymbol{\eta}_p^0$. Then, the parameters are estimated by minimizing

$$R(\boldsymbol{\eta}_p) = \sum_{t=1}^{n}\left[y(t) - \sum_{k=1}^{p}\left(A_k \cos(\alpha_k t + \beta_k t^2) + B_k \sin(\alpha_k t + \beta_k t^2)\right)\right]^2$$

with respect to $\boldsymbol{\xi}_1, \ldots, \boldsymbol{\xi}_p$. We also assume that $(\alpha_k^0, \beta_k^0) \neq (\alpha_j^0, \beta_j^0)$ for $k \neq j = 1, \ldots, p$ and $(\boldsymbol{\xi}_1^0, \ldots, \boldsymbol{\xi}_p^0)$ is an interior point of its parameters space. Let N_k be a neighborhood of $\boldsymbol{\xi}_k^0$. Then, the unknown parameters can also be estimated by minimizing $R(\boldsymbol{\eta}_p)$ locally, in the sense that, estimate $\boldsymbol{\xi}_k$ as

$$\widehat{\boldsymbol{\xi}}_k = \arg\min_{\boldsymbol{\xi} \in N_k} \sum_{t=1}^{n}\left[y(t) - A\cos(\alpha t + \beta t^2) - B\sin(\alpha t + \beta t^2)\right]^2,$$

where $y(t)$ is given in (10.12). It can be shown that the parameters estimated in this way, are strongly consistent. The asymptotic distribution of $\widehat{\boldsymbol{\xi}}_k$ for $k = 1, \ldots, p$ is as follows:

$$\left(\sqrt{n}(\widehat{A}_k - A_k^0), \sqrt{n}(\widehat{B}_k - B_k^0), n\sqrt{n}(\widehat{\alpha}_k - \alpha_k^0), n^2\sqrt{n}(\widehat{\beta}_k - \beta_k^0)\right)$$

$$\xrightarrow{d} \mathcal{N}_4\left(\mathbf{0}, \boldsymbol{\Sigma}_k^{-1}\boldsymbol{\Gamma}_k\boldsymbol{\Sigma}_k^{-1}\right),$$

where $\boldsymbol{\Sigma}_k$ and $\boldsymbol{\Gamma}_k$ are same as $\boldsymbol{\Sigma}$ and $\boldsymbol{\Gamma}$ with A^0, B^0, σ_δ and ρ^0, replaced by $A_k^0, B_k^0, \sigma_{k\delta}$ and ρ_k^0, respectively. Also, $\left(\sqrt{n}(\widehat{A}_k - A_k^0), \sqrt{n}(\widehat{B}_k - B_k^0), n\sqrt{n}(\widehat{\alpha}_k - \alpha_k^0), n^2\sqrt{n}(\widehat{\beta}_k - \beta_k^0)\right)$ and $\left(\sqrt{n}(\widehat{A}_j - A_j^0), \sqrt{n}(\widehat{B}_j - B_j^0), n\sqrt{n}(\widehat{\alpha}_j - \alpha_j^0), n^2\sqrt{n}(\widehat{\beta}_j - \beta_j^0)\right)$ for $k \neq j$ are independently distributed for large n.

In the class of real-valued random amplitude chirp models, one can also consider a model of the following form

$$y(t) = a(t)\cos(\alpha^0 t + \beta^0 t^2) + b(t)\sin(\alpha^0 t + \beta^0 t^2) + e(t), \tag{10.13}$$

where $\{a(t)\}$ and $\{b(t)\}$ are sequences of i.i.d. random variables with nonzero means and finite variances. Additionally, $\{a(t)\}$ and $\{b(t)\}$ are independently distributed. It will be interesting to estimate the unknown parameters, study the properties of the estimators and explore real-life applications.

10.4   Discussion

In this chapter, the random amplitude sinusoidal/chirp model has been considered. The random amplitude chirp model generalizes the basic sinusoidal model in two ways. The first one is that the frequency is not constant and changes linearly; the second one is that the amplitude is random. The most general model in this class is the random amplitude polynomial phase signal which belongs to a wide class of nonstationary signals used for modeling in engineering applications. In this case, the frequency in the sinusoidal model is replaced by a polynomial in index t, see Remark 10.3. Morelande and Zoubir [14] considered a polynomial phase signal modulated by a nonstationary random process. They used nonlinear least squares type phase and amplitude parameter estimators with sequential generalizations of the Bonferroni test. Fourt and Benidir [7] proposed a robust estimation method of parameters of random amplitude polynomial phase signal. The estimation of parameters of random amplitude polynomial phase signal under different error (both multiplicative and additive) assumptions and studying the properties of estimators are important problems. Other problems of interest are the estimation of the degree of the polynomial and developing some efficient algorithms to estimate the parameters. The reason is that random amplitude polynomial phase signal is a highly nonlinear model in its parameters and the estimators which have nice theoretical properties, may not work well in practice.

There are many other open problems. How to estimate σ_α^2 in the case of model (10.6) and $\sigma_{k\alpha}^2$, $k = 1, \ldots, p$ for model (10.9). Then, the problem of estimation of the variance of the additive error in all the models considered in the chapter. To use the asymptotic distribution in practice, these need to be estimated. Similarly for model (10.11), we need to estimate $\sigma_{\delta0}^2$ as well as σ^2, which has not been discussed here and are open problems. Estimation of the number of components p in both complex and real-valued multicomponent models is another important problem. The problem of detection of random amplitude is the most crucial decision in studying random amplitude models. Tests can be formulated based on asymptotic distributions of constant amplitudes case and random amplitudes case. The likelihood ratio test can be an option in this case.

Appendix A

In this Appendix, the proof of the consistency of $\widehat{\theta}$, defined in Sect. 10.2, is given. Write $z(t) = y^2(t) = z_R(t) + iz_I(t)$, then

$$z_R(t) = \alpha^2(t) \cos(2(\theta_1^0 t + \theta_2^0 t^2)) + (e_R^2(t) - e_I^2(t))$$
$$+2\alpha(t) e_R(t) \cos(\theta_1^0 t + \theta_2^0 t^2) - 2\alpha(t) e_I(t) \sin(\theta_1^0 t + \theta_2^0 t^2), \quad (10.14)$$
$$z_I(t) = \alpha^2(t) \sin(2(\theta_1^0 t + \theta_2^0 t^2)) + 2\alpha(t) e_I(t) \cos(\theta_1^0 t + \theta_2^0 t^2)$$
$$+2\alpha(t) e_R(t) \sin(\theta_1^0 t + \theta_2^0 t^2) + 2e_R(t) e_I(t). \quad (10.15)$$

The following lemmas will be required to prove the result.

Lemma 10.1 *(Lahiri, Kundu and Mitra [13]) If $(\theta_1, \theta_2) \in (0, \pi) \times (0, \pi)$, then except for a countable number of points, the following results are true.*

$$\lim_{n \to \infty} \frac{1}{n} \sum_{t=1}^{n} \cos(\theta_1 t + \theta_2 t^2) = \lim_{n \to \infty} \frac{1}{n} \sum_{t=1}^{n} \sin(\theta_1 t + \theta_2 t^2) = 0,$$

$$\lim_{n \to \infty} \frac{1}{n^{k+1}} \sum_{t=1}^{n} t^k \cos^2(\theta_1 t + \theta_2 t^2) = \lim_{n \to \infty} \frac{1}{n^{k+1}} \sum_{t=1}^{n} t^k \sin^2(\theta_1 t + \theta_2 t^2) = \frac{1}{2(k+1)},$$

$$\lim_{n \to \infty} \frac{1}{n^{k+1}} \sum_{t=1}^{n} t^k \cos(\theta_1 t + \theta_2 t^2) \sin(\theta_1 t + \theta_2 t^2) = 0, \quad k = 0, 1, 2.$$

Lemma 10.2 *Let $\widehat{\boldsymbol{\theta}} = (\widehat{\theta}_1, \widehat{\theta}_2)$ be an estimate of $\boldsymbol{\theta}^0 = (\theta_1^0, \theta_2^0)$ that maximizes $Q(\boldsymbol{\theta})$, defined in (10.8) and for any $\varepsilon > 0$, let $S_\varepsilon = \{\boldsymbol{\theta} : |\boldsymbol{\theta} - \boldsymbol{\theta}^0| > \varepsilon\}$ for some fixed $\boldsymbol{\theta}^0 \in (0, \pi) \times (0, \pi)$. If for any $\varepsilon > 0$,*

$$\overline{\lim}_{n \to \infty} \sup_{S_\varepsilon} \frac{1}{n} \big[Q(\boldsymbol{\theta}) - Q(\boldsymbol{\theta}^0) \big] < 0, \quad a.s. \quad (10.16)$$

then as $n \to \infty$, $\widehat{\boldsymbol{\theta}} \to \boldsymbol{\theta}^0$ a.s., that is, $\widehat{\theta}_1 \to \theta_1^0$ and $\widehat{\theta}_2 \to \theta_2^0$ a.s.

Proof of Lemma 10.2 We write $\widehat{\boldsymbol{\theta}}$ by $\widehat{\boldsymbol{\theta}}_n$ and $Q(\boldsymbol{\theta})$ by $Q_n(\boldsymbol{\theta})$ to emphasize that these quantities depend on n. Suppose (10.16) is true but $\widehat{\boldsymbol{\theta}}_n$ does not converge to $\boldsymbol{\theta}^0$ as $n \to \infty$. Then, there exists an $\varepsilon > 0$ and a subsequence $\{n_k\}$ of $\{n\}$ such that $|\widehat{\boldsymbol{\theta}}_{n_k} - \boldsymbol{\theta}^0| > \varepsilon$ for $k = 1, 2, \ldots$. Therefore, $\widehat{\boldsymbol{\theta}}_{n_k} \in S_\varepsilon$ for all $k = 1, 2, \ldots$. By definition, $\widehat{\boldsymbol{\theta}}_{n_k}$ is the estimator of $\boldsymbol{\theta}^0$ that maximizes $Q_{n_k}(\boldsymbol{\theta})$ when $n = n_k$. This implies that

$$Q_{n_k}(\widehat{\boldsymbol{\theta}}_{n_k}) \geq Q_{n_k}(\boldsymbol{\theta}^0) \Rightarrow \frac{1}{n_k} \Big[Q_{n_k}(\widehat{\boldsymbol{\theta}}_{n_k}) - Q_{n_k}(\boldsymbol{\theta}^0) \Big] \geq 0.$$

Therefore, $\overline{\lim}_{n \to \infty} \sup_{\boldsymbol{\theta}_{n_k} \in S_\varepsilon} \frac{1}{n_k} \big[Q_{n_k}(\widehat{\boldsymbol{\theta}}_{n_k}) - Q_{n_k}(\boldsymbol{\theta}^0) \big] \geq 0$, which contradicts inequality (10.16). Hence, the result follows.   ∎

Lemma 10.3 *(Nandi and Kundu [15]) Let $\{e(t)\}$ be a sequence of i.i.d. real-valued random variables with mean zero and finite variance $\sigma^2 > 0$, then as $n \to \infty$,*

$$\sup_{a,b} \left| \frac{1}{n} \sum_{t=1}^{n} e(t) \cos(at) \cos(bt^2) \right| \xrightarrow{a.s.} 0.$$

The result is true for all combinations of sine and cosine functions.

Lemma 10.4 *(Grover, Kundu, and Mitra [11]) If $(\beta_1, \beta_2) \in (0, \pi) \times (0, \pi)$, then except for a countable number of points, the following results hold.*

$$\lim_{n \to \infty} \frac{1}{n^{\frac{2m+1}{2}}} \sum_{t=1}^{n} t^m \cos(\beta_1 t + \beta_2 t^2) = \lim_{n \to \infty} \frac{1}{n^{\frac{2m+1}{2}}} \sum_{t=1}^{n} t^m \sin(\beta_1 t + \beta_2 t^2) = 0, \quad m = 0, 1, 2.$$

Lemma 10.5 *Under the same set-up as in Sect. 10.2, the following results are true for model (10.6).*

$$\frac{1}{n^{m+1}} \sum_{t=1}^{n} t^m z_R(t) \cos(2\theta_1^0 t + 2\theta_2^0 t^2) \xrightarrow{a.s} \frac{1}{2(m+1)}(\sigma_\alpha^2 + \mu_\alpha^2), \quad (10.17)$$

$$\frac{1}{n^{m+1}} \sum_{t=1}^{n} t^m z_I(t) \cos(2\theta_1^0 t + 2\theta_2^0 t^2) \xrightarrow{a.s} 0, \quad (10.18)$$

$$\frac{1}{n^{m+1}} \sum_{t=1}^{n} t^m z_R(t) \sin(2\theta_1^0 t + 2\theta_2^0 t^2) \xrightarrow{a.s} 0, \quad (10.19)$$

$$\frac{1}{n^{m+1}} \sum_{t=1}^{n} t^m z_I(t) \sin(2\theta_1^0 t + 2\theta_2^0 t^2) \xrightarrow{a.s} \frac{1}{2(m+1)}(\sigma_\alpha^2 + \mu_\alpha^2), \quad (10.20)$$

for $m = 0, 1, \ldots, 4$.

Proof of Lemma 10.5: Note that $E[e_R(t)e_I(t)] = 0$ and $\text{Var}[e_R(t)e_I(t)] = \frac{\sigma^4}{2}$ because $e_R(t)$ and $e_I(t)$ are each independently distributed with mean 0, variance $\frac{\sigma^2}{2}$ and fourth moment γ. Therefore, $\{e_R(t)e_I(t)\} \overset{i.i.d.}{\sim} (0, \frac{\sigma^4}{2})$. Similarly, we can show that

$$\{e_R^2(t) - e_I^2(t)\} \overset{i.i.d.}{\sim} (0, 2\gamma - \frac{\sigma^4}{2}), \quad \{\alpha(t)e_R(t)\} \overset{i.i.d.}{\sim} (0, (\sigma_\alpha^2 + \mu_\alpha^2)\frac{\sigma^2}{2}),$$

$$\{\alpha(t)e_I(t)\} \overset{i.i.d.}{\sim} (0, (\sigma_\alpha^2 + \mu_\alpha^2)\frac{\sigma^2}{2}). \quad (10.21)$$

This is due to the assumptions that $\alpha(t)$ is i.i.d. with mean μ_α and variance σ_α^2 and $\alpha(t)$ and $e(t)$ are independently distributed. Now, consider

$$\frac{1}{n^{m+1}} \sum_{t=1}^{n} t^m z_R(t) \cos(2\theta_1^0 t + 2\theta_2^0 t^2)$$

$$= \frac{1}{n^{m+1}} \sum_{t=1}^{n} t^m \alpha^2(t) \cos^2(2\theta_1^0 t + 2\theta_2^0 t^2)$$

$$+ \frac{1}{n^{m+1}} \sum_{t=1}^{n} t^m (e_R^2(t) - e_I^2(t)) \cos(2\theta_1^0 t + 2\theta_2^0 t^2)$$

$$+ \frac{2}{n^{m+1}} \sum_{t=1}^{n} t^m \alpha(t) e_R(t) \cos(\theta_1^0 t + \theta_2^0 t^2) \cos(2\theta_1^0 t + 2\theta_2^0 t^2)$$

$$+ \frac{2}{n^{m+1}} \sum_{t=1}^{n} t^m \alpha(t) e_I(t) \sin(\theta_1^0 t + \theta_2^0 t^2) \cos(2\theta_1^0 t + 2\theta_2^0 t^2).$$

The second term converges to zero as $n \to \infty$ using Lemma 10.3 as $(e_R^2(t) - e_I^2(t))$ is a mean zero and finite variance process. Similarly, the third and fourth terms also converge to zero as $n \to \infty$ using (10.21). Now the first term can be written as

$$\frac{1}{n^{m+1}} \sum_{t=1}^{n} t^m \alpha^2(t) \cos^2(2\theta_1^0 t + 2\theta_2^0 t^2)$$

$$= \frac{1}{n^{m+1}} \left[\sum_{t=1}^{n} t^m \{\alpha^2(t) - E[\alpha^2(t)]\} \cos^2(2\theta_1^0 t + 2\theta_2^0 t^2) \right.$$

$$\left. + \sum_{t=1}^{n} t^m E[\alpha^2(t)] \cos^2(2\theta_1^0 t + 2\theta_2^0 t^2) \right]$$

$$\xrightarrow{a.s.} 0 + \frac{1}{2(m+1)} E[\alpha^2(t)]$$

$$= \frac{1}{2(m+1)} (\sigma_\alpha^2 + \mu_\alpha^2),$$

using Lemmas 10.1 and 10.3. Note that here we have used the fact that the fourth moment of $\alpha(t)$ exists. In a similar way, (10.18), (10.19) and (10.20) can be proved. ∎

Proof of Consistency of $\widehat{\theta}$:

Expanding $Q(\theta)$ and using $y^2(t) = z(t) = z_R(t) + i z_I(t)$

$$\frac{1}{n}[Q(\boldsymbol{\theta}) - Q(\boldsymbol{\theta}^0)] = \left[\frac{1}{n}\sum_{t=1}^{n}\left\{z_R(t)\cos(2\theta_1 t + 2\theta_2 t^2) + z_I(t)\sin(2\theta_1 t + 2\theta_2 t^2)\right\}\right]^2$$

$$+ \left[\frac{1}{n}\sum_{t=1}^{n}\left\{-z_R(t)\sin(2\theta_1 t + 2\theta_2 t^2) + z_I(t)\cos(2\theta_1 t + 2\theta_2 t^2)\right\}\right]^2$$

$$- \left[\frac{1}{n}\sum_{t=1}^{n}\left\{z_R(t)\cos(2\theta_1^0 t + 2\theta_2^0 t^2) + z_I(t)\sin(2\theta_1^0 t + 2\theta_2^0 t^2)\right\}\right]^2$$

$$- \left[\frac{1}{n}\sum_{t=1}^{n}\left\{-z_R(t)\sin(2\theta_1^0 t + 2\theta_2^0 t^2) + z_I(t)\cos(2\theta_1^0 t + 2\theta_2^0 t^2)\right\}\right]^2$$

$$= S_1 + S_2 + S_3 + S_4 \quad (say)$$

Using Lemma 10.5, we have,

$$\frac{1}{n}\sum_{t=1}^{n} z_R(t)\cos(2\theta_1^0 t + 2\theta_2^0 t^2) \xrightarrow{a.s.} \frac{1}{2}(\sigma_\alpha^2 + \mu_\alpha^2),$$

$$\frac{1}{n}\sum_{t=1}^{n} z_I(t)\cos(2\theta_1^0 t + 2\theta_2^0 t^2) \xrightarrow{a.s.} 0,$$

$$\frac{1}{n}\sum_{t=1}^{n} z_R(t)\sin(2\theta_1^0 t + 2\theta_2^0 t^2) \xrightarrow{a.s.} 0,$$

$$\frac{1}{n}\sum_{t=1}^{n} z_I(t)\sin(2\theta_1^0 t + 2\theta_2^0 t^2) \xrightarrow{a.s.} \frac{1}{2}(\sigma_\alpha^2 + \mu_\alpha^2).$$

Then

$$\lim_{n\to\infty} S_3 = -(\sigma_\alpha^2 + \mu_\alpha^2)^2 \quad \text{and} \quad \lim_{n\to\infty} S_4 = 0.$$

Now

$$\overline{\lim}_{n\to\infty} \sup_{S_\varepsilon} S_1$$

$$= \overline{\lim}_{n\to\infty} \sup_{S_\varepsilon}\left[\frac{1}{n}\sum_{t=1}^{n}\left\{z_R(t)\cos(2\theta_1 t + 2\theta_2 t^2) + z_I(t)\sin(2\theta_1 t + 2\theta_2 t^2)\right\}\right]^2$$

$$= \overline{\lim}_{n\to\infty} \sup_{S_\varepsilon}\left[\frac{1}{n}\sum_{t=1}^{n}\left\{\alpha^2(t)\cos[2(\theta_1^0 - \theta_1)t + 2(\theta_2^0 - \theta_2)t^2]\right.\right.$$

$$+ 2e_R^2(t)e_I^2(t)\sin(2\theta_1 t + 2\theta_2 t^2) + (e_R^2(t) - e_I^2(t))\cos(2\theta_1 t + 2\theta_2 t^2)$$

$$+ 2\alpha(t)e_R(t)\cos[(2\theta_1^0 - \theta_1)t + (2\theta_2^0 - \theta_2)t^2]$$

$$\left.\left. + 2\alpha(t)e_I(t)\sin[(2\theta_1^0 - \theta_1)t + (2\theta_2^0 - \theta_2)t^2]\right\}\right]^2$$

$$= \overline{\lim}_{n\to\infty} \sup_{|\theta^0-\theta|>\varepsilon} \left[\frac{1}{n} \sum_{t=1}^{n} \left\{ (\alpha^2(t) - (\sigma_\alpha^2 + \mu_\alpha^2)) \cos[2(\theta_1^0 - \theta_1)t + 2(\theta_2^0 - \theta_2)t^2] \right. \right.$$

$$+ 2\alpha(t)e_R(t)\cos[(2\theta_1^0 - \theta_1)t + (2\theta_2^0 - \theta_2)t^2]$$

$$+ 2\alpha(t)e_I(t)\sin[(2\theta_1^0 - \theta_1)t + (2\theta_2^0 - \theta_2)t^2]$$

$$\left. \left. + (\sigma_\alpha^2 + \mu_\alpha^2)\cos[2(\theta_1^0 - \theta_1)t + 2(\theta_2^0 - \theta_2)t^2] \right\} \right]^2$$

(Second and third terms are independent of θ^0 and vanish using Lemma 10.3)

$$\longrightarrow 0, \quad \text{a.s.}$$

using (10.21) and Lemmas 10.1 and 10.3. Similarly, we can show that $\overline{\lim}_{n\to\infty} \sup_{S_\varepsilon}$ $S_2 \xrightarrow{a.s.} 0$. Therefore,

$$\overline{\lim}_{n\to\infty} \sup_{S_\varepsilon} \frac{1}{n}\left[Q(\theta) - Q(\theta^0)\right] = \overline{\lim}_{n\to\infty} \sup_{S_\varepsilon}\left[\sum_{i=1}^{4} S_i\right] \longrightarrow -(\sigma_\alpha^2 + \mu_\alpha^2)^2 < 0 \text{ a.s.}$$

and using Lemma 10.2, $\widehat{\theta}_1$ and $\widehat{\theta}_2$ which maximize $Q(\theta)$ are strongly consistent estimators of θ_1^0 and θ_2^0.   ∎

Appendix B

In this Appendix, we derive the asymptotic distribution of the estimator, discussed in Sect. 10.2, of the unknown parameters of model (10.6). The first order derivatives of $Q(\theta)$ with respect to θ_k, $k = 1, 2$ are as follows:

$$\frac{\partial Q(\theta)}{\partial \theta_k} = \frac{2}{n} \left\{ \sum_{t=1}^{n} z_R(t)\cos(2\theta_1 t + 2\theta_2 t^2) + \sum_{t=1}^{n} z_I(t)\sin(2\theta_1 t + 2\theta_2 t^2) \right\} \times$$

$$\left\{ \sum_{t=1}^{n} z_I(t)2t^k\cos(2\theta_1 t + 2\theta_2 t^2) - \sum_{t=1}^{n} z_R(t)2t^k\sin(2\theta_1 t + 2\theta_2 t^2) \right\}$$

$$+ \frac{2}{n}\left\{ \sum_{t=1}^{n} z_I(t)\cos(2\theta_1 t + 2\theta_2 t^2) - \sum_{t=1}^{n} z_R(t)\sin(2\theta_1 t + 2\theta_2 t^2) \right\} \times$$

$$\left\{ -\sum_{t=1}^{n} z_I(t)2t^k\sin(2\theta_1 t + 2\theta_2 t^2) - \sum_{t=1}^{n} z_R(t)2t^k\cos(2\theta_1 t + 2\theta_2 t^2) \right\}$$

$$= \frac{2}{n} f_1(\theta)g_1(k; \theta) + \frac{2}{n} f_2(\theta)g_2(k; \theta), \quad \text{(say)} \qquad (10.22)$$

where

$$f_1(\theta) = \sum_{t=1}^{n} z_R(t) \cos(2\theta_1 t + 2\theta_2 t^2) + \sum_{t=1}^{n} z_I(t) \sin(2\theta_1 t + 2\theta_2 t^2),$$

$$g_1(k; \theta) = \sum_{t=1}^{n} z_I(t) 2t^k \cos(2\theta_1 t + 2\theta_2 t^2) - \sum_{t=1}^{n} z_R(t) 2t^k \sin(2\theta_1 t + 2\theta_2 t^2),$$

$$f_2(\theta) = \sum_{t=1}^{n} z_I(t) \cos(2\theta_1 t + 2\theta_2 t^2) - \sum_{t=1}^{n} z_R(t) \sin(2\theta_1 t + 2\theta_2 t^2),$$

$$g_2(k; \theta) = -\sum_{t=1}^{n} z_I(t) 2t^k \sin(2\theta_1 t + 2\theta_2 t^2) - \sum_{t=1}^{n} z_R(t) 2t^k \cos(2\theta_1 t + 2\theta_2 t^2).$$

Observe that using Lemma 10.5, it immediately follows that

$$(a) \quad \lim_{n \to \infty} \frac{1}{n} f_1(\theta^0) = (\sigma_\alpha^2 + \mu_\alpha^2) \quad \text{and} \quad (b) \quad \lim_{n \to \infty} \frac{1}{n} f_2(\theta^0) = 0 \quad \text{a.s.} \quad (10.23)$$

Therefore, for large n, $\dfrac{\partial Q(\theta)}{\partial \theta_k} = \dfrac{2}{n} f_1(\theta) g_1(k; \theta)$, ignoring the second term in (10.22) which involves $f_2(\theta)$. The second order derivatives with respect to θ_k for $k = 1, 2$ are

$$\frac{\partial^2 Q(\theta)}{\partial \theta_k^2} = \frac{2}{n} \left\{ -\sum_{t=1}^{n} z_R(t) 2t^k \sin(2\theta_1 t + 2\theta_2 t^2) + \sum_{t=1}^{n} z_I(t) 2t^k \cos(2\theta_1 t + 2\theta_2 t^2) \right\}^2$$

$$+ \frac{2}{n} \left\{ \sum_{t=1}^{n} z_R(t) \cos(2\theta_1 t + 2\theta_2 t^2) + \sum_{t=1}^{n} z_I(t) \sin(2\theta_1 t + 2\theta_2 t^2) \right\} \times$$

$$\left\{ -\sum_{t=1}^{n} z_R(t) 4t^{2k} \cos(2\theta_1 t + 2\theta_2 t^2) - \sum_{t=1}^{n} z_I(t) 4t^{2k} \sin(2\theta_1 t + 2\theta_2 t^2) \right\}$$

$$+ \frac{2}{n} \left\{ -\sum_{t=1}^{n} z_R(t) 2t^k \cos(2\theta_1 t + 2\theta_2 t^2) - \sum_{t=1}^{n} z_I(t) 2t^k \sin(2\theta_1 t + 2\theta_2 t^2) \right\}^2$$

$$+ \frac{2}{n} \left\{ -\sum_{t=1}^{n} z_R(t) \sin(2\theta_1 t + 2\theta_2 t^2) + \sum_{t=1}^{n} z_I(t) \cos(2\theta_1 t + 2\theta_2 t^2) \right\} \times$$

$$\left\{ \sum_{t=1}^{n} z_R(t) 4t^{2k} \sin(2\theta_1 t + 2\theta_2 t^2) - \sum_{t=1}^{n} z_I(t) 4t^{2k} \cos(2\theta_1 t + 2\theta_2 t^2) \right\},$$

$$\frac{\partial^2 Q(\theta)}{\partial \theta_1 \partial \theta_2} = \frac{2}{n} \left\{ \sum_{t=1}^{n} z_R(t) \cos(2\theta_1 t + 2\theta_2 t^2) + \sum_{t=1}^{n} z_I(t) \sin(2\theta_1 t + 2\theta_2 t^2) \right\} \times$$

$$\left\{ -\sum_{t=1}^{n} z_R(t) 4t^3 \cos(2\theta_1 t + 2\theta_2 t^2) - \sum_{t=1}^{n} z_I(t) 4t^3 \sin(2\theta_1 t + 2\theta_2 t^2) \right\}$$

$$+ \frac{2}{n} \left\{ -\sum_{t=1}^{n} z_R(t) 2t^2 \sin(2\theta_1 t + 2\theta_2 t^2) + \sum_{t=1}^{n} z_I(t) 2t^2 \cos(2\theta_1 t + 2\theta_2 t^2) \right\} \times$$

$$\left\{ -\sum_{t=1}^{n} z_R(t) 2t \sin(2\theta_1 t + 2\theta_2 t^2) + \sum_{t=1}^{n} z_I(t) 2t \cos(2\theta_1 t + 2\theta_2 t^2) \right\}$$

$$+ \frac{2}{n} \left\{ -\sum_{t=1}^{n} z_R(t) \sin(2\theta_1 t + 2\theta_2 t^2) + \sum_{t=1}^{n} z_I(t) \cos(2\theta_1 t + 2\theta_2 t^2) \right\} \times$$

$$\left\{ \sum_{t=1}^{n} z_R(t) 4t^3 \sin(2\theta_1 t + 2\theta_2 t^2) - \sum_{t=1}^{n} z_I(t) 4t^3 \cos(2\theta_1 t + 2\theta_2 t^2) \right\}$$

$$+ \frac{2}{n} \left\{ -\sum_{t=1}^{n} z_R(t) 2t^2 \cos(2\theta_1 t + 2\theta_2 t^2) - \sum_{t=1}^{n} z_I(t) 2t^2 \sin(2\theta_1 t + 2\theta_2 t^2) \right\} \times$$

$$\left\{ -\sum_{t=1}^{n} z_R(t) 2t \cos(2\theta_1 t + 2\theta_2 t^2) - \sum_{t=1}^{n} z_I(t) 2t \sin(2\theta_1 t + 2\theta_2 t^2) \right\}.$$

Now, we can show the following using Lemma 10.5,

$$\lim_{n\to\infty} \frac{1}{n^3} \frac{\partial^2 Q(\theta)}{\partial \theta_1^2} \bigg|_{\theta^0} = -\frac{2}{3} (\sigma_\alpha^2 + \mu_\alpha^2)^2, \tag{10.24}$$

$$\lim_{n\to\infty} \frac{1}{n^5} \frac{\partial^2 Q(\theta)}{\partial \theta_2^2} \bigg|_{\theta^0} = -\frac{32}{45} (\sigma_\alpha^2 + \mu_\alpha^2)^2, \tag{10.25}$$

$$\lim_{n\to\infty} \frac{1}{n^4} \frac{\partial^2 Q(\theta)}{\partial \theta_1 \partial \theta_2} \bigg|_{\theta^0} = -\frac{2}{3} (\sigma_\alpha^2 + \mu_\alpha^2)^2. \tag{10.26}$$

Write $Q'(\theta) = \left(\frac{\partial Q(\theta)}{\partial \theta_1}, \frac{\partial Q(\theta)}{\partial \theta_2} \right)$ and $Q''(\theta) = \begin{pmatrix} \frac{\partial^2 Q(\theta)}{\partial \theta_1^2} & \frac{\partial^2 Q(\theta)}{\partial \theta_1 \partial \theta_2} \\ \frac{\partial^2 Q(\theta)}{\partial \theta_1 \partial \theta_2} & \frac{\partial^2 Q(\theta)}{\partial \theta_2^2} \end{pmatrix}$. Define a diagonal

matrix $\mathbf{D} = \text{diag} \left\{ \frac{1}{n^{\frac{3}{2}}}, \frac{1}{n^{\frac{5}{2}}} \right\}$. Expand $Q'(\widehat{\theta})$ using bivariate Taylor series expansion around θ^0,

$$Q'(\widehat{\theta}) - Q'(\theta^0) = (\widehat{\theta} - \theta^0) Q''(\bar{\theta}),$$

where $\bar{\theta}$ is a point on the line joining $\widehat{\theta}$ and θ^0. As $\widehat{\theta}$ maximizes $Q(\theta)$, $Q'(\widehat{\theta}) = 0$, the above equation can be written as

$$-[Q'(\theta^0)\mathbf{D}] = (\widehat{\theta} - \theta^0) \mathbf{D}^{-1} \mathbf{D} Q''(\bar{\theta}) \mathbf{D}$$
$$\Rightarrow (\widehat{\theta} - \theta^0) \mathbf{D}^{-1} = -[Q'(\theta^0)\mathbf{D}][\mathbf{D} Q''(\bar{\theta})\mathbf{D}]^{-1},$$

provided $[\mathbf{D} Q''(\bar{\theta})\mathbf{D}]$ is an invertible matrix a.s. Because $\widehat{\theta} \to \theta^0$ a.s. and $Q''(\theta)$ is a continuous function of θ, using continuous mapping theorem, we have

$$\lim_{n\to\infty} [\mathbf{D} Q''(\bar{\theta})\mathbf{D}] = \lim_{n\to\infty} [\mathbf{D} Q''(\theta^0)\mathbf{D}] = -\Sigma,$$

where $\boldsymbol{\Sigma}$ can be obtained using (10.24)-(10.26) as $\boldsymbol{\Sigma} = \dfrac{2(\sigma_\alpha^2 + \mu_\alpha^2)^2}{3} \begin{pmatrix} 1 & 1 \\ 1 & \frac{16}{15} \end{pmatrix}$. Using (10.23), the elements of $Q'(\boldsymbol{\theta}^0)\mathbf{D}$ are

$$\frac{1}{n^{\frac{3}{2}}} \frac{\partial Q(\boldsymbol{\theta}^0)}{\partial \theta_1} = 2\frac{1}{n} f_1(\boldsymbol{\theta}^0) \frac{1}{n^{\frac{3}{2}}} g_1(1; \boldsymbol{\theta}^0) \text{ and } \frac{1}{n^{\frac{5}{2}}} \frac{\partial Q(\boldsymbol{\theta}^0)}{\partial \theta_2} = 2\frac{1}{n} f_1(\boldsymbol{\theta}^0) \frac{1}{n^{\frac{5}{2}}} g_1(2; \boldsymbol{\theta}^0),$$

for large n. Therefore, to find the asymptotic distribution of $Q'(\boldsymbol{\theta}^0)\mathbf{D}$, we need to study the large sample distribution of $\frac{1}{n^{\frac{3}{2}}} g_1(1; \boldsymbol{\theta}^0)$ and $\frac{1}{n^{\frac{5}{2}}} g_1(2; \boldsymbol{\theta}^0)$. Replacing $z_R(t)$ and $z_I(t)$ in $g_1(k; \boldsymbol{\theta}^0)$, $k = 1, 2$, we have

$$\frac{1}{n^{\frac{2k+1}{2}}} g_1(k; \boldsymbol{\theta}^0)$$

$$= \frac{2}{n^{\frac{2k+1}{2}}} \sum_{t=1}^{n} t^k z_I(t) \cos(2\theta_1^0 t + 2\theta_2^0 t^2) - \frac{2}{n^{\frac{2k+1}{2}}} \sum_{t=1}^{n} t^k z_R(t) \sin(2\theta_1^0 t + 2\theta_2^0 t^2)$$

$$= \frac{4}{n^{\frac{2k+1}{2}}} \sum_{t=1}^{n} t^k e_R(t) e_I(t) \cos(2\theta_1^0 t + 2\theta_2^0 t^2) + \frac{4}{n^{\frac{2k+1}{2}}} \sum_{t=1}^{n} t^k \alpha(t) e_I(t) \cos(\theta_1^0 t + \theta_2^0 t^2)$$

$$- \frac{4}{n^{\frac{2k+1}{2}}} \sum_{t=1}^{n} t^k \alpha(t) e_R(t) \sin(\theta_1^0 t + \theta_2^0 t^2)$$

$$- \frac{2}{n^{\frac{2k+1}{2}}} \sum_{t=1}^{n} t^k (e_R^2(t) - e_I^2(t)) \sin(2\theta_1^0 t + 2\theta_2^0 t^2).$$

The random variables $e_R(t)e_I(t)$, $\alpha(t)e_R(t)$, $\alpha(t)e_I(t)$ and $(e_R^2(t) - e_I^2(t))$ are all mean zero finite variance random variables. Therefore, $E[\frac{1}{n^{\frac{3}{2}}} g_1(1; \boldsymbol{\theta}^0)] = 0$ and $E[\frac{1}{n^{\frac{5}{2}}} g_1(2; \boldsymbol{\theta}^0)] = 0$ for large n and all the terms above satisfy the Lindeberg-Feller's condition. So, $\frac{1}{n^{\frac{3}{2}}} g_1(1; \boldsymbol{\theta}^0)$ and $\frac{1}{n^{\frac{5}{2}}} g_1(2; \boldsymbol{\theta}^0)$ converge to normal distributions with zero mean and finite variances. In order to find the large sample covariance matrix of $Q'(\boldsymbol{\theta}^0)\mathbf{D}$, we first find the variances of $\frac{1}{n^{\frac{3}{2}}} g_1(1; \boldsymbol{\theta}^0)$ and $\frac{1}{n^{\frac{5}{2}}} g_1(2; \boldsymbol{\theta}^0)$ and their covariance.

$$\mathrm{Var}\left[\frac{1}{n^{\frac{3}{2}}} g_1(1; \boldsymbol{\theta}^0) \right]$$

$$= \mathrm{Var}\left[\frac{4}{n^{\frac{3}{2}}} \sum_{t=1}^{n} t e_R(t) e_I(t) \cos(2\theta_1^0 t + 2\theta_2^0 t^2) + \frac{4}{n^{\frac{3}{2}}} \sum_{t=1}^{n} t \alpha(t) e_I(t) \cos(\theta_1^0 t + \theta_2^0 t^2) \right.$$

$$\left. - \frac{4}{n^{\frac{3}{2}}} \sum_{t=1}^{n} t \alpha(t) e_R(t) \sin(\theta_1^0 t + \theta_2^0 t^2) - \frac{2}{n^{\frac{3}{2}}} \sum_{t=1}^{n} t (e_R^2(t) - e_I^2(t)) \sin(2\theta_1^0 t + 2\theta_2^0 t^2) \right]$$

$$= E\left[\frac{16}{n^3} \sum_{t=1}^{n} t^2 e_R^2(t) e_I^2(t) \cos^2(2\theta_1^0 t + 2\theta_2^0 t^2) + \frac{16}{n^3} \sum_{t=1}^{n} t^2 \alpha^2(t) e_I^2(t) \cos^2(\theta_1^0 t + \theta_2^0 t^2) \right.$$

$$+\frac{16}{n^3}\sum_{t=1}^{n}t^2\alpha^2(t)e_R^2(t)\sin^2(\theta_1^0 t + \theta_2^0 t^2)$$

$$+\frac{4}{n^3}\sum_{t=1}^{n}t^2(e_R^2(t) - e_I^2(t))^2\sin^2(2\theta_1^0 t + 2\theta_2^0 t^2)\Big]$$

(The cross-product terms vanish due to Lemma 10.1 and independence of $\alpha(t)$, $e_R(t)$ and $e_I(t)$.)

$$\longrightarrow 16.\frac{\sigma^2}{2}.\frac{\sigma^2}{2}.\frac{1}{6} + 16.\frac{\sigma^2}{2}.(\sigma_\alpha^2 + \mu_\alpha^2).\frac{1}{6} + 16.\frac{\sigma^2}{2}.(\sigma_\alpha^2 + \mu_\alpha^2).\frac{1}{6} + 4.(2\gamma - \frac{\sigma^4}{2}).\frac{1}{6}$$

$$= \frac{8}{3}\Big[(\sigma_\alpha^2 + \mu_\alpha^2)\sigma^2 + \frac{1}{2}\gamma + \frac{1}{8}\sigma^4\Big].$$

Similarly, we can show that for large n

$$\mathrm{Var}\Big[\frac{1}{n^{\frac{5}{2}}}g_1(2;\boldsymbol{\theta}^0)\Big]\longrightarrow \frac{8}{5}\Big[(\sigma_\alpha^2 + \mu_\alpha^2)\sigma^2 + \frac{1}{2}\gamma + \frac{1}{8}\sigma^4\Big],$$

$$\mathrm{Cov}\Big[\frac{1}{n^{\frac{3}{2}}}g_1(1;\boldsymbol{\theta}^0), \frac{1}{n^{\frac{5}{2}}}g_1(2;\boldsymbol{\theta}^0)\Big]\longrightarrow 2\Big[(\sigma_\alpha^2 + \mu_\alpha^2)\sigma^2 + \frac{1}{2}\gamma + \frac{1}{8}\sigma^4\Big].$$

Now, note that $Q'(\boldsymbol{\theta}^0)\mathbf{D}$ can be written as

$$Q'(\boldsymbol{\theta}^0)\mathbf{D} = \frac{2}{n}f_1(\boldsymbol{\theta}^0)\Big[\frac{1}{n^{\frac{3}{2}}}g_1(1;\boldsymbol{\theta}^0), \frac{1}{n^{\frac{5}{2}}}g_1(2;\boldsymbol{\theta}^0)\Big].$$

Then, as $n \to \infty$, $\frac{2}{n}f_1(\boldsymbol{\theta}^0) \xrightarrow{a.s.} 2(\sigma_\alpha^2 + \mu_\alpha^2)$ and

$$\Big[\frac{1}{n^{\frac{3}{2}}}g_1(1;\boldsymbol{\theta}^0), \frac{1}{n^{\frac{5}{2}}}g_1(2;\boldsymbol{\theta}^0)\Big] \xrightarrow{d} \mathcal{N}_2(\mathbf{0}, \boldsymbol{\Gamma}),$$

where

$$\boldsymbol{\Gamma} = 8\Big[(\sigma_\alpha^2 + \mu_\alpha^2)\sigma^2 + \frac{1}{2}\gamma + \frac{1}{8}\sigma^4\Big]\begin{bmatrix}\frac{1}{3} & \frac{1}{4}\\\frac{1}{4} & \frac{1}{5}\end{bmatrix}.$$

Therefore, using Slutsky's theorem, as $n \to \infty$,

$$Q'(\boldsymbol{\theta}^0)\mathbf{D} \xrightarrow{d} \mathcal{N}_2(\mathbf{0}, 4(\sigma_\alpha^2 + \mu_\alpha^2)^2\boldsymbol{\Gamma}),$$

and hence

$$(\widehat{\boldsymbol{\theta}} - \boldsymbol{\theta}^0)\mathbf{D}^{-1} \xrightarrow{d} \mathcal{N}_2(\mathbf{0}, 4(\sigma_\alpha^2 + \mu_\alpha^2)^2\boldsymbol{\Sigma}^{-1}\boldsymbol{\Gamma}\boldsymbol{\Sigma}^{-1}).$$

This is the asymptotic distribution stated in Sect. 10.2. ∎

References

1. Besson, O. (1995). Improved detection of a random amplitude sinusoid by constrained least-squares technique. *Signal Processing*, *45*(3), 347–356.
2. Besson, O., & Castanie, F. (1993). On estimating the frequency of a sinusoid in autoregressive multiplicative noise. *Signal Processing*, *3*(1), 65–83.
3. Besson, O., Ghogho, M., & Swami, A. (1999). Parameter estimation for random amplitude chirp signals. *IEEE Transactions on Signal Processing*, *47*(12), 3208–3219.
4. Besson, O., & Stoica, P. (1995). Sinusoidal signals with random amplitude: Least-squares estimation and their statistical analysis. *IEEE Transactions on Signal Processing*, *43*, 2733–2744.
5. Besson, O., & Stoica, P. (1999). Nonlinear least-squares approach to frequency estimation and detection for sinusoidal signals with arbitrary envelope. *Digital Signal Processing*, *9*(1), 45–56.
6. Ciblata, P., Ghoghob, M., Forsterc, P., & Larzabald, P. (2005). Harmonic retrieval in the presence of non-circular Gaussian multiplicative noise: Performance bounds. *Signal Processing*, *85*, 737–749.
7. Fourt, O., & Benidir, M. (2009). Parameter estimation for polynomial phase signals with a fast and robust algorithm. In *17th European Signal Processing Conference (EUSIPCO 2009)* (pp. 1027–1031).
8. Francos, J. M., & Friedlander, B. (1995). Bounds for estimation of multicomponent signals with random amplitudes and deterministic phase. *IEEE Transactions on Signal Processing*, *43*, 1161 – 1172.
9. Gabor, D. (1946). Theory of communication. Part 1: The analysis of information. *Journal of the Institution of Electrical Engineers - Part III: Radio and Communication Engineering*, *93*, 429–441.
10. Ghogho, M., Nandi, A. K., & Swami, A. (1999). Crame-Rao bounds and maximum likelihood estimation for random amplitude phase-modulated signals. *IEEE Transactions on Signal Processing*, *47*(11), 2905–2916.
11. Grover, R., Kundu, D., & Mitra, A. (2018). On approximate least squares estimators of parameters on one dimensional chirp signal. *Statistics*, *52*, 1060–1085.
12. Kundu, D., & Nandi, S. (2012). *Statistical Signal Processing: Frequency Estimation*. New Delhi: Springer.
13. Lahiri, A., Kundu, D., & Mitra, A. (2015). Estimating the parameters of multiple chirp signals. *Journal of Multivariate Analysis*, *139*, 189–205.
14. Morelande, M. R., & Zoubir, A. M. (2002). Model selection of random amplitude polynomial phase signals. *IEEE Transactions on Signal Processing*, *50*(3), 578–589.
15. Nandi, S., & Kundu, D. (2004). Asymptotic properties of the least squares estimators of the parameters of the chirp signals. *Annals of the Institute of Statistical Mathematics*, *56*, 529–544.
16. Nandi, S., & Kundu, D. (2020). Estimation of parameters in random amplitude chirp signal. *Signal Processing*, *168*, Art. 107328.
17. Stoica, P., & Moses, R. (2005). *Spectral analysis of signals*. Upper Saddle River: Prentice Hall.
18. Zhou, G., & Giannakis, G. B. (1994). On estimating random amplitude modulated harmonics using higher order spectra. *IEEE Transactions on Oceanic Engineering*, *19*(4), 529–539.

Chapter 11
Related Models

11.1 Introduction

The sinusoidal frequency model is a well-known model in different fields of science and technology and as has been observed in previous chapters, is a very useful model in explaining nearly periodical data. There are several other models which are practically some form of the multiple sinusoidal model, but also exploit some extra features in the data. In most of such cases, the parameters satisfy some additional conditions other than the assumptions required for the sinusoidal model. For example, in the complex-valued sinusoidal model, depending on the form of the terms in the exponent, β_j, it defines some related models. If β_j is real-valued, it is a *real compartment model*. Sometimes different form of modulation in amplitudes give rise to *amplitude modulated signals*. If the usual sinusoidal model is observed in a linear or higher order polynomial trend function, then we have a *partially sinusoidal model*. The *burst type model* is another special form of amplitude modulated model which has burst like features. Both *static and dynamic sinusoidal models* belong to the class of damped sinusoidal models. The fundamental frequency model where the frequencies are at $\lambda, 2\lambda, \ldots, p\lambda$, the harmonics of a fundamental frequency λ and have been discussed thoroughly in Chap. 6. The generalization of the fundamental frequency model to the class of chirp models is the *harmonic chirp model or fundamental chirp model* where the frequencies are $\lambda, 2\lambda, \ldots, p\lambda$ and frequency rates are $\mu, 2\mu, \ldots, p\mu$ for a p-component model. The *3-D chirp model for colored textures* is another extension of the 3-D sinusoidal model discussed in Chap. 8. The *chirp-like model* combines the sinusoidal frequency model with the chirp rate model. In the chirp rate part, the initial frequency is zero and it changes linearly. *Multi-channel sinusoidal model* is useful when data are generated from the same system with the same inherent frequencies but channel-wise different amplitudes.

This chapter is organized in the following way. The damped sinusoid and the amplitude modulated model are discussed in Sects. 11.2 and 11.3, respectively. These are complex-valued models. The rest of the models, discussed here, are real-valued. The partial sinusoidal model is in Sect. 11.4 and the burst type model in Sect. 11.5.

© Springer Nature Singapore Pte Ltd. 2020
S. Nandi and D. Kundu, *Statistical Signal Processing*,
https://doi.org/10.1007/978-981-15-6280-8_11

The static and the dynamic sinusoidal models are discussed in Sect. 11.6 whereas the harmonic chirp model in Sect. 11.7. The 3-D chirp model for colored textures is introduced in Sect. 11.8. The chirp-like model is discussed in Sect. 11.9 and the multi-channel sinusoidal model in Sect. 11.10. Finally, the chapter is concluded by a brief discussion of some more related models in Sect. 11.11.

11.2 Damped Sinusoidal Model

The superimposed damped exponential signal in presence of noise is an important model in signal processing literature. It is a complex-valued model in general form and can be written as

$$
y(t) = \mu(\boldsymbol{\alpha}, \boldsymbol{\beta}, t) + \varepsilon(t) = \sum_{j=1}^{p} \alpha_j \exp\{\beta_j t\} + \varepsilon(t), \quad t = t_i, \tag{11.1}
$$

where $\boldsymbol{\alpha} = (\alpha_1, \ldots, \alpha_p)$, $\boldsymbol{\beta} = (\beta_1, \ldots, \beta_p)$; $t_i, i = 1, \ldots, n$, are equidistant; $\alpha_j \ j = 1, \ldots, p$ are unknown complex-valued amplitudes; p is the total number of sinusoids present; $\alpha_j, j = 1, \ldots, p$ are assumed to be distinct; $\{\varepsilon(t_1), \ldots, \varepsilon(t_n)\}$ are complex-valued random variables with mean zero and finite variance.

Model (11.1) represents general complex-valued sinusoidal model and has three special cases; (i) undamped sinusoidal model, (ii) damped sinusoidal model, and (iii) real compartment model, depending on the form of $\beta_j, j = 1, \ldots, p$. When $\beta_j = i\omega_j, \omega_j \in (0, \pi)$, model (11.1) is an undamped sinusoid; if $\beta_j = -\delta_j + i\omega_j$ with $\delta_j > 0$ and $\omega_j \in (0, \pi)$ for all j, it is a damped sinusoid where δ_j is the damping factor and ω_j is the frequency corresponding to the jth component; if for all j, α_j and β_j are real numbers, model (11.1) represents a real compartment model. All the three models are quite common among engineers and scientists. For applications of damped and undamped models, the readers are referred to Kay [11] and for the real compartment model, see Seber and Wild [31] and Bates and Watts [1].

Tufts and Kumaresan in a series of papers considered model (11.1) with $t_i = i$. Some modifications of Prony's method was suggested by different authors. See Kumaresan [12], Kumaresan and Tufts [13], Kannan and Kundu [10], Kundu and Mitra [16, 17] and the references therein. It is pointed out by Rao [29], that solutions obtained by these methods may not be consistent. Moreover, Wu [35], showed that any estimator of the unknown parameters of model (11.1) is inconsistent with $t_i = i$. Due to this reason Kundu [15], consider the following alternative model. Write model (11.1) as

$$
y_{ni} = \mu(\boldsymbol{\alpha}^0, \boldsymbol{\beta}^0, t_{ni}) + \varepsilon_{ni}, \quad i = 1, \ldots, n, \tag{11.2}
$$

where $t_{ni} = i/n, i = 1, \ldots, n$ take values in the unit interval; $\boldsymbol{\theta} = (\boldsymbol{\alpha}, \boldsymbol{\beta})^T = (\alpha_1, \ldots, \alpha_p, \beta_1, \ldots, \beta_p)^T$ be the parameter vector. Least norm squares estimator is the most natural estimator in this case. Kundu [14] extends the results of Jennrich [8] for the LSEs to be consistent and asymptotically normal to the complex parameter case. But the damped sinusoid does not satisfy Kundu's condition. It is necessary for the LSEs to be consistent that t_{ni}, $i = 1, \ldots, n$, $n = 1, 2, \ldots$ are bounded. It is also assumed that $\{\varepsilon_{ni}\}$, $i = 1, \ldots, n$, $n = 1, 2, \ldots$ is a double array sequence of complex-valued random variables. Each row $\{\varepsilon_{n1}, \ldots, \varepsilon_{nn}\}$ are i.i.d. with mean zero. The real and imaginary parts of ε_{ni} are independently distributed with finite fourth moments. The parameter space Θ is a compact subset of $\mathcal{C}^p$ and the true parameter vector $\boldsymbol{\theta}^0$ is an interior point of Θ. Further the function

$$\int_0^1 |\mu(\boldsymbol{\alpha}^0, \boldsymbol{\beta}^0, t) - \mu(\boldsymbol{\alpha}, \boldsymbol{\beta}, t)|^2 dt \tag{11.3}$$

has a unique minimum at $(\boldsymbol{\alpha}, \boldsymbol{\beta}) = (\boldsymbol{\alpha}^0, \boldsymbol{\beta}^0)$. Under these assumptions, the LSEs of the unknown parameters of model (11.1) are strongly consistent and with proper normalization the LSEs become asymptotically normally distributed.

11.3 Amplitude Modulated (AM) Model

This is a special type of amplitude modulated undamped signal model and naturally complex-valued like damped or undamped model. The discrete-time complex random process $\{y(t)\}$ consisting of p single-tone AM signals is given by

$$y(t) = \sum_{k=1}^{p} A_k \left[1 + \mu_k e^{i\nu_k t}\right] e^{i\omega_k t} + X(t); \quad t = 1, \ldots, n, \tag{11.4}$$

where for $k = 1, \ldots, p$, A_k is the complex-valued carrier amplitude of the constituent signal, μ_k is the modulation index and can be real or complex-valued, ω_k is the carrier angular frequency, and ν_k is the modulating angular frequency. The sequence of additive errors $\{X(t)\}$ is a complex-valued stationary linear process.

The model was first proposed by Sircar and Syali [33]. Nandi and Kundu [20] and Nandi, Kundu, and Iyer [24], proposed LSEs and ALSEs and studied their theoretical properties for large n. The method, proposed by Sircar and Syali [33], is based on accumulated correlation functions, power spectrum and Prony's difference type equations but applicable if $\{X(t)\}$ is a sequence of complex-valued i.i.d. random variables. Readers are referred to Sircar and Syali [33], for the physical interpretation of different model parameters. Model (11.4) was introduced for analyzing some special types of nonstationary signals in steady-state analysis. If $\mu_k = 0$ for all k, model (11.4) coincides with the sum of complex exponential, that is, the undamped model, discussed in Sect. 11.2. The undamped model is best suited for transient nonstationary signal but it may lead to large order when the signal is not decaying over time. Sircar and Syali [33] argued that complex-valued AM model is more suitable for steady-state nonstationary signal. This model was proposed to analyze some short duration speech data. Nandi and Kundu [20] and Nandi, Kundu, and Iyer [24] analyzed two such datasets.

The following restrictions are required on the true values of the model parameters: for all k, $A_k \neq 0$ and $\mu_k \neq 0$ and they are bounded. Also $0 < \nu_k < \pi, 0 < \omega_k < \pi$ and

$$\omega_1 < \omega_1 + \nu_1 < \omega_2 < \omega_2 + \nu_2 < \cdots < \omega_M < \omega_M + \nu_M. \tag{11.5}$$

A complex-valued stationary linear process implies that $\{X(t)\}$ has the following representation

$$X(t) = \sum_{k=0}^{\infty} a(k)e(t-k),$$

where $\{e(t)\}$ is a sequence of i.i.d. complex-valued random variables with mean zero and variance $\sigma^2 < \infty$, that is, the real and imaginary parts are sequences of i.i.d. random variables with mean zero and variance $\frac{\sigma^2}{2}$ and are uncorrelated. The sequence $\{a(k)\}$ of arbitrary complex-valued constants is such that

$$\sum_{k=0}^{\infty} |a(k)| < \infty.$$

The least (norm) squares estimators of the unknown parameters of model (11.4) minimize

$$Q(\mathbf{A}, \boldsymbol{\mu}, \boldsymbol{\nu}, \boldsymbol{\omega}) = \sum_{t=1}^{n} \left| y(t) - \sum_{k=1}^{p} A_k (1 + \mu_k e^{i\nu_k t}) e^{i\omega_k t} \right|^2, \tag{11.6}$$

with respect to $\mathbf{A} = (A_1, \ldots, A_p)^T$, $\boldsymbol{\mu} = (\mu_1, \ldots, \mu_p)^T$, $\boldsymbol{\nu} = (\nu_1, \ldots, \nu_p)^T$, $\boldsymbol{\omega} = (\omega_1, \ldots, \omega_p)^T$, subject to the restriction (11.5).

Similarly, the ALSEs of the unknown parameters are obtained by maximizing

$$I(\boldsymbol{\nu}, \boldsymbol{\omega}) = \sum_{k=1}^{p} \left\{ \frac{1}{n} \left| \sum_{t=1}^{n} y(t) e^{-i\omega_k t} \right|^2 + \frac{1}{n} \left| \sum_{t=1}^{n} y(t) e^{-i(\omega_k + \nu_k)t} \right|^2 \right\}, \tag{11.7}$$

with respect to $\boldsymbol{\nu}$ and $\boldsymbol{\omega}$ under restriction (11.5). Write $(\widetilde{\omega}_k, \widetilde{\nu}_k)$ as the ALSE of (ω_k, ν_k), for $k = 1, \ldots, p$. Then the corresponding ALSEs of the linear parameters of A_k and μ_k are estimated as

$$\widetilde{A}_k = \frac{1}{n} \sum_{t=1}^{n} y(t) e^{-i\widetilde{\omega}_k t}, \qquad \widetilde{A}_k \widetilde{\mu}_k = \frac{1}{n} \sum_{t=1}^{n} y(t) e^{-i(\widetilde{\omega}_k + \widetilde{\nu}_k)t}. \tag{11.8}$$

Minimization of $Q(\mathbf{A}, \boldsymbol{\mu}, \boldsymbol{\nu}, \boldsymbol{\omega})$ is a $5p$ or $6p$ dimensional optimization problem depending on whether μ_k is real or complex-valued. But, as in case of sequential method of multiple sinusoidal model, it can be reduced to $2p$, 2-D optimization problem. Maximization of $I(\nu, \omega)$ involves a $2p$-dimensional maximization problem, but it can be implemented sequentially, that is a $2p$-dimensional maximization problem can be reduced to $2p$, one-dimensional maximization problems.

Nandi and Kundu [20] and Nandi, Kundu, and Iyer [24], established the strong consistency of the LSEs and ALSEs and obtained their asymptotic distributions for large n under restriction (11.5) and the assumption that the sequence of the error random variables follow the assumption of stationary complex-valued linear process. Additionally, it also required that $A_k \neq 0$, $\mu_k \neq 0$, and $\nu_k, \omega_k \in (0, \pi)$. LSEs and ALSEs have the same asymptotic distribution similar to the multiple sinusoidal model. LSEs, as well as ALSEs are asymptotically component-wise independent. It has been found that the linear parameters, that is, the real and imaginary parts of the amplitudes A_k and modulation index μ_k for $k = 1, \ldots, p$ are estimated with a rate $O_p(n^{-1/2})$, whereas the frequencies are estimated with rate $O_p(n^{-3/2})$.

11.4 Partially Sinusoidal Frequency Model

The Partially Sinusoidal Frequency Model is proposed by Nandi and Kundu [22], in the aim of analyzing data with periodic nature superimposed with a polynomial trend component. The model in its simplest form, in presence of a linear trend, is written as

$$y(t) = a + bt + \sum_{k=1}^{p} [A_k \cos(\omega_k t) + B_k \sin(\omega_k t)] + X(t), \quad t = 1, \ldots, n + 1.$$

(11.9)

Here, $\{y(t), t = 1, \ldots, n + 1\}$ are the observed data and a and b, unknown real numbers, are parameters of the linear trend component. The specifications in the sinusoidal part in model (11.9), are same as the multiple sinusoidal model (4.36), described in Sect. 4.6. That is, $A_k, B_k \in \mathbb{R}$ are unknown amplitudes, $\omega_k \in (0, \pi)$ are unknown frequencies. The sequence of noise $\{X(t)\}$ satisfies the following assumptions;

Assumption 11.1 $\{X(t)\}$ is a stationary linear process with the following form

$$X(t) = \sum_{j=-\infty}^{\infty} a(j)e(t - j),$$

(11.10)

where $\{e(t); t = 1, 2, \ldots\}$, are i.i.d. random variables with $E(e(t)) = 0$, $V(e(t)) = \sigma^2$, and $\sum_{j=-\infty}^{\infty} |a(j)| < \infty$.

Assumption 11.1 is similar to Assumption 3.2 in Chap. 3. The number of sinusoidal components present is p and it is assumed to be known in advance. The initial sample size is taken as $n + 1$ instead of the usual convention as n due to the fact that we work with first difference of the observed series. Therefore, eventually, we have n observations for estimation and inference.

If b is zero, model (11.9) is nothing but model (4.36) with a constant mean term. A more general model in the class of partially sinusoidal frequency model includes a polynomial of degree q, $2 < q \ll n$ instead of the linear contribution $a + bt$. This model was motivated by some real datasets.

Consider model (11.9) with $p = 1$ and write the model as

$$y(t) = a + bt + A\cos(\omega t) + B\sin(\omega t) + X(t), \quad t = 1, \ldots, n + 1. \quad (11.11)$$

Define $z(t) = y(t + 1) - y(t)$, say for $t = 1, \ldots n$, then

$$z(t) = b + A[\cos(\omega t + \omega) - \cos(\omega t)] + B[\sin(\omega t + \omega) - \sin(\omega t)] + x_d(t), (11.12)$$

where $x_d(t) = X(t + 1) - X(t)$ is the first difference of $\{X(t)\}$. It can be written as

$$x_d(t) = \sum_{k=-\infty}^{\infty} \beta(k) e(t - k)$$

where $\{\beta(k)\} = \{a(k + 1) - a(k)\}$. Since $\{a(k)\}$ is absolutely summable, so is $\{\beta(k)\}$ and $\{x_d(t)\}$ satisfies Assumption 11.1, the assumption of a stationary linear process. The constant term in linear trend part is not part of the differenced series $\{z(t)\}$. In matrix notation, Eq. (11.12) is written as

$$\mathbf{Z} = b\mathbf{1} + \mathbf{X}(\omega)\boldsymbol{\eta} + \mathbf{E}, \quad (11.13)$$

with $\mathbf{Z} = (z(1), z(2), \ldots, z(n))^T$, $\mathbf{1} = (1, 1, \ldots, 1)^T$, $\mathbf{E} = (x_d(1), \ldots, x_d(n))^T$, $\boldsymbol{\eta} = (A, B)^T$ and

$$\mathbf{X}(\omega) = \begin{bmatrix} \cos(2\omega) - \cos(\omega) & \sin(2\omega) - \sin(\omega) \\ \cos(3\omega) - \cos(2\omega) & \sin(3\omega) - \sin(2\omega) \\ \vdots & \vdots \\ \cos((n+1)\omega) - \cos(n\omega) & \sin((n+1)\omega) - \sin(n\omega) \end{bmatrix}. \quad (11.14)$$

Observe that $\frac{1}{n}\sum_{t=1}^{n} x_d(t) = O_p(n^{-1})$, since $\{\beta(k)\}$ is absolutely summable. Therefore, averaging over t, we have

$$\frac{1}{n}\sum_{t=1}^{n} z(t) = b + O(n^{-1}) + O_p(n^{-1}).$$

The results of Mangulis [18] (See Result 2.2 in Chap. 2) has been used in second approximation. Then for large n, b is estimated as $\widehat{b} = \dfrac{1}{n} \sum\limits_{t=1}^{n} z(t)$ and is a consistent estimator of b, see Nandi and Kundu [22]. Plug $\widehat{b}$ in (11.13) and write

$$\mathbf{Z}^* = \mathbf{Z} - \widehat{b}\mathbf{1} = \mathbf{X}(\omega)\boldsymbol{\eta} + \mathbf{E}. \tag{11.15}$$

Then, based on observations $\mathbf{Z}^*$, the LSEs of $\boldsymbol{\eta}$ and ω minimize the residual sum of squares

$$Q(\boldsymbol{\eta}, \omega) = \mathbf{E}^T \mathbf{E} = (\mathbf{Z}^* - \mathbf{X}(\omega)\boldsymbol{\eta})^T (\mathbf{Z}^* - \mathbf{X}(\omega)\boldsymbol{\eta}). \tag{11.16}$$

For a given ω, $\boldsymbol{\eta}$ is estimated as a function of ω

$$\widehat{\boldsymbol{\eta}}(\omega) = \left[\mathbf{X}(\omega)^T \mathbf{X}(\omega)\right]^{-1} \mathbf{X}(\omega)^T \mathbf{Z}^* = (\widehat{A}(\omega), \widehat{B}(\omega))^T. \tag{11.17}$$

Now use $\widehat{\boldsymbol{\eta}}(\omega)$ in (11.16) as

$$R(\omega) = Q(\widehat{\boldsymbol{\eta}}(\omega), \omega) = \mathbf{Z}^{*T} \left(\mathbf{I} - \mathbf{X}(\omega) \left[\mathbf{X}(\omega)^T \mathbf{X}(\omega)\right]^{-1} \mathbf{X}(\omega)^T \right) \mathbf{Z}^*. \tag{11.18}$$

The LSE of ω, $\widehat{\omega}$ is obtained by maximizing $R(\omega)$ and then plugging in $\widehat{\omega}$ into (11.17), $\boldsymbol{\eta}$ is estimated as $\widehat{\boldsymbol{\eta}}(\widehat{\omega})$. This two stage estimation process basically uses separable regression technique, discussed in Sect. 2.2.1 of Chap. 2.

Another interesting problem is testing whether b is zero, which has not been addressed so far.

Remark 11.1 In case of a model of q degree polynomial superimposed with a single sinusoid, one is required to take differences of the observed data q times to define $z^*(t)$. In such a scenario, ω can be obtained by maximizing a similar criterion function as $R(\omega)$, defined in (11.18). Then the entries of $\mathbf{X}(\omega)$ matrix will be more complicated functions of ω and one requires $n + q$ data points to start with.

Nandi and Kundu [22] studied the strong consistency and obtained the distribution of the estimators for large sample size. Denote the true values of A, B, and ω as A^0, B^0, and ω^0, respectively. It is proved that under the condition $A^{0^2} + B^{0^2} > 0$, the estimators are consistent and for large n,

$$(\sqrt{n}(\widehat{A} - A^0), \sqrt{n}(\widehat{B} - B^0), n\sqrt{n}(\widehat{\omega} - \omega^0) \xrightarrow{d} \mathcal{N}_3(\mathbf{0}, \boldsymbol{\Sigma}(\boldsymbol{\theta}^0))$$

where $\boldsymbol{\theta}^0 = (A^0, B^0, \omega^0)^T$ and

$$\Sigma(\theta^0) = \frac{\sigma^2 c_{par}(\omega^0)}{(1 - \cos(\omega^0))(A^{0^2} + B^{0^2})} \begin{bmatrix} A^{0^2} + 4B^{0^2} & -3A^0 B^0 & -6B^0 \\ -3A^0 B^0 & 4A^{0^2} + B^{0^2} & 6A^0 \\ -6B^0 & 6A^0 & 12 \end{bmatrix},$$

$$c_{par}(\omega) = \left| \sum_{k=-\infty}^{\infty} (a(k+1) - a(k))e^{-i\omega k} \right|^2. \tag{11.19}$$

Note that $c_{par}(\omega)$ takes the same form as $c(\omega)$ presented in (4.10) defined for the differenced error process $\{x_d(t)\}$. Also,

$$\mathrm{Var}(\widehat{\theta}^1) = \frac{c_{par}(\omega^0)}{2(1 - \cos(\omega^0))c(\omega^0)} \mathrm{Var}(\widehat{\theta}^2),$$

where $\widehat{\theta}^1 = (\widehat{A}^1, \widehat{B}^1, \widehat{\omega}^1)^T$ denote the LSE of $(A, B, \omega)^T$ in the case of model (11.11) and $\widehat{\theta}^2 = (\widehat{A}^2, \widehat{B}^2, \widehat{\omega}^2)^T$ is the LSE of the same model without any trend component.

In case of model (11.9), the estimation technique is the same as the one component frequency plus linear trend model (11.11). One needs to estimate the frequencies and the amplitudes using the differenced data. The coefficient b is estimated as the average of the differenced data. The corresponding design matrix $\mathbf{X}$ is of the order $n \times 2p$. Because $\mathbf{X}(\omega_j)$ and $\mathbf{X}(\omega_k)$, $k \neq j$, are orthogonal matrices, $\mathbf{X}(\omega_j)^T \mathbf{X}(\omega_k)/n = \mathbf{0}$ for large n. Therefore, the parameters corresponding to each sinusoidal component can be estimated sequentially. For sequential estimation, we assume that

$$A_1^{0^2} + B_1^{0^2} > A_2^{0^2} + B_2^{0^2} > \cdots > A_p^{0^2} + B_p^{0^2} > 0. \tag{11.20}$$

Nandi and Kundu [22] observe that the estimators are consistent and the parameters of the jth frequency component have similar asymptotic distribution as model (11.11) and the estimators corresponding to jth component is asymptotically independently distributed as the estimators corresponding to kth estimators, for $j \neq k$. Denote $\widehat{\theta}_j = (\widehat{A}_j, \widehat{B}_j, \widehat{\omega}_j)^T$ as the LSE of $\theta_j^0 = (A_j^0, B_j^0, \omega_j^0)^T$ corresponding to jth component of model (11.9). Then, for large n,

$$(\sqrt{n}(\widehat{A}_j - A_j^0), \sqrt{n}(\widehat{B}_j - B_j^0), n\sqrt{n}(\widehat{\omega}_j - \omega_j^0)) \overset{d}{\to} \mathcal{N}_3(\mathbf{0}, \Sigma(\theta_j^0)),$$

where $\Sigma(\theta)$ is same as defined above and $\widehat{\theta}_k$ and $\widehat{\theta}_j$, $j \neq k$ are asymptotically independently distributed.

Remark 11.2 In order to implement the asymptotic theoretical results in practice, it is required to estimate σ^2, as well as $c_{par}(\omega)$ at the frequency. We cannot estimate them separately, but it is possible to estimate $\sigma^2 c_{par}(\omega)$ because

$$\sigma^2 c_{par}(\omega) = E\left(\frac{1}{n}\left|\sum_{t=1}^{n} x_d(t)e^{-i\omega t}\right|^2\right),$$

the expected value of the periodogram function of $\{x_d(t)\}$, the error variables present in $z^*(t)$. This is actually needed to form confidence intervals based on asymptotic results. For numerical calculations, the method of averaging the periodogram function of the estimated noise over a two-sided window around the estimated frequency is generally used.

11.5 Burst Type Model

The burst type signal is proposed by Sharma and Sircar [32], to describe a segment of an ECG signal. This model is a generalized model of the multiple sinusoidal model with time-dependent amplitudes of certain form. The model exhibits occasional burst and is expressed as

$$y(t) = \sum_{j=1}^{p} A_j \exp[b_j\{1 - \cos(\alpha t + c_j)\}]\cos(\theta_j t + \phi_j) + X(t), \qquad t = 1,\ldots,n,$$

(11.21)

where for $j = 1,\ldots,p$, A_j is the amplitude of the carrier wave, $\cos(\theta_j t + \phi_j)$; b_j, and c_j are the gain part and the phase part of the exponential modulation signal; α is the modulation angular frequency; θ_j is the carrier angular frequency and ϕ_j is the phase corresponding to the carrier angular frequency θ_j; $\{X(t)\}$ is the sequence of the additive error random variables. The number of components, p, denotes the number of carrier frequencies present. The modulation angular frequency α is assumed to be same through all components. This ensures the occasional burst at regular intervals. Nandi and Kundu [21] studies the LSEs of the unknown parameters under the i.i.d. assumption of $\{X(t)\}$ and known p. Model (11.21) can be viewed as a sinusoidal model with a time-dependent amplitude $\sum_{j=1}^{p} A_j s_j(t) \cos(\theta_j t + \phi_j) + X(t)$ where $s_j(t)$ takes the particular exponential form $\exp[b_j\{1 - \cos(\alpha t + c_j)\}]$.

Mukhopadhyay and Sircar [19] proposed a similar model to analyze an ECG signal with a different representation of parameters. They assumed that θ_js are integer multiple of α. Nandi and Kundu [21] shows the strong consistency of the LSEs and obtain the asymptotic distribution as normal under the assumption of i.i.d. error and $\exp\{|b^0|\} < J$, where b^0 is the true value of b without any constraint on θ_j, like integer multiple of α. It is observed as in case of other related models that the carrier angular frequencies θ_j, $j = 1,\ldots,p$ and the modulation angular frequency α are estimated with a rate $O_p(n^{-3/2})$ and rest of the other parameters with rate equal to $O_p(n^{-1/2})$. When $p = 1$, the estimators of the pairs of parameters (A, b), (α, c) and (θ, ϕ) are asymptotically independent of each other, whereas the estimators of the parameters in each pair are asymptotically dependent. In the case $p > 1$, the estimator of α depends on the parameters of all the components. This is quite natural as α does not change over different components. The modular angular frequency can be taken different corresponding to different carrier angular frequency θ_j which a general model and is being considered by Nandi and Kundu [23].

11.6   Static and Dynamic Sinusoidal Model

A different form of damped sinusoidal model is given by

$$y(t) = \sum_{l=1}^{L} \rho_l^t \left(i_l \cos(\omega_l t) + q_l \sin(\omega_l t) \right) + e(t), \qquad (11.22)$$

where $t = 1, \ldots, n$, denote the uniformly sampled data. The unknown parameters $i_l, q_l, \omega_l \in [0, \pi]$ denote the undamped in-phase component, the undamped quadrature component and the frequency and $\rho_l \in (0, 1)$ denote the damping coefficient of the lth sinusoidal component, respectively; $\{e(t)\}$ is a sequence of random variables with mean zero and finite variance $\sigma_e^2 > 0$. Nielsen et al. [25] assumed normality on the additive error $\{e(t)\}$. This model can also be written as

$$y(t) = \sum_{l=1}^{L} \rho_l^t \alpha_l \cos(\omega_l t - \phi_l) + e(t) \qquad (11.23)$$

where $\alpha_l = \sqrt{i_l^2 + q_l^2}$ and $\phi_l = \tan^{-1}\left(q_l / i_l \right)$ are the undamped amplitude and phase of the lth sinusoidal component, respectively. Nielsen et al. [25] named this model as static sinusoidal model in contrast with the dynamic sinusoidal model which will be discussed below.

The static sinusoidal model and its variant is an important model and have drawn interest of many researchers for quite some time, particularly for its wide applicability in different real-life situation. Mathematically, the damping coefficient ρ_l and the frequency appear in the model in a nonlinear way and that complicates the estimation of unknown parameters. In model (11.22), the in-phase and quadrature components, i_l and q_l and in model (11.23), the undamped amplitude α_l and the phase ϕ_l are assumed to be constant over a segment of n sample points. These two models are widely used to represent the damped periodical behavior in nature. But many real-life signals do not fit the set-up of models (11.22) and (11.23) and violates the model assumptions. Therefore, Nielsen et al. [25], use a dynamic sinusoidal model where the in-phase and quadrature components i_l and q_l evolve as a first-order Gauss–Markov process. Relaxing the assumption of constant i_l and q_l, a dynamic sinusoidal model is given by

$$\text{(Observation equation):} \qquad y(t) = \mathbf{b}^T \mathbf{s}(t) + e(t) \qquad (11.24)$$

$$\text{(State equation):} \qquad \mathbf{s}(t+1) = \mathbf{A}\mathbf{s}(t) + \mathbf{v}(t), \qquad (11.25)$$

where $\mathbf{s}(t)$ is a state vector of order $2L$ at time point t and $\mathbf{v}(t)$ is a zero-mean Gaussian state noise vector with covariance matrix

$$\mathbf{\Sigma_v} = \text{diag}\left\{\sigma_{v,1}^2 \mathbf{I}_2, \dots, \sigma_{v,l}^2 \mathbf{I}_2, \dots \sigma_{v,L}^2 \mathbf{I}_2\right\}.$$

The state noise vectors corresponding to different components are mutually independent and independent of the observation noise $\{e(t)\}$. Also,

$$\mathbf{b} = \begin{bmatrix} 1 & 0 & \cdots & 1 & 0 \end{bmatrix}^T, \qquad (11.26)$$

$$\mathbf{A} = \text{diag}\left\{\mathbf{A}_1, \dots, \mathbf{A}_l, \dots, \mathbf{A}_L\right\}, \qquad (11.27)$$

$$\mathbf{A}_l = \rho_l \begin{bmatrix} \cos\omega_l & \sin\omega_l \\ -\sin\omega_l & \cos\omega_l \end{bmatrix}. \qquad (11.28)$$

Due to the block-diagonal structure of $\mathbf{A}$ and $\mathbf{\Sigma_v}$, the state equation in (11.25) can be written as

$$\mathbf{s}^l(t+1) = \mathbf{A}_l \mathbf{s}^l(t) + \mathbf{v}^l(t).$$

Here $\mathbf{s}^l(t)$ and $\mathbf{v}^l(t)$ denote vectors of order two comprising $(2l-1)$-st and $2l$th components of $\mathbf{s}(t)$ and $\mathbf{v}(t)$, respectively. In absence of state noise, the dynamic sinusoidal model reduces to static sinusoidal model. But, in the usual dynamic model, in-phase and quadrature components are like first-order Gauss–Markov processes. Insert the state equation into the observation equation in order to observe this.

$$
\begin{aligned}
y(t) &= \mathbf{b}^T \mathbf{s}(t) + e(t) \\
&= \mathbf{b}^T \left(\mathbf{A}\mathbf{s}(t-1) + \mathbf{v}(t-1) \right) + e(t) \\
&= \mathbf{b}^T \mathbf{A}^t \left(\mathbf{A}^{-1}\mathbf{s}(1) + \mathbf{A}^{-t} \sum_{k=1}^{t-1} \mathbf{A}^{k-1}\mathbf{v}(t-k) \right) + e(t) \\
&= \sum_{l=1}^{L} \left[i_{t,l}\cos(\omega_l t) + q_{t,l}\cos(\omega_l t) \right] + e(t), \qquad (11.29)
\end{aligned}
$$

where $i_{t,l}$ and $q_{t,l}$ are defined as follows:

$$\begin{bmatrix} i_{t,l} \\ q_{t,l} \end{bmatrix} \overset{def}{=} \rho_l^t \left(\mathbf{A}_l^{-1}\mathbf{s}^l(1) + \mathbf{A}_l^{-t} \sum_{k=1}^{t-1} \mathbf{A}_l^{k-1}\mathbf{v}^l(t-k) \right). \qquad (11.30)$$

Model (11.29) has the same form as model (11.22) with an important difference that the in-phase and quadrature components are time-varying. Therefore, the amplitudes and phases in the form (11.23) are also time-varying.

Define a stochastic process $\mathbf{z}^l(t) \overset{def}{=} [i_{t,l}\ q_{t,l}]^T$. Then, $\mathbf{z}^l(t)$ can be written as a recursive way

$$\mathbf{z}^l(t+1) = \rho_l \mathbf{z}^l(t) + \left(\rho^{-1}\mathbf{A}_l\right)^{-(t+1)}\mathbf{v}^l(t), \tag{11.31}$$

with $\mathbf{z}^l(1) = \rho_l \mathbf{A}_l^{-1}\mathbf{s}^l(1)$. If a Gaussian distribution is selected for the initial state, $\mathbf{s}^l(1)$, that is, $\mathbf{s}^l(1) \sim \mathcal{N}_2(\boldsymbol{\mu}_{\mathbf{s}^l(1)}, \sigma_{\mathbf{s}^l(1)}^2 \mathbf{I}_2)$, then $\mathbf{z}^l(t)$ is a finite-order Gauss–Markov process. Write $\widetilde{\mathbf{v}}^l(t) = \left(\rho_l^{-1}\mathbf{A}_l\right)^{-(n+1)}\mathbf{v}^l(t)$, then this is an orthogonal transformation and the distribution of $\widetilde{\mathbf{v}}^l(t)$ is same as the distribution of $\mathbf{v}^l(t)$ which is $\mathcal{N}(\mathbf{0}, \sigma_{v,l}^2 \mathbf{I}_2)$. Thus, $\mathbf{z}^l(t+1) = \rho_l \mathbf{z}^l(t) + \widetilde{\mathbf{v}}^l(t)$ have the same statistical behavior as (11.31). Therefore, $z^l(t)$ is a simple first-order Gauss–Markov process, independent of the frequency parameter. In case, $\boldsymbol{\mu}_{\mathbf{s}^l(1)} = \mathbf{0}$ and $\sigma_{\mathbf{s}^l(1)}^2 = \sigma_{v,l}^2/(1-\rho_l^2)$, $z^l(t)$ represents a stationary AR(1) process. Also, if $e(t) = 0$, the dynamic sinusoidal model (11.29) reduces to an AR(1) process. In general, considering above discussions, the dynamic sinusoidal model is equivalent to

$$\widetilde{y}(t) = \sum_{l=1}^{L}\left[\cos(\omega_l t)\ \sin(\omega_l t)\right]\begin{bmatrix}\widetilde{i}_{t,l}\\\widetilde{q}_{t,l}\end{bmatrix} + e(t),$$

$$\begin{bmatrix}\widetilde{i}_{t+1,l}\\\widetilde{q}_{t+,l}\end{bmatrix} = \rho_l\begin{bmatrix}\widetilde{i}_{t,l}\\\widetilde{q}_{t,l}\end{bmatrix} + \mathbf{v}^l(t). \tag{11.32}$$

In econometrics, this class of dynamic models is referred to as stochastic cyclical models, see Harvey [6] and Harvey and Jaeger [7]. Bayesian inference methods for dynamic sinusoidal models are considered by recent papers, Dubois and Davy [3] and Vincent and Plumbley [34]. Nielsen et al. [25] used MCMC technique and used Gibbs sampler. They assumed that only vague prior information is available and considered the joint prior distribution as products of priors of individual unknown parameters. The case of missing data is also considered by Nielsen et al. [25]. The computational complexity of the proposed algorithm is quite high and heavily depends on proper initialization.

11.7 Harmonic Chirp Model/Fundamental Chirp Model

One generalization the standard periodic signal is the harmonic chirp model and is given by

$$y(t) = \sum_{k=1}^{p}\left[A_k \cos(k(\lambda t + \mu t^2)) + B_k \sin(k(\lambda t + \mu t^2))\right] + X(t),$$
$$t = 1, \ldots, n. \tag{11.33}$$

Here, for $k = 1, \ldots, p$, A_k, and B_k are unknown real-valued amplitudes; λ is the fundamental frequency; μ is the fundamental chirp/frequency rate; p is the total number of harmonics. The harmonic chirp model is a generalization of harmonic sinusoidal model (fundamental frequency model discussed in Chap. 6) where the fundamental frequency of the harmonics changes linearly as a function of time. The harmonic chirp model is more accurate than the harmonic sinusoidal model in many practical scenarios, see Pantazis, Rosec, and Stylianou [27], Doweck, Amar, and Cohen [2], Nørholm, Jensen, and Christensen [26], and Jensen et al. [9]. Grover [4] considered the harmonic chirp model and studied the theoretical properties of the proposed estimators.

The nonlinear least squares (NLS) method can be used to estimate the unknown parameters of the harmonic chirp model (11.33), and this is identical to the MLEs under i.i.d. normal error assumption. A number of approximate methods were discussed by Doweck, Amar, and Cohen [2]. Jensen et al. [9] presented an algorithm exploiting the matrix structures of the LS optimization criterion function and employing fast Fourier transform on a uniform grid. Some of the methods and results discussed for the fundamental frequency model in Chap. 6 can be extended for the harmonic chirp model. Observe that the harmonic chirp model is a special case of the usual multiple chirp model discussed in Chap. 9 where frequency $\lambda_k = k\lambda$ and chirp rate $\mu_k = k\mu, k = 1, \ldots, p$. If the special structure of model (11.33) is suitable for a particular dataset, then the complexity of methods of estimation reduces to a large extent. This is because the total number of nonlinear parameters reduces to two from $2p$. Therefore, methods employed for multiple chirp models in Chap. 9 can be modified using the special structure of the harmonic chirp model.

11.8 Three-Dimensional Chirp Model for Colored Textures

In the same line as the 3-D colored sinusoidal model discussed in (8.3) of Chap. 8, we propose the following 3-D chirp model for colored texture.

$$y(m, n, s) = \sum_{k=1}^{p} \Big[A_k \cos(m\alpha_k + m^2\beta_k + n\gamma_k + n^2\delta_k + s\lambda_k + s^2\mu_k)$$

$$+ B_k \sin(m\alpha_k + m^2\beta_k + n\gamma_k + n^2\delta_k + s\lambda_k + s^2\mu_k) \Big] + X(m, n, s);$$

$$\text{for } m = 1, \ldots, M, \ n = 1, \ldots, N, \ s = 1, \ldots, S. \tag{11.34}$$

Here, for $k = 1, \ldots, p$, A_k, B_k are real-valued amplitudes; $\{\alpha_k, \gamma_k, \lambda_k\}$ are unknown frequencies; $\{\beta_k, \delta_k, \mu_k\}$ are frequency rates; $\{X(m, n, s)\}$ is a 3-D sequence of stationary (weak) error random variables with mean zero and finite variance; "p" the number of 3-D chirp components.

Model (11.34) can be used in modeling colored texture similarly as Prasad and Kundu [28]. The third dimension represents the intensity of different colors. In the digital representation of RGB type color pictures, the color at each pixel is described by red, green, and blue color intensities. One needs three values, the RGB triplets, to determine the color at each pixel, and therefore, $S = 3$ is fixed in case of colored texture. A large sample size corresponds to a bigger colored picture which implies M and N goes to infinity but the third dimension always remains fixed at 3.

The unknown parameters can be obtained by the LS method or sequential method along with separable regression technique of Richards [30]. The theoretical results will be asymptotic, that is, for large M and N. It will be interesting to develop the consistency and asymptotic normality properties of the estimators under the assumption that the frequencies and frequency rates belong to $(0, \pi)$ and the vector of true parameter values is an interior point of the parameter space.

11.9   Frequency Plus Chirp Rate Model (Chirp-Like Model)

The chirp-like model is proposed by Grover, Kundu, and Mitra [5], recently to somehow simplify the chirp model and is given by

$$y(t) = \sum_{k=1}^{p}\left[A_k^0 \cos(\alpha_k^0 t) + B_k^0 \sin(\alpha_k^0 t)\right] + \sum_{k=1}^{q}\left[C_k^0 \cos(\beta^0 t^2) + D_k^0 \sin(\beta^0 t^2)\right] + X(t),$$

(11.35)

where A_k^0s, B_k^0s, C_k^0s, and D_k^0s are unknown amplitudes and α_k^0s and β_k^0s are frequencies and frequency rates, respectively. Note that if $C_k^0 = D_k^0 = 0; k = 1, \ldots, q$, then model (11.35) reduces to a sinusoidal model, defined in (3.1), whereas if $A_k^0 = B_k^0 = 0; k = 1, \ldots, p$, them model (11.35) becomes a chirp rate model.

The form of the model is motivated by the fact that a vowel sound data "aaa" which was discussed by Grover, Kundu, and Mitra [5], can also be analyzed using model (11.35). Therefore, the data which can be analyzed using the chirp model can also be modeled using the chirp-like model. Grover [4] and Grover, Kundu, and Mitra [5] discuss the method of LS as the estimation technique. It is observed by Grover, Kundu, and Mitra [5], that computation of the LSEs of the chirp-like model is much simpler compared to the chirp model. To reduce the computation in obtaining the LSEs, the sequential estimation procedure is proposed. The consistency property of the estimators is established and large sample distribution is obtained under Assumption 11.1 on $\{X(t)\}$ along with the following ordering assumption.

$$\infty > A_1^{0^2} + B_1^{0^2} > A_2^{0^2} + B_2^{0^2} > \cdots > A_p^{0^2} + B_p^{0^2} > 0,$$

$$\infty > C_1^{0^2} + D_1^{0^2} > C_2^{0^2} + D_2^{0^2} > \cdots > C_q^{0^2} + D_q^{0^2} > 0.$$

The rates of convergence of the parameter estimates of chirp-like model are the same as those for the chirp model for amplitudes estimators, as well as frequency and frequency rate estimators. When $p = q = 1$, we need to solve two 1-D optimization problems to estimate the unknown parameters, instead of solving a 2-D optimization problem, as required to obtain the LSEs. For the multicomponent chirp-like model, if the number of components, p and q are very large, finding the LSEs becomes computationally challenging. Using the sequential procedure, the $(p + q)$-dimensional optimization problem can be reduced to $p + q$, 1-D optimization problems.

Chirp signals are ubiquitous in many areas of science and engineering and hence their parameter estimation is of great significance in signal processing. But it has been observed that parameter estimation of this model, particularly using the LS method is computationally complex. It seems that the chirp-like model can serve as an alternate model to the chirp model. It is observed that the data that have been analyzed using chirp models can also be analyzed using the chirp-like model and estimating its parameters using sequential LSEs is simpler than that for the chirp model. Therefore, it has a great potential to be used for real-life applications.

11.10 Multi-channel Sinusoidal Model

The single-component multi-channel sinusoidal model with two channels takes the following form

$$\begin{pmatrix} y_1(t) \\ y_2(t) \end{pmatrix} = \begin{pmatrix} A_1^0 & B_1^0 \\ A_2^0 & B_2^0 \end{pmatrix} \begin{pmatrix} \cos(\omega^0 t) \\ \sin(\omega^0 t) \end{pmatrix} + \begin{pmatrix} e_1(t) \\ e_2(t) \end{pmatrix} \tag{11.36}$$

$$\mathbf{y}(t) = \mathbf{A}^0 \boldsymbol{\theta}(\omega^0, t) + \mathbf{e}(t) \tag{11.37}$$

$$= \boldsymbol{\mu}(t) + \mathbf{e}(t), \quad \boldsymbol{\mu}(t) = \begin{pmatrix} \mu_1(t) \\ \mu_2(t) \end{pmatrix}, \tag{11.38}$$

where A_1^0 and B_1^0 are amplitudes corresponding to the first channel, whereas A_2^0 and B_2^0 are from the second channel; ω^0 is the frequency which is common for both the channels. The bivariate random vector $\mathbf{e}(t)$ is the error for the bivariate data $\mathbf{y}(t)$, $t = 1, \ldots, n$. The aim is to estimate the entries of the matrix of amplitudes $\mathbf{A}^0$ and the frequency ω^0 given a sample of size n under certain assumptions on $\mathbf{e}(t)$. Assume that $\{\mathbf{e}(t), t = 1, \ldots, n\}$ are i.i.d. with mean vector $\mathbf{0}$ and dispersion matrix

$$\Sigma = \begin{pmatrix} \sigma_1^2 & \sigma_{12} \\ \sigma_{21} & \sigma_2^2 \end{pmatrix},$$

where $\sigma_1^2, \sigma_2^2 > 0$ and $\sigma_{12} = \sigma_{21} \neq 0$.

Note that one can add the components of $\mathbf{y}(t)$, that is, $y_1(t) + y_2(t)$ and estimate ω^0. In that case the variance of the efftive error process will increase and also the linear parameters A_1^0, A_2^0, B_1^0 and B_2^0 are not identifiable. One can estimate $A_1^0 + A_2^0$ and $B_1^0 + B_2^0$, but cannot estimate them individually. An alternative way is to minimize

$$\sum_{t=1}^{n} [y_1(t) - \mu_1(t)]^2 + \sum_{t=1}^{n} [y_2(t) - \mu_2(t)]^2$$

with respect to the unknown parameters.

Let $\boldsymbol{\xi} = (A_1, B_1, A_2, B_2, \omega)^T$ and $\boldsymbol{\xi}^0$ denote the true value of $\boldsymbol{\xi}$. The LSE $\widehat{\boldsymbol{\xi}}$ of $\boldsymbol{\xi}$ in this case minimizes the residual sum of squares with respect to $\boldsymbol{\xi}$, defined as follows:

$$
\begin{aligned}
R(\boldsymbol{\xi}) &= \sum_{t=1}^{n} \mathbf{e}^T(t)\mathbf{e}(t) \\
&= \sum_{t=1}^{n} [e_1^2(t) + e_2^2(t)] \\
&= \sum_{t=1}^{n} [y_1(t) - A_1 \cos(\omega t) - B_1 \sin(\omega t)]^2 + \sum_{t=1}^{n} [y_2(t) - A_2 \cos(\omega t) - B_2 \sin(\omega t)]^2.
\end{aligned}
$$

Using matrix notation $R(\boldsymbol{\xi})$ can be written as follows:

$$R(\boldsymbol{\xi}) = (\mathbf{Y}_1 - \mathbf{X}(\omega)\boldsymbol{\beta})^T (\mathbf{Y}_1 - \mathbf{X}(\omega)\boldsymbol{\beta}_1) + (\mathbf{Y}_2 - \mathbf{X}(\omega)\boldsymbol{\beta}_2)^T (\mathbf{Y}_2 - \mathbf{X}(\omega)\boldsymbol{\beta}_2),$$

(11.39)

where $\mathbf{Y}_k = (y_k(1), \ldots, y_k(n))^T$, $\boldsymbol{\beta}_k = (A_k, B_k)^T$ for $k = 1, 2$ and

$$\mathbf{X}^T(\omega) = \begin{pmatrix} \cos(\omega) & \cos(2\omega) & \cdots & \cos(n\omega) \\ \sin(\omega) & \sin(2\omega) & \cdots & \sin(n\omega) \end{pmatrix}.$$

Minimizing (11.39) with respect to $\boldsymbol{\beta}_1$ and $\boldsymbol{\beta}_2$ for a given ω, we obtain

$$\widehat{\boldsymbol{\beta}}_k(\omega) = (\mathbf{X}^T(\omega)\mathbf{X}(\omega))^{-1}\mathbf{X}^T(\omega)\mathbf{Y}_k, \quad k = 1, 2.$$

(11.40)

Replacing $\boldsymbol{\beta}_k$ by $\boldsymbol{\beta}_k(\omega)$ in $R(\boldsymbol{\xi}) = R(\boldsymbol{\beta}_1, \boldsymbol{\beta}_2, \omega)$, we have

$$
\begin{aligned}
&R(\widehat{\boldsymbol{\beta}}_1(\omega), \widehat{\boldsymbol{\beta}}_2(\omega), \omega) \\
&= (\mathbf{Y}_1 - P_{\mathbf{X}(\omega)}\mathbf{Y}_1)^T (\mathbf{Y}_1 - P_{\mathbf{X}(\omega)}\mathbf{Y}_1) + (\mathbf{Y}_2 - P_{\mathbf{X}(\omega)}\mathbf{Y}_2)^T (\mathbf{Y}_2 - P_{\mathbf{X}(\omega)}\mathbf{Y}_2) \\
&= \mathbf{Y}_1^T (\mathbf{I} - P_{\mathbf{X}(\omega)})\mathbf{Y}_1 + \mathbf{Y}_2^T (\mathbf{I} - P_{\mathbf{X}(\omega)})\mathbf{Y}_2,
\end{aligned}
$$

where $P_{\mathbf{X}(\omega)} = \mathbf{X}(\omega)(\mathbf{X}^T(\omega)\mathbf{X}(\omega))^{-1}\mathbf{X}^T(\omega)$. We can write similarly if we have two channels and p frequencies, same for each channel. Only, $\boldsymbol{\beta}_1, \boldsymbol{\beta}_2$ and $\mathbf{X}(\omega)$ will be changed accordingly. If we have m channels with single frequency, say ω, then one can minimize

$$Q(\omega) = R(\widehat{\boldsymbol{\beta}}_1(\omega), \widehat{\boldsymbol{\beta}}_2(\omega), \ldots, \widehat{\boldsymbol{\beta}}_m(\omega), \omega) = \sum_{k=1}^{m} \mathbf{Y}_k^T (\mathbf{I} - P_{\mathbf{X}(\omega)})\mathbf{Y}_k,$$

where $\widehat{\boldsymbol{\beta}}_k(\omega)$ is same as defined in (11.40).

It can be shown that the LSE $\widehat{\boldsymbol{\xi}}$ of $\boldsymbol{\xi}^0$ is a strongly consistent estimator and the large sample distribution can be obtained as follows:

$$\mathbf{D}_1(\widehat{\boldsymbol{\xi}} - \boldsymbol{\xi}^0) \xrightarrow{d} \mathcal{N}_5(\mathbf{0}, \boldsymbol{\Gamma}^{-1}\boldsymbol{\Sigma}\boldsymbol{\Gamma}^{-1}). \tag{11.41}$$

The matrix $\mathbf{D}_1 = \text{diag}\{\sqrt{n}, \sqrt{n}, \sqrt{n}, \sqrt{n}, n\sqrt{n}\}$ and the $\boldsymbol{\Sigma} = ((\Sigma_{ij}))$ has the following form

$$\Sigma_{11} = 2\sigma_1^2, \quad \Sigma_{12} = 0, \quad \Sigma_{13} = 2\sigma_{12}, \quad \Sigma_{14} = 0, \quad \Sigma_{15} = B_1\sigma_1^2 + B_2\sigma_{12},$$
$$\Sigma_{22} = 2\sigma_1^2, \quad \Sigma_{23} = 0, \quad \Sigma_{24} = 2\sigma_{12}, \quad \Sigma_{25} = -A_1\sigma_1^2 - A_2\sigma_{12},$$
$$\Sigma_{33} = 2\sigma_2^2, \quad \Sigma_{34} = 0, \quad \Sigma_{35} = B_2\sigma_2^2 + B_1\sigma_{12},$$
$$\Sigma_{44} = 2\sigma_2^2, \quad \Sigma_{45} = -A_2\sigma_2^2 - A_1\sigma_{12},$$
$$\Sigma_{55} = \frac{2}{3}\Big[\sigma_1^2(A_1^2 + B_1^2) + \sigma_2^2(A_2^2 + B_2^2) + \sigma_{12}(A_1A_2 + B_1B_2)\Big],$$

and with $\rho_s = (A_1^2 + B_1^2 + A_2^2 + B_2^2)$, the matrix $\boldsymbol{\Gamma}$ is equal to

$$\begin{pmatrix} 1 + \frac{3B_1^2}{\rho_s} & -\frac{3A_1B_1}{\rho_s} & \frac{3B_1B_2}{\rho_s} & -\frac{3A_2B_1}{\rho_s} & -\frac{6B_1}{\rho_s} \\ -\frac{3A_1B_1}{\rho_s} & 1 + \frac{3A_1^2}{\rho_s} & -\frac{3A_1B_2}{\rho_s} & \frac{3A_1A_2}{\rho_s} & \frac{6A_1}{\rho_s} \\ \frac{3B_1B_2}{\rho_s} & -\frac{3A_1B_2}{\rho_s} & 1 + \frac{3B_2^2}{\rho_s} & -\frac{3A_2B_2}{\rho_s} & -\frac{6B_2}{\rho_s} \\ -\frac{3A_2B_1}{\rho_s} & \frac{3A_1A_2}{\rho_s} & -\frac{3A_2B_2}{\rho_s} & 1 + \frac{3A_2^2}{\rho_s} & \frac{6A_2}{\rho_s} \\ -\frac{6B_1}{\rho_s} & \frac{6A_1}{\rho_s} & -\frac{6B_2}{\rho_s} & \frac{6A_2}{\rho_s} & \frac{12}{\rho_s} \end{pmatrix}.$$

The multicomponent two channels sinusoidal model with p frequencies can be defined as the sum of p summands similar to model (11.36). The LSEs of the unknown parameters can be obtained similarly as above by minimizing sum of $\mathbf{e}(t)^T\mathbf{e}(t)$. If the diagonal entries of $\boldsymbol{\Sigma}$, σ_1^2 and σ_2^2 are known, then we can consider weighted least squares method and the theoretical properties can be established. Instead of known σ_1^2 and σ_2^2, if we assume that consistent estimates of σ_1^2 and σ_2^2 are available, then also the consistency goes along the same line and the asymptotic distribution can be derived. If the variance structure of $\mathbf{e}(t)$ is entirely known (that is, σ_1^2, σ_2^2 and σ_{12} for two channels model), then generalized least squares estimator can be obtained.

11.11 Discussion

Apart from the models discussed in this chapter, there are many other models which have important real-life applications and can be categorized as related models of the sinusoidal frequency model. In array signal processing, the signals recorded at the sensors contain information about the structure of the generating signals including the frequency and amplitudes of the underlying sources. It is basically the problem of identifying multiple superimposed exponential signals in noise. In two-dimensional sinusoidal and chirp models can be generalized with an interaction term in the arguments of sine and cosine functions. Such models can be used in many real-life image processing problems, for example, in fingerprints. In statistical signal processing literature, there are many variations of the models discussed in this monograph. Those models, which will be of interest to statisticians as per the understanding of the authors, are discussed in the monograph.

References

1. Bates, D. M., & Watts, D. G. (1988). *Nonlinear regression and its applications*. New York: Wiley.
2. Doweck, Y., Amar, A., & Cohen, I. (2015). Joint model order selection and parameter estimation of chirps with harmonic components. *IEEE Transactions on Signal Processing, 63*, 1765–1778.
3. Dubois, C., & Davy, M. (2007). Joint detection and tracking of time-varying harmonic components: A flexible Bayesian approach. *IEEE Transactions on Audio, Speech, and Language Processing, 15*, 1283–1295.
4. Grover, R. (2018). *Frequency and frequency rate estimation of some non-stationary signal processing models*. Ph.D. thesis, IIT Kanpur.
5. Grover, R., Kundu, D., & Mitra, A. (2018). Chirp-like model and its parameters estimation. arXiv:1807.09267v1.
6. Harvey, A. C. (1989). *Forecasting, structural time series models and the Kalman filter*. Cambridge: Cambridge University Press.
7. Harvey, A. C., & Jaeger, A. (1993). Detrending, sylized facts and the business cycle. *Journal of Applied Econometrics, 8*, 231–247.
8. Jennrich, R. I. (1969). Asymptotic properties of the non linear least squares estimators. *The Annals of Mathematical Statistics, 40*, 633–643.
9. Jensen, T. L., Nielsen, J. K., Jensen, J. R., Christensen, M.G., & Jensen, S. H. (2017). A fast algorithm for maximum-likelihood estimation of harmonic chirp parameters. *IEEE Transactions on Signal Processing, 65*, 5137–5152.
10. Kannan, N., & Kundu, D. (2001). Estimating parameters of the damped exponential models. *Signal Processing, 81*, 2343–2352.
11. Kay, S. M. (1988). *Fundamentals of statistical signal processing, Volume II - Detection theory*. New York: Prentice Hall.
12. Kumaresan, R. (1982). *Estimating the parameters of exponential signals*, Ph.D. thesis, University of Rhode Island.
13. Kumaresan, R., & Tufts, D. W. (1982). Estimating the parameters of exponentially damped sinusoids and pole-zero modeling in noise. *IEEE Transactions on Acoustics, Speech, and Signal Processing, 30*, 833–840.
14. Kundu, D. (1991). Asymptotic properties of the complex valued non-linear regression model. *Communications in Statistics-Theory and Methods, 20*, 3793–3803.
15. Kundu, D. (1994). A modified Prony algorithm for sum of damped or undamped exponential signals. *Sankhya, 56*, 524–544.
16. Kundu, D., & Mitra, A. (1995). Estimating the parameters of exponentially damped/undamped sinusoids in noise; a non-iterative approach. *Signal Processing, 46*, 363–368.
17. Kundu, D., & Mitra, A. (2001). Estimating the number of signals of the damped exponential models. *Computational Statistics & Data Analysis, 36*, 245–256.
18. Mangulis, V. (1965). *Handbook of series for scientists and engineers*. New York: Academic.
19. Mukhopadhyay, S., & Sircar, P. (1996). Parametric modeling of ECG signal. *Medical & Biological Engineering & Computing, 34*, 171–174.

20. Nandi, S., & Kundu, D. (2002). Complex AM model for non stationary speech signals. *Calcutta Statistical Association Bulletin*, *52*, 349–370.
21. Nandi, S., & Kundu, D. (2010). Estimating the parameters of Burst-type signals. *Statistica Sinica*, *20*, 733–746.
22. Nandi, S., & Kundu, D. (2013). Estimation of parameters of partially sinusoidal frequency model. *Statistics*, *47*, 45–60.
23. Nandi, S. and Kundu, D. (2020). Asymptotic properties of the least squares estimators of the parameters of burst-type signals in stationary noise. Preprint.
24. Nandi, S., Kundu, D., & Iyer, S. K. (2004). Amplitude modulated model for analyzing non stationary speech signals. *Statistics*, *38*, 439–456.
25. Nielsen, J. K., Christensen, M. G., Cemgil, A. T., Godsill, S. J., & Jensen, S. H. (2011). Bayesian interpolation and parameter estimation in a dynamic sinusoidal model. *IEEE Transactions on Audio, Speech, and Language Processing*, *19*, 1986–1998.
26. Nørholm, S. M., Jensen, J. R., & Christensen, M. G. (2016). Instantaneous pitch estimation with optimal segmentation for non-stationary voiced speech. *IEEE Transactions on Audio, Speech, and Language Processing*, *24*, 2354–2367.
27. Pantazis, Y., Rosec, O. & Stylianou, Y. (2009). Chirp rate estimation of speech based on a time-varying quasi-harmonic model. *Proceedings of IEEE International Conference on Acoustics, Speech, Signal Processing* (pp. 3985–3988).
28. Prasad, A., & Kundu, D. (2009). Modeling and estimation of symmetric color textures. *Sankhya, Series B*, *71*, 30–54.
29. Rao, C. R. (1988). Some results in signal detection. In S. S. Gupta & J. O. Berger (Eds.), *Decision theory and related topics, IV* (Vol. 2, pp. 319–332). New York: Springer.
30. Richards, F. S. G. (1961). A method of maximum likelihood estimation, *Journal of the Royal Statistical Society: Series B*, 469–475.
31. Seber, A., & Wild, B. (1989). *Nonlinear regression*. New York: Wiley.
32. Sharma, R. K., & Sircar, P. (2001). Parametric modeling of burst-type signals. *Journal of the Franklin Institute*, *338*, 817–832.
33. Sircar, P., & Syali, M. S. (1996). Complex AM signal model for non-stationary signals. *Signal Processing*, *53*, 35–45.
34. Vincent, E., & Plumbley, M. D. (2008). Efficient Bayesian inference for harmonic models via adaptive posterior factorization. *Neurocomputing*, *72*, 79–87.
35. Wu, C. F. J. (1981). Asymptotic theory of non linear least squares estimation. *The Annals of Statistics*, *9*, 501–513.

Index

© Springer Nature Singapore Pte Ltd. 2020
S. Nandi and D. Kundu, *Statistical Signal Processing*,
https://doi.org/10.1007/978-981-15-6280-8